Studies in Computational Intelligence

Volume 1118

Series Editor

Janusz Kacprzyk, Polish Academy of Sciences, Warsaw, Poland

The series "Studies in Computational Intelligence" (SCI) publishes new developments and advances in the various areas of computational intelligence—quickly and with a high quality. The intent is to cover the theory, applications, and design methods of computational intelligence, as embedded in the fields of engineering, computer science, physics and life sciences, as well as the methodologies behind them. The series contains monographs, lecture notes and edited volumes in computational intelligence spanning the areas of neural networks, connectionist systems, genetic algorithms, evolutionary computation, artificial intelligence, cellular automata, self-organizing systems, soft computing, fuzzy systems, and hybrid intelligent systems. Of particular value to both the contributors and the readership are the short publication timeframe and the world-wide distribution, which enable both wide and rapid dissemination of research output.

Indexed by SCOPUS, DBLP, WTI Frankfurt eG, zbMATH, SCImago.

All books published in the series are submitted for consideration in Web of Science.

Kevin Daimi · Abeer Alsadoon · Luis Coelho
Editors

Cutting Edge Applications of Computational Intelligence Tools and Techniques

Springer

Editors
Kevin Daimi
University of Detroit Mercy
Detroit, MI, USA

Abeer Alsadoon
Asia Pacific International College (APIC)
Sydney, NSW, Australia

Luis Coelho
Polytechnic Institute of Porto
Porto, Portugal

ISSN 1860-949X ISSN 1860-9503 (electronic)
Studies in Computational Intelligence
ISBN 978-3-031-44126-4 ISBN 978-3-031-44127-1 (eBook)
https://doi.org/10.1007/978-3-031-44127-1

This Springer imprint is published by the registered company Springer Nature Switzerland AG
The registered company address is: Gewerbestrasse 11, 6330 Cham, Switzerland

Preface

The main objective of computational intelligence is to recognize the rules, theories, dogmas, and assumptions that result in conceivable intelligent performance in applications and systems. The Institute of Electrical and Electronics Engineers (IEEE) defines Computational Intelligence (CI) to be "the theory, design, application, and development of biologically and linguistically motivated computational paradigms. Traditionally the three main pillars of CI have been Neural Networks, Fuzzy Systems and Evolutionary Computation." Computational intelligence is composed of various computational techniques and methods to tackle complex real-life problems when other traditional approaches fail to solve such problems.

To contribute to the objectives of computational intelligence (CI), this book presents six areas of cutting-edge applications of CI tools and techniques. The first area focuses on CI in Human-Machine Interaction. CI in Robotics and Automation and CI in Manufacturing, Engineering, and Industry are covered in Areas 2 and 3, respectively. Area 4 introduces CI in Recognition and Processing and Area 5 depicts CI in Finance, Business, Economics, and Education. Finally, Area 6 concentrates on applications of CI in Vehicles, Smart Cities/Energy, and Networking.

Area 1 presents a comprehensive overview of the history, science, technology, and ethics of Brain-Computer Interfaces (BCIs), and discusses the current state and prospects of BCIs, as well as their potential impact on society and humanity. Using artificial NN to predict the structural response of regular domains (such as rectangular, circular plates) submitted to uniform loads is a straightforward application. However, for more demanding problems, such as irregular bio-structures under complex load cases, it is necessary to explore and investigate the performance and efficiency of such machine learning frameworks. This is discussed in this area. In addition, this area reviews major psycholinguistic subclasses of conversational gestures, and the application of machine learning and deep learning techniques to conversational gesture recognition. It then describes a Synchronous Colored Petri Net (SCPN) model for automated training of social robots to detect and learn major subclasses of conversational gestures based upon synchronization of organ-motions and speech. Then Area 1 presents an approach to integrate spatial-temporal information to enhance user identification accuracy by using kernel density estimation to measure the similarity of users' trajectories, taking both spatial and temporal information into account, assigning weights to each check-in record to prioritize discriminative ones, and utilizing inconsistencies among check-in records to compute penalties for trajectory similarity.

Area 2 deals with a hybrid model, ATIAS, which combines AI ethics variables with technology acceptance model (TAM) variables. It is applied to the healthcare field to examine the impact of known technology acceptance factors and AI ethical factors on users' trust in and positive attitudes toward CI. This area then moves to a strategic overview of applications in the computer vision domain. Initially the etymology of computer vision, main tasks, key techniques, and algorithms are introduced. Then, traditional feature extraction methods and deep learning techniques, including prominent algorithms like Region-Based Convolutional Neural Network (R-CNN) and You Only Look Once (YOLO), are explored.

In Area 3, a random forest-based out-of-bag permutation feature importance study was conducted by assessing the influence of temperature, pH and organic loading rate (OLR), chemical oxygen demand (COD), total solids (TS), biological oxygen demand (BOD), suspended solids (SS), and hydraulic retention time (HRT) on methane, carbon dioxide, and hydrogen sulphide emission. Another chapter in this area elaborates on an algorithm that can predict autonomously the optimal positioning of a component through the innovative techniques of deep learning, employed for their ability to draw information from a set of examples and build complex models. A convolutional neural network (CNN) is developed to predict the rotation angle pair that leads to the optimal printing configuration starting at the tri-dimensional representation of an object.

Area 4 examines SINATRA, a novel multi-label classifier of music genres of songs. By following an iterative procedure that continuously reduces the dimensional space of the genres, SINATRA can tag a song with multiple and complementary genres. In addition, this area enhances a usual face recognition model with additional task capability without any additional storage cost by formulating the underlying relationship between the two tasks, and seamlessly embedding this relationship in a distance ranking deep model, which directly works on features rather than classification labels to make the system well generalized on unseen data.

Area 5 stresses that conceptual intelligence helps leaders to adopt a long-term perspective and forecast upcoming trends and opportunities by thinking outside the box and using data-driven decision-making in their business. Leaders must prioritize ongoing learning for themselves and their employees, which can lead to growth. The area also introduces a novel explainable artificial intelligent (AI) visualization system for the neuro-fuzzy architecture. The proposed explainable AI visualization system is designed in a form of graphical user interface to assist users to better understand inner function mechanism on how rules are generated and how conclusions are drawn from the data fed into the neuro-fuzzy system.

The book is concluded with Area 5 forming a daily living environment of smart cities in which CI applications play a dominant role. Autonomous (pilotless) vehicles present a shining example of the application of CI exploring which vehicles are allowed to move autonomously in both residential and rural areas. This environment examines the robustness of detecting adversarial attacks in the physical layer. This area further demonstrates the vast potential of CI in addressing complex challenges in modern urban and energy systems. It suggests that CI holds the promise of opening avenues for developing more effective, secure, and sustainable solutions in the dynamic realm of smart cities and smart energy systems. Finally, Area 5 analyzes main types of dirty data processed by computational intelligence, criteria of their classification, and means of their detection. The results of this analysis are represented by ontological model that contains taxonomy of classical and non-classical data and knowledge-oriented methods of their transformation.

Detroit, USA — Kevin Daimi
Sydney, Australia — Abeer Alsadoon
Porto, Portugal — Luis Coelho

Acknowledgements

It gives us great pleasure to admire the authors of the chapters of the book who invested their expertise in the field to furnish cutting-edge applications chapters. We are also indebted to the chapters' reviewers for their dedication, hard work, and valuable comments and corrections which significantly participated in improving the chapters. The editors would like to express their gratefulness and gratitude to Dr. Thomas Ditzinger, the Editorial Director of Interdisciplinary Applied Sciences, for his patience, encouragement, dedication, and support. Finally, we would also like to thank Naveen Kumar Madhan Mohan, the Project Coordinator of Cutting-Edge Applications of Computational Intelligence Tools and Techniques book, for his help and cooperation.

Nadire Cavus, Near East University, Cyprus
Somenath Chakraborty, West Virginia University Institute of Technology, USA
Luis Coelho, Polytechnic of Porto, Portugal
Kevin Daimi, University of Detroit Mercy, USA
Valentina Furlan, Università degli Studi di Pavia, Italy
Anatoly Gladun, National Academy of Sciences of Ukraine, Ukraine
Dimitris Glotsos, University of West Attica, Greece
Petr Hajek, University of Pardubice, Czech Republic
Gahangir Hossain, University of North Texas, USA
Eenjun Hwang, Korea University, South Korea
Vasyltsov Ihor, Samsung Advanced Institute of Technology, South Korea
Dariusz Jacek Jakobczak, Koszalin University of Technology, Poland
Spiros Kostaupolos, University of West Attica, Greece
Yongjun Li, Northwestern Polytechnical University, China
Luís Mourão, Polytechnic of Porto, Portugal
Mais Nijim, Texas A&M University-Kingsville, USA
George A. Papakostas, International Hellenic University, Greece
Marcin Paprzycki, Polish Academy of Sciences, Poland
Cao Qi, University of Glasgow Singapore, Singapore
Ricardo Queiros, Escola Superior de Media Arts, Portugal

Sunny Raj, Oakland University, USA
Sara Reis, Polytechnic of Porto, Portugal
Carlos Ramos, Polytechnic of Porto, Portugal
Julia Rogushina, National Academy of Science of Ukraine, Ukraine
Abeer Alsadoon, Asia Pacific International College, Australia
Fernando Terroso-Saenz, Catholic University of Murcia, Spain
Mario Versaci, Mediterranea University, Italy
Gina Vilão, Polytechnic of Porto, Portugal
Yi Wang, Manhattan College, USA

Contents

CI in Human-Machine Interaction

Brain-Computer Interfaces: High-Tech Race to Merge Minds and Machines 3
Nadire Cavus, Oluwafemi Ayotunde Oke, and Jamilu Maipan-uku Yahaya

Using Artificial Neural Networks to Predict Critical Displacement and Stress Values in the Proximal Femur for Distinct Geometries and Load Cases 21
Ana Pais, Jorge Lino Alves, and Jorge Belinha

An Integrated Model for Automated Identification and Learning of Conversational Gestures in Human–Robot Interaction 33
Aditi Singh and Arvind K. Bansal

Computational Intelligence Methods for User Matching 63
Yongjun Li, Shuai Yang, and Wenqiang He

CI in Robotics and Automation

ATIAS: A Model for Understanding Intentions to Use AI Technology 85
Farhana Faruqe, Larry Medsker, and Ryan Watkins

Electronics Engineering Perspectives on Computer Vision Applications: An Overview of Techniques, Sub-areas, Advancements and Future Challenges 113
Yu Xun Zheng, K.-W. (G. H.) A. Chee, Anand Paul, Jeonghong Kim, and H. Lv

CI in Manufacturing, Engineering, and Industry

Feature Importance Study for Biogas Production from POME Treatment Plants Using Out-of-Bag Permutation 145
Kishaan Ragu, Ong Qian Yee, Kiew Xin Yun, Hia Hung Yi, Kashwin Selvanathan, Sara Kazemi Yazdi, Chen Zhi Yuan, Chan Yi Jing, and Reza Godary

Convolutional Neural Networks for Part Orientation in Additive Manufacturing 165
Valentina Furlan, Kevin Castelli, Luca Scaburri, and Hermes Giberti

CI in Recognition and Processing

SINATRA: A Music Genre Classifier Based on Clustering and Graph Analysis 185
Fernando Terroso-Saenz, Jesús Soto, and Andres Muñoz

Towards an Enhanced and Lightweight Face Authentication System ... 211
Ying Zhang, Roger Zimmermann, Zhiwen Yu, and Bin Guo

CI in Finance, Business, Economics, and Education

Conceptual Intelligence, Digital Transformation, and Leadership Skills: Key Concepts for Modern Business Success 231
Nadire Cavus and Seyedali Aghamiri

GEMM-SaFIN(FRIE)++: Explainable Artificial Intelligence Visualisation System with Episodic Memory 255
Nelson Mingwei Ko, Chen Xie, Qi Cao, and Chai Quek

CI in Vehicles, Smart Cities/Energy, and Networking

Traffic Sign Recognition Robustness in Autonomous Vehicles Under Physical Adversarial Attacks 287
Kyriakos D. Apostolidis, Emmanouil V. Gkouvrikos, Eleni Vrochidou, and George A. Papakostas

Computational Intelligence in Smart Cities and Smart Energy Systems 305
Yi Wang and Jian Fu

Ontology-Based Similarity Estimates for Fuzzy Data: Semantic Wiki Approach 327
Julia Rogushina and Anatoly Gladun

About the Editors

Kevin Daimi received his Ph.D. from the University of Cranfield, England. He has a long academic and industry experience. His research interests include Computer and Network Security with emphasis on vehicle network security, Software Engineering, Data Science, Computational Intelligence, and Computer Science and Software Engineering Education. He has published several papers on vehicle security. He is the Editor of seven books in Cybersecurity and Data Science; Computer and Network Security Essentials, Innovation in Cybersecurity Education, Advances in Cybersecurity Management, Principles of Data Science, Breakthroughs in Digital Biometrics and Forensics, Principles and Practice of Blockchains, and Emerging Trends in Cybersecurity Applications, which were published by Springer. He is also the Editor of the Proceedings of the ICR'22 and ICR'23 International Conference on Innovations in Computing Research book published by Springer Book Series: Advances in Intelligent Systems and Computing, Lecture Notes in Networks and Systems, and ACR'23 International Conference on Advances in Computing Research book published by Springer Book Series: Lecture Notes in Networks and Systems. He has been chairing the annual International Conference on Security and Management (SAM) since 2012. He is also Program Chair of the annual International Conference on Innovations in Computing Research for the years 2022–2024, and the annual International Conference on Advances in Computing Research for the years 2023–2024. He is a Fellow of the British Computer

Society (BCS), a Senior Member of the Association for Computing Machinery (ACM), and a Senior Member of the Institute of Electrical and Electronic Engineers (IEEE). He is the recipient of the Outstanding Achievement Award from the 2010 World Congress in Computer Science, Computer Engineering, and Applied Computing (WORLDCOMP'10) in Recognition and Appreciation of his Leadership, Service and Research Contributions to the Field of Network Security. He is currently Professor Emeritus of Computer Science and Software Engineering at the University of Detroit Mercy.

Abeer Alsadoon received her Ph.D. from the University of Technology, Baghdad. She published more than 130 papers in A and B ranking journals and has more than 100 conference papers published in different peer-review IEEE conferences, and she has received four (4) best paper awards. She edited an Emerging Trends in Cybersecurity Applications book published by Springer. She received twenty (20) awards for teaching and research excellence from different Australian Universities. She has been recognized nationally as one out of five finalists for the prestigious 2019 Australian Women's Agenda Leadership Awards in the category of Emerging Female Leader in the Government and Public Sector. She chaired the 2022 International Conference on Health Informatics and Medical System (HIMS'22), Las Vegas, USA. She is the Conference Chair of HIMS'23 conference too. She is the Program Chair of the 2022 International Conference on Innovations in Computing Research (ICR'22), Athens, Greece; the 2023 International Conference on Advances in Computing Research (ACR'23), Orlando, Florida, USA; the ICR'23, Madrid, Spain; and the ACR'24, Madrid, Spain. She has been chairing the Program Committee of the annual IEEE International Conference on Innovative Technologies in Intelligent Systems & Industrial Application (CITISIA'20), Sydney, since 2020. She also chaired special sessions at different conferences. She is a Senior Member of the Institute of Electrical and Electronic Engineers (IEEE). She has many years of academic experience. She is currently an Associate Professor at the Education Centre of Australia (ECA)-Asia Pacific International College (APIC). She was a Dean of Scholarship and Research at Kent Institute Australia. In addition, she is an Adjunct Associate Professor at Charles Sturt University (CSU), Australia.

Luis Coelho received his Ph.D. from the University of Vigo, Spain. He has published more than 80 scientific articles in conferences and journals. He actively promotes publishing as editor of special edition books. He has received two best paper awards and was awarded with a national innovation prize. He has been a member of the organization committee for several conferences including the Sixth World Congress on Nature and Biologically Inspired Computing (NaBIC2014), Porto, Portugal; the 2014 Congress on Computational Aspects of Social Networks (CASON2014), Porto, Portugal; the 2018 Global Medical Engineering Physics Exchanges (GMEPE), Porto, Portugal; and the Pacific HealthCare Engineering (PAHCE), Porto, Portugal. He served as the Sessions/Workshops Chair for the 2022 International Conference on Health Informatics and Medical System (HIMS'22), Las Vegas, USA and the 2022 International Conference on Innovations in Computing Research (ICR'22), Athens, Greece. He is currently the Sessions/Workshops Chair of the 2023 International Conference on Innovations in Computing Research (ICR'23), Madrid, Spain. Furthermore, he is a program committee member of a number of conferences including the 2023 International Conference on Advances in Computing Research (ACR'23), Orlando, USA; the 2023 International Conference on Health Informatics and Medical Systems (HIMS'23), Las Vegas, USA; and the 32nd International Conference on Flexible Automation and Intelligent Manufacturing (FAIM'23), Porto, Portugal. He has been actively involved with the coordination of several healthcare-related degrees, and in the promotion of bridges between industry and academia.

CI in Human-Machine Interaction

Brain-Computer Interfaces: High-Tech Race to Merge Minds and Machines

Nadire Cavus, Oluwafemi Ayotunde Oke, and Jamilu Maipan-uku Yahaya

Abstract Brain-Computer Interfaces (BCIs) are a cutting-edge technology that allows for direct brain and external device communication, such as a computer. BCIs use advanced neuroscience techniques to measure and interpret brain activity, making it possible for users to control devices and software with their thoughts alone. This emerging technology has the potential to revolutionize many aspects of our lives, from healthcare to gaming to military operations. However, the development of BCIs also raises ethical concerns, such as privacy invasion, loss of autonomy, and human enhancement. As the race to merge minds and machines continues, it is crucial for researchers, policymakers, and society as a whole to consider the implications of this technology and ensure that it is developed and used responsibly. BCIs have various applications, such as restoring lost functions, augmenting existing abilities, or creating new experiences. BCIs also pose various challenges and risks, such as ethical, social, and security issues. This study gives a comprehensive overview of the history, science, technology, and ethics of BCIs. It also discusses the current state and prospects of BCIs, as well as their potential impact on society and humanity. This study is intended for anyone interested in learning more about BCIs and their implications.

Keywords Cutting-edge technology · Brain-computer interface · Human brain · EEG · Brain implants

---

N. Cavus (✉) · O. A. Oke · J. M. Yahaya
Department of Computer Information Systems, Near East University, Nicosia, Cyprus
e-mail: nadire.cavus@neu.edu.tr

Computer Information Systems Research and Technology Centre, Nicosia, Turkey

O. A. Oke
e-mail: 20206831@std.neu.edu.tr

J. M. Yahaya
e-mail: jymaipanuku@ibbu.edu.ng; 20205996@std.neu.edu.tr

J. M. Yahaya
Department of Computer Science, Ibrahim Badamasi Babangida University, Lapai, Nigeria

K. Daimi et al. (eds.), *Cutting Edge Applications of Computational Intelligence Tools and Techniques*, Studies in Computational Intelligence 1118,
https://doi.org/10.1007/978-3-031-44127-1_1

1 Introduction

Envision the idea of having abilities to control a computer, a robot, or even a spaceship with just your thoughts. Imagine being able to communicate with another person or an artificial intelligence without speaking or typing. Imagine being able to enhance your memory, creativity, or intelligence by connecting your brain to a device. These are some of the possibilities that brain-computer interfaces offer. Brain-Computer Interfaces (BCIs), as depicted in Fig. 1, shows how the brain interacts with systems and applications via the use of commands [1]. They can be used for various uses, such as restoring lost functions, augmenting existing abilities, or creating new experiences. BCIs can also enable users to access and manipulate information in novel ways, such as through virtual reality or telepresence.

BCIs are not science fiction; they are already being developed and tested by researchers and entrepreneurs around the world. In this study, this paper will explore the history, science, technology, and ethics of BCIs. Technologies such as BCIs and neural interfaces, make it possible for the brain to communicate directly with an external device like a computer or prosthetic limb. The technology works by picking up electrical signals from the brain's neurons using electrodes that are implanted or placed on the brain surface. A computer algorithm can then interpret these signals and issue commands or actions [2]. Neuromarketing and advertising, education and self-regulation, security and authentication, and games and entertainment are just a few of the many fields in which BCIs could be used [3].

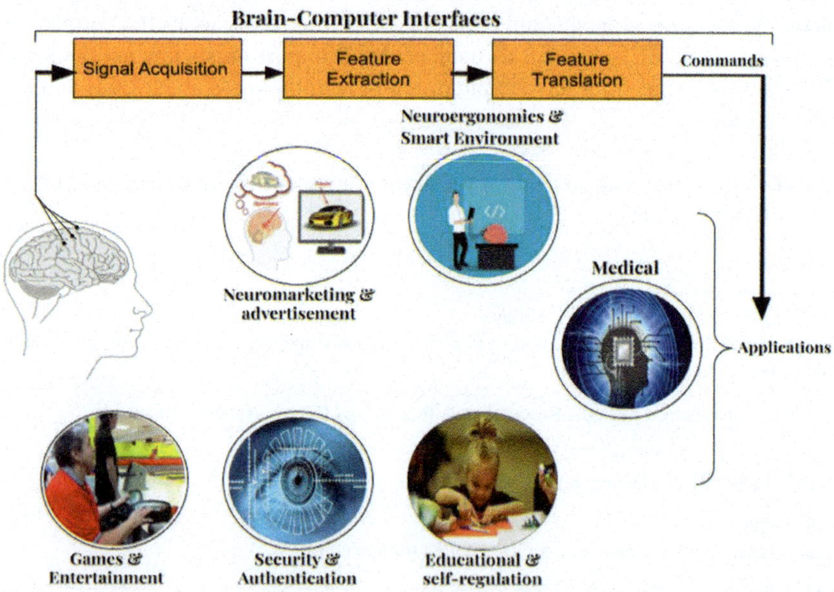

Fig. 1 Brain-computer interfaces transition

BCIs are utilized in a variety of fields such as medical, gaming, educational, communication, military, and business applications are just examples [4]. For example, people with disabilities can control and communicate with the help of noninvasive BCIs which derives intents from EEG activity of scalp recordings [5]. There are four main sections in this study: BCIs' history, science, technology, and ethics. This paper will provide an overview of the major ideas, developments, and issues associated with BCIs in each section and also outline some examples, case studies, and exercises to help understand BCIs better. This paper will also examine the current challenges and prospects of BCIs, as well as their potential impact on society and humanity. This study is intended for anyone interested in learning more about BCIs and their implications. Whether you are a student, a professional, a hobbyist, or a curious reader, you will find something in this study that will spark your imagination and inspire you to think about the possibilities and limitations of BCIs [6]. This paper in this study will not only inform you but also challenge you to ask questions and engage in discussions about BCIs.

1.1 History of BCIs

From ancient times to the present, this paper will examine the history and development of BCIs in this section. This paper will investigate the ways in which neuroscience, psychology, engineering, computer science, and other fields have contributed to the development of BCIs. This paper will also discuss about some of the firsts and significant achievements in the field of BCIs. The development of electroencephalography (EEG) and the discovery of the electrical activity of the human brain can be traced back to the beginnings of BCIs. Hans Berger, a German psychiatrist, was the first to use EEG to record human brain activity in 1924. As electrodes, he utilized silver wires that were inserted beneath the patient's scalp or silver foil electrodes that were attached to the head [7]. In addition, he discovered the brain's alpha and beta waves and coined the term "electroencephalogram". American composer Alvin Lucier demonstrated the use of a BCI for the first time in 1965. He controlled acoustic instruments like piano, gongs, and percussion with the help of EEG data and techniques for analog signal processing [8]. He composed a piece of music titled "Music for Solo Performer" in which he used the opening and closing of his eyes to control his brain waves, which were then used to turn on loudspeakers that made various objects vibrate. Furthermore, Jacques Vidal first used the term "brain-computer interface" in 1973 [9]. He came up with a system that would use EEG signals to let people talk to computers. He experimented on human subjects who were instructed to perform mental tasks like mental arithmetic, visual imagery, or relaxation to alter their EEG activity. Additionally, he created algorithms for EEG signal feature extraction, classification, and feedback. Rehabilitation of Cognitive and Perceptual Skills [10] was established by monitoring and identifying the magnitude of steady-state visual evoked processing (SSVER) with advanced signal processing technology and non-invasive scalp electrodes; An electrically driven hand orthosis was developed by Pfurtscheller et al. [11] to restore hand grasp function; the device's operation is directly based on the brain's bioelectrical signals; According to Millan [12], researchers have been fascinated over the past few years by the concept of moving robots or prosthetic devices without the use of manual control and simply through thought. Initially, intracranial electrodes implanted in the motor cortex of a monkey served as the basis for demonstrations of the approach's viability. Non-invasive methods based on EEG signals are preferred for humans, but measurements on the scalp result in decreased spatial resolution and increased noise. For the time being, signals from an EEG-based brain-machine interface are sufficient to continuously control a small mobile robot inside. Zander [13] demonstrated the significance of passive brain computer interfaces and how they can be used to obtain useful information about human–machine systems that are difficult to detect through external factors.

Phillip Kennedy built the first intracortical BCI in 1987. He embedded neurotrophic-cone cathodes into monkeys that could keep single-unit movement from neurons in the engine cortex. He demonstrated that neural activity could be used to teach monkeys to control a screen cursor. Later, he used his method on people who had severe motor impairments like quadriplegia or locked-in syndrome ([14], see Fig. 2). Cyberkinetics created the first BCI device that was approved by the FDA in 2004. It was known as BrainGate, and it made use of a collection of micro-electrodes that were inserted into the motor cortex of people who had tetraplegia. They were able to use their neural activity to control a wheelchair, a robotic arm, or a computer cursor. Through electrical stimulation of the somatosensory cortex, it also provided sensory feedback. The history section shows how BCIs have grown rapidly and how far it has come since inception till date with it now existing in a wide range of forms, modalities, applications, and domains. BCIs have been used to restore lost functions like speech, vision, hearing, and movement; enhancing abilities that are already in place, like memory, attention, or learning; or inventing novel experiences like entertainment, gaming, or art. BCIs have also been integrated with other technologies like artificial intelligence, nanotechnology, and virtual reality.

2 Science of BCIs

In this section, the fundamentals of BCIs and how they work will be explored. This paper will explain how BCIs measure and analyze various kinds of brain signals. Additionally, this paper will talk about the classification and evaluation of various kinds of BCIs based on their features and functions. The idea that brain activity can be extracted, measured, and converted into outputs that can replace, restore, enhance, supplement, or improve human functions is the foundation of the science of BCIs (Fig. 3). BCIs can make use of electrical, magnetic, metabolic, optical, and other kinds of brain signals. In addition, BCIs can measure brain signals using invasive, partially invasive, or noninvasive techniques. The electrical signal is the brain

Fig. 2 Brain computer-interfaces evolution (Adapted from [15])

signal that BCIs use the most frequently. When neurons in the brain communicate with one another through chemical and electrical synapses, they produce electrical signals. Electrodes can be implanted into the brain tissue (invasive, such as electrocorticography (ECoG)), inserted into the skull (non-invasive, such as EEG), or attached to the scalp (non-invasive, such as EEG). EEG, which records the voltage fluctuations that occur along the scalp, is the method that is most commonly used to measure electrical signals. Computer output is the type of output that BCIs use the most frequently. Algorithms process and interpret brain signals to turn them into commands or actions that can be used to control external devices like computers, robotic arms, wheelchairs, and prosthetics. These outputs come from computers. The user can also receive feedback from computer outputs through electrical, visual, auditory, or tactile stimulation. Visual feedback, which displays the user's performance or state on a screen, is the most common form of feedback [16].

The communication function is the type of function that BCIs use the most frequently. Using brain signals, communication functions enable individuals with severe motor or speech impairments to express their thoughts or needs [18]. Evoked potentials (EPs), event-related potentials (ERPs), steady-state potentials (SSPs), slow cortical potentials (SCPs), sensorimotor rhythms (SMRs), and P300 potentials are examples of specific brain signals that communication functions can use to elicit from the user [18]. Neuroscience, psychology, engineering, computer science, mathematics, and medicine all play a role in the interdisciplinary field of BCI science. Its objective is to comprehend how the brain functions and how it interacts with machines [17]. Additionally, it aims to develop novel technologies and approaches that have the potential to enhance the well-being and quality of life of people with disabilities or diseases.

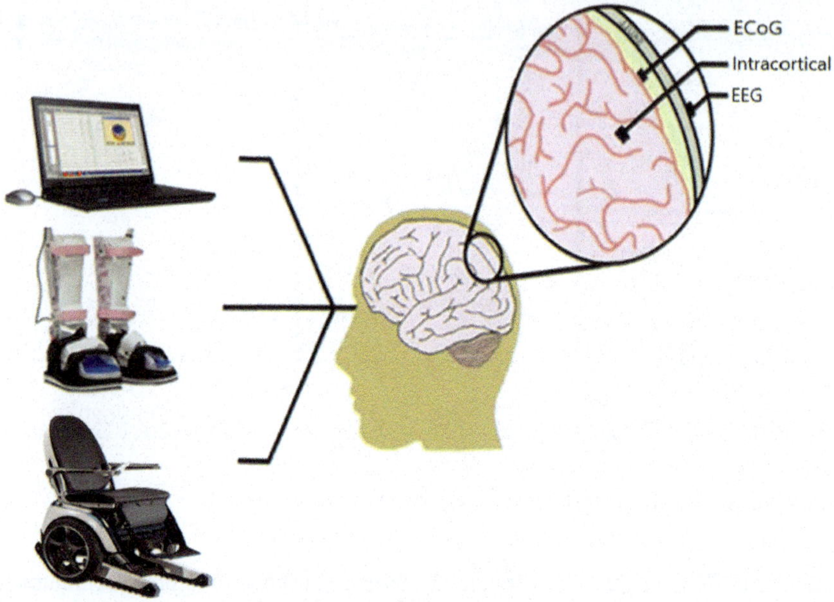

Fig. 3 Kind of brain computer-interfaces (Adapted from [17])

3 Technology of BCIs

In this section, some of the new and existing technologies that make BCIs possible will be discussed such as some of the hardware and software parts that go into making and running BCIs. In addition, this paper will discuss a few of the domains and applications that make use of BCIs for a variety of purposes. The concept that brain signals can be recorded, processed, and translated into outputs that can control external devices like computers, robotic arms, wheelchairs, or prosthetics underpins the technology of BCIs ([16], see Fig. 4). BCIs can record, process, and translate brain signals using a variety of tools and techniques. The implantable sensor is the kind of device that BCIs use the most. Small electrodes called implantable sensors are inserted surgically into the skull or brain tissue to measure electrical signals from specific brain regions. Although implantable sensors have the potential to deliver signals to the brain that are of a high resolution and quality, they also carry the potential for infection, inflammation, or rejection. Technologies known as BCIs make it possible for the brain to communicate with a computer or other external device. There are many different kinds of BCIs on the market today, each with its own set of capabilities and features. Some examples of BCIs and the companies that make them are as follows:

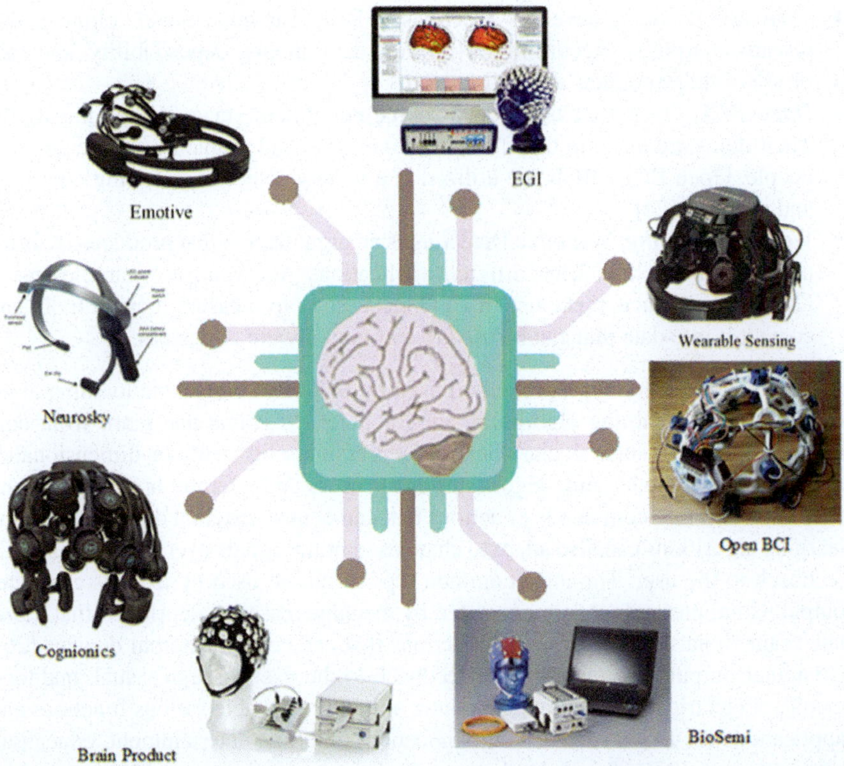

Fig. 4 BCIs technologies

- ***Emotive***: BCIs are manufactured by Emotiv for use in gaming, research, and other fields. EEG technology is used in their products to measure brain electrical activity and convert it into commands that can be used to control software or devices [19].
- ***EGI***: Another company that makes BCIs based on EEG is EGI. In clinical, educational, and research settings, their products are used to measure brain activity and analyze data. EGI's BCIs are renowned for their user-friendliness and high-quality data [20].
- ***Neurosky***: Neurosky manufactures BCIs that convert EEG data into data that can be used to control devices and measure brain activity. Their products are utilized in a variety of fields, including gaming and education [21].
- ***Cognionics***: High-quality EEG-based BCIs for neuroscience research, clinical applications, and other uses are manufactured by Cognionics. Their goods are well-known for being accurate, dependable, and simple to use.
- ***Brain Product***: A number of EEG-based BCIs, including systems for research, clinical applications, and other uses, are manufactured by Brain Products. Cgxsystems [22] says that their products are well-known for their high-quality data and ease of use.

- ***SemiBio***: BioSemi develops EEG-based BCIs for educational, clinical, and research purposes. According to [23], the data quality, dependability, and ease of use of their products are well-known.
- ***Open BCI***: Open BCI is an open-source equipment stage for building BCIs. Their items are made to be affordable, adaptable, and usable by a wide range of people. Open BCI's BCIs are utilized in a variety of fields, including education and research [24]
- ***Wearable Sensing***: Wearable Detecting is an organization that produces BCIs for use in research, clinical, and different applications. According to Wearablesensing [25], their products make use of EEG technology to measure brain activity and convert it into data that can be used to control software or devices.

The most common type of method used by BCIs is the machine learning algorithm. Machine learning algorithms are computer programs that learn from data and perform tasks such as classification, regression, clustering, or dimensionality reduction. Machine learning algorithms can process and interpret brain signals and translate them into commands or actions that can control external devices. Machine learning algorithms can also adapt to changes in brain signals over time and provide feedback to the user. The most common type of output used by BCIs is computer output. Computer outputs are generated by machine learning algorithms that translate brain signals into commands or actions that can control external devices [26]. Computer outputs can also provide feedback to the user through visual, auditory, tactile, or electrical stimulation. Computer outputs can enable various functions and applications for BCIs, such as communication, mobility, entertainment, education, or health care [27]. The technology of BCIs is an innovative field that involves engineering, computer science, mathematics, and medicine. It aims to develop new devices and methods that can enable direct neural control between the human brain and machines. It also aims to create new opportunities and challenges for the future of work and society.

4 Ethics of BCIs

In this section, this paper will examine some of the ethical, social, and legal implications of BCIs and will identify some of the benefits and risks that BCIs pose for individuals and society. Some of the values and principles that should guide the development and use of BCIs will also be explored. The ethics of BCIs is based on the principle that brain signals can be used to control external devices, such as computers, robotic arms, wheelchairs, or prosthetics. BCIs can raise various ethical issues that surround their use as depicted in Fig. 5, such as privacy, consent, autonomy, responsibility, identity, and social justice [28]. One of the main ethical issues of BCIs is privacy. Privacy refers to the right of individuals to control their personal information and to protect it from unauthorized access or misuse [29]. BCIs invasion of privacy is one major ethical concern as it can provide access to intimate and personal

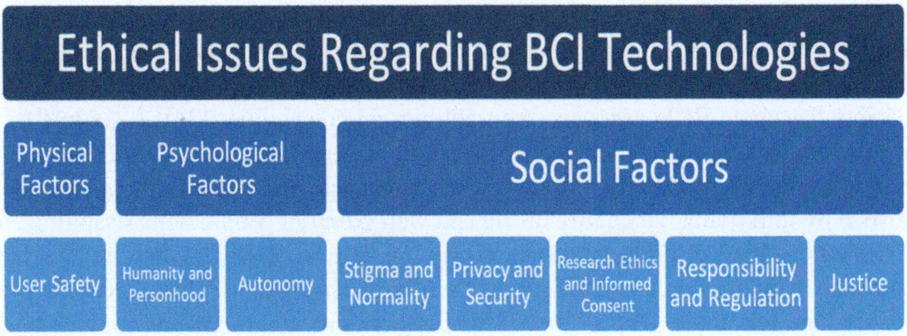

Fig. 5 BCI technologies ethical issues [33]

information about an individual's thoughts, emotions, and behaviors thereby potentially violating privacy by exposing sensitive brain data. This information could be collected without the individual's knowledge or consent, leading to serious privacy concerns and exposure to third parties such as researchers, companies, hackers, or governments. Brain data can reveal personality traits, preferences, emotions, intentions, memories, or thoughts that the user may not want to share or disclose [30]. Another ethical issue of BCIs is consent. Consent refers to the voluntary agreement of individuals to participate in research or treatment involving BCIs. The use of BCIs raises questions about informed consent. Participants must be fully informed about the potential risks and benefits of BCIs before they can make an informed decision about whether to use them or not [31]. However, there may be individuals who are unable to provide informed consent due to cognitive or communicative disabilities such as awareness, understanding, reasoning, or decision-making. Thereby affecting the user's decisions and also influencing the user's preferences, values, beliefs, or goals by providing feedback or stimulation that may alter their brain states or behavior [17]. A third ethical issue of BCIs is autonomy. Autonomy refers to the ability of individuals to act according to their own will and interests. BCIs can enhance autonomy by restoring or augmenting lost or impaired functions for people with disabilities or diseases [28]. However, ethical concern is the potential for BCIs to interfere with an individual's autonomy. If BCIs are used to control an individual's behavior or decisions, it could result in a loss of self-autonomy and freedom which could raise a concern if individuals are forced to use BCIs or have them implanted against their will thereby overriding self-autonomy by interfering with the user's sense of agency, control, or ownership over their actions and outcomes [32].

A fourth ethical issue of BCIs is responsibility. Responsibility refers to the moral obligation of individuals to account for their actions and their consequences. BCIs can complicate responsibility by creating ambiguity about who is accountable for the user's actions and outcomes: The user, the BCIs system, the BCIs provider, or a combination of them [29]. Responsibility also depends on the reliability and accuracy of the BCI system and its potential for errors or malfunctions. A fifth ethical issue of BCIs is identity. Identity refers to the sense of self that individuals have based on their personal characteristics and social roles. BCIs can affect identity by changing how the user perceives themselves and how others perceive them concerning their BCI system. Identity can also be influenced by the expectations and norms that society has about BCIs users and their abilities and disabilities [34]. A sixth ethical issue of BCIs is social justice. Social justice refers to the fair distribution of benefits and burdens among individuals and groups in society. BCIs can promote social justice by improving the quality of life and well-being of people with disabilities or diseases who face discrimination or exclusion. However, BCIs can also create social injustice by increasing inequality or disparity between those who have access to BCI technology and those who do not [15]. Furthermore, the use of BCIs for human enhancement raises ethical questions about the boundaries of human nature and the potential for inequality. BCIs could be used to enhance cognitive abilities, such as memory or learning, or physical abilities, such as strength or endurance. However, this could lead to a situation where those who can afford BCIs have a significant advantage over those who cannot [34]. In addition, the use of BCIs raises serious cybersecurity concerns as hackers may be able to gain access to an individual's brain signals or implantable devices, potentially causing harm or accessing personal information. This could be especially concerning if BCIs are used for sensitive applications, such as military or government operations [35]. The ethics of BCIs is an important field that involves philosophy, law, sociology, and medicine. It aims to identify and address the moral values and principles that guide the development and use of BCIs technology. It also aims to balance the risks and benefits of BCIs technology for individuals and society.

5 Application of BCIs

BCIs has been integrated into different fields and aspects of fields and numerous applications for BCIs have positively impacted the various fields. Some of which are:

i. ***Medicine***: In the medical field, BCIs are frequently used to treat and diagnose diseases or injuries that affect the nervous system, such as Amyotrophic Lateral Sclerosis (ALS), stroke, brain injury, and spinal cord injuries [36]. People with these conditions can use BCIs to communicate, control prosthetics, and enhance their quality of life. BCIs touches on every aspect of health-related phases in the medical field, including;

   - Smoking, alcoholism, and motion sickness prevention [37]
   - Brain structure (such as a brain tumor) [38]
   - Seizures (such as epilepsy) [39]
   - Disorders of sleep [37]
   - Brain swelling [40]
   - Stroke rehabilitation and restoration [41]

ii. ***Gaming and Entertainment***: In applications for entertainment and gaming, BCIs are used to give players an immersive experience such that the player's brain activity can be tracked with BCIs, and this data can be used to control the game or device [42]. For example; BrainArea implements BCIs in entertainment and games [43].
iii. ***Educational Training***: BCIs are used to improve learning and skill development in education and training applications as it can be used to enhance learning outcomes thereby measuring and monitoring student engagement [44]. According to Wegemer [45], BCI also makes conducting virtual reality training more interactive. Furthermore, according to [46], BCIs impacts education in terms of self-regulation and skill acquisition in education through Neurofeedback and functional Magnetic Resonance Imaging (fMRI).
iv. ***Communication***: BCIs are used in the development of communication software to help people with disabilities communicate more effectively. BCIs can help people communicate more effectively by converting their thoughts into actions [47]. This can also be helpful in promoting healthy green-living lifestyle by implementation of Neuroerganomics and smart environments such as Brain-computer interface-based smart living environmental auto-adjustment control systems [48].
v. ***Military Application***: The military uses BCIs for a number of different tactical and technical things, like controlling unmanned aerial vehicles (UAVs), increasing situational awareness, and performing better in high-stress environments [49]. Furthermore, according to [50], BCIs could offer an important means to expand and improve human–machine teaming by incorporating artificial intelligence (AI) and semiautonomous systems into its operations. It has also shown to impact security and authentication such as cognitive biometrics, which makes use of bio-signals (brain signals).
vi. ***Business***: In companies, offices and workplace generally, BCIs are used to boost productivity and boost employee well-being by increasing efficiencies in terms of workflows, monitor employee stress levels, and pinpoint areas for improvement [51]. It also the fields of marketing and advertising in business, such as the use of neuromarketing TV advertisements that use EEG for political and commercial purposes [52].

6 Discussion

BCIs are gadgets that empower direct correspondence between the cerebrum and outside gadgets, like PCs, robots, prosthetics, or computer-generated reality. BCIs are used in a variety of fields, including work, entertainment, education, and medicine. BCIs are being investigated by leading neuroscience researchers who are competing to unlock the secrets of the mind and merge the human brain with machines [27]. The current state and prospects of BCIs are shown in the High-Tech Race to Merge Minds and Machines. Instead of the normal output pathways of the brain's peripheral nerves and muscles, users of BCIs are able to communicate with or control external devices using brain signals. According to Mak and Wolpaw [53], researchers are looking into novel applications of BCIs to assist people with disabilities, such as paralysis or limb loss. BCIs' potential for use in the treatment of mental health conditions like depression and anxiety is also the subject of research [6].

According to Gonfalonieri [27], BCIs also have the potential to change how we live, work, and interact with one another, as evidenced by their potential effects on humanity and society. According to Thomas [54], they may result in novel means of communication, novel approaches to education, and novel perspectives on the world around us. In areas like defense and space, BCIs have the potential to speed up and simplify interactions between humans and machines. Additionally, some researchers have proposed that BCIs-controlled robots might be able to assist people in potentially hazardous settings like coal mines [18]. However, there are also concerns regarding the technology's ethical implications, including privacy, security, and autonomy concerns [6]. Furthermore, Gonfalonieri [27] stated that experts say 15–30% of people are inherently incapable of producing brain signals strong enough to operate a BCIs. This situation has the potential to result in incorrect outcomes and severe consequences. The High-Tech Race to Merge Minds and Machines is a fascinating and rapidly developing field that has the potential to alter our way of life and our interactions with one another. However, it is essential to carefully consider the ethical implications of this technology and to make certain that it is developed and utilized in an ethical manner. Ideas, techniques, approaches not yet implemented from the study (Table 1).

These are few of the ides from the study. If these ideas are followed-up and accepted, they would lead to BCIs evolution and innovation. Hence, making it a useful and sought-after technology for humanity.

Table 1 Prospective BCIs ideas and techniques

No	Field	Ideas	References
1.	Health	Disorders of sleep, and brain swelling	Sharanreddy and Kulkami [55]
2.	Health	Stroke rehabilitation and restoration	Ang et al. [41]
3.	IoT	Brain-computer interface-based Smart living environmental auto-adjustment control systems	Lin et al. [48]
4.	Business	Neuromarketing and advertising	Vecchiato et al. [52]
5.	Education	Neurofeedback and functional magnetic resonance imaging (fMRI)	Birbaumer et al. [46]

7 Conclusion

In this study, this paper focuses and explored the fascinating world of BCIs. BCIs have shown how to use brain activity to control and communicate with machines. Which can be used for a variety of things, like improving abilities, or making new experiences. This study found out about the set of experiences, science, innovation, and morals of BCIs and have looked at BCIs' potential effects on humanity and society, as well as their current challenges and future prospects. Hence, the benefit of this study is the informative impact on increasing awareness about brain interfaces: The emotional innovative competition to combine brains and machines as well as its implications which buttresses that fact that BCIs are not just devices for only improving our capacities, but also for observing and enhancing our natural world. However, the use and development of BCIs bring up a number of ethical issues that require careful consideration and resolution. While BCIs have the potential to enhance our lives in numerous ways, it is essential to ensure that they are developed and utilized ethically, taking into account human enhancement, informed consent, privacy, autonomy, and cybersecurity. This study proffers no new discovery. However, it is a bibliographical research compilation carried out to analyze BCIs information and importance. The current status of BCIs in terms of the state of the art is the use of EEG modalities that are noninvasive with some level of control thanks to some control signals. Furthermore, numerous ongoing researches being carried out in BCIs also amount for its current status which is aimed at improving the reliability and efficiency of the technology. Hence leading to a great innovation which would ensure its positive impact on society and humanity. Especially for people with disabilities. For example, Rao et al. [56] combined transcranial magnetic stimulation (TMS) with EEG to record brain signals and deliver information to the brain so as to evaluate the effectiveness of the brain-brain interface. BCIs will see an increase in demand across a variety of sectors as technology continues to advance. Hence this study recommends that more people should get involved in exploring BCIs, its applications, and devices as BCIs are not just tailored to engineers and scientists but to everyone who cares about future of humanity.

References

1. C. Chen, "All you need to Know About the Brain-Computer Interface, the Technique Elon Musk Wants to Use to Merge Man and Machines," 2019, https://www.scmp.com/tech/start-ups/article/3019288/all-you-need-know-about-brain-computer-interface-technique-elon-musk, [Retrieved: March, 2023].
2. G. Malcolm, "Brain-Computer Interfaces, Biomedical Engineering - Trends, Man-Machine Systems, Research Subjects," Farrar, Straus and Giroux: USA, New York, 2015.
3. S. N. Abdulkader, A. Atia, and M. M. Mostafa-Sami. "Brain Computer Interfacing: Applications and Challenges," Egyptian Informatics Journal, vol. 16, issue 2, pp. 213–230, 2015. https://doi.org/10.1016/j.eij.2015.06.002.

4. Z. M. Hanafiah, M. N. Taib, and N. Hamid, "EEG Pattern of Smokers for Theta, Alpha and Beta Band Frequencies," in Proc. the IEEE Student Conference on Research and Development, IEEE, 2010, pp. 320–23.
5. D. J. McFarland and J. R. Wolpaw, "Brain-Computer Interface Operation of Robotic and Prosthetic Devices," Computer, vol. 41, issue 10, pp. 52–56, 2008, https://doi.org/10.1109/MC.2008.409.
6. M. Norris, and A. Youngblood, "Brain-Computer Interfaces Are Coming. Will We Be Ready?," 2020, https://www.rand.org/blog/articles/2020/08/brain-computer-interfaces-are-coming-will-we-be-ready.html, [Retrieved: March, 2023].
7. R. Hajare and S. Kadam, "Comparative Study Analysis of Practical EEG Sensors in Medical Diagnoses," Global Transitions Proceedings, vol. 2, issue 2, pp. 467–475, 2021, https://doi.org/10.1016/j.gltp.2021.08.009.
8. V. Straebel, and W. Thoben, "Alvin Lucier's Music for Solo Performer: Experimental Music Beyond Sonification," Organised Sound. Vol. 19. Issue 1, pp. 17–29, 2014, https://doi.org/10.1017/S135577181300037X.
9. H. Si-Mohammed, G. Casiez, F. Argelaguet, N. Roussel and A. Lécuyer. "Defining Brain-Computer Interfaces: A Human-Computer Interaction Perspective," in Proc. the 8th Graz Brain-Computer Interface Conference, 2019, https://doi.org/10.3217/978-3-85125-682-6-58.
10. G. McMillan, M. Middendorf, G. Calhoun, G. Calhoun, and K. S. Jones, "Brain-Computer Interfaces Based on the Steady-State Visual-Evoked Response," IEEE Transactions on Rehabilitation Engineering, vol 8, issue 2, pp. 211–214, 2000, https://doi.org/10.1109/86.847819.
11. G. Pfurtscheller, C. Guger, G. Müller, G. Krausz, and C. Neuper, "Brain Oscillations Control Hand Orthosis in a Tetraplegic," Neuroscience Letters, vol. 292, issue 3, pp. 211–214, 2000. https://doi.org/10.1016/s0304-3940(00)01471-3.
12. J. R. Millan, F. Renkens, J. Mourino and W. Gerstner, "Noninvasive Brain-Actuated Control of a Mobile Robot by Human EEG" IEEE Transactions on Biomedical Engineering, vol. 51, issue 6, pp. 1026-1033, June 2004, https://doi.org/10.1109/TBME.2004.827086.
13. T. O. Zander, C. Kothe, S. Welke, M. Roetting, "Utilizing Secondary Input from Passive Brain-Computer Interfaces for Enhancing Human-Machine Interaction," in Proc. the Foundations of Augmented Cognition. Neuroergonomics and Operational Neuroscience: 5th International Conference, FAC 2009 Held as Part of HCI International 2009 San Diego, CA, USA, 2009, pp. 759–771. https://doi.org/10.1007/978-3-642-02812-0_86.
14. A. Kübler, "The History of BCI: From a Vision for the Future to Real Support for Personhood in People with Locked-in Syndrome," Neuroethics, vol. 13, pp.163–180, 2020, https://doi.org/10.1007/s12152-019-09409-4.
15. S. Saha, K. A. Mamun, K. Ahmed, R. Mostafa, G. R Naik, S. Darvishi, A. H. Khandok, and M. Baumert, "Progress in Brain Computer Interface: Challenges and Opportunities," Frontiers in Systems Neuroscience, vol. 15, article no. 578875, 2021, https://doi.org/10.3389/fnsys.2021.578875.
16. Brown University, "Toward Next-Generation Brain-Computer Interface Systems," 2021, https://www.sciencedaily.com/releases/2021/08/210812135910.htm, [Retrieved: March, 2023].
17. S. Khan, and T. Aziz, "Transcending the Brain: Is There a Cost to Hacking the Nervous System?," Brain Communications, vol. 1, issue 1, 2019, fcz015, https://doi.org/10.1093/braincomms/fcz015.
18. K. Howard, "Science & Tech Spotlight: Brain-Computer Interfaces," 2022, https://www.gao.gov/products/gao-22-106118, [Retrieved: March, 2023].
19. Emotiv, "Brain Data Measuring Hardware and Software Solutions," 2023, https://www.emotiv.com, [Retrieved: March, 2023].
20. Egi, "Electrical Geodesics," 2023, https://www.egi.com, [Retrieved: March, 2023].
21. Neurosky, "EEG & ECG Biosensors," 2023, https://neurosky.com, [Retrieved: March, 2023].
22. Cgxsystems, "Dry EEG Headsets," 2023, https://www.cgxsystems.com/, [Retrieved: March, 2023].

23. Biosemi, "EEG ECG EMG BSPM NEURO Amplifier Electrodes," 2023, https://www.biosemi.com, [Retrieved: March, 2023].
24. Openbci, "Galea: Biosensing + Spatial Computing," 2023, https://openbci.com, [Retrieved: March, 2023].
25. Wearablesensing, "Wearable Sensing Dry EEG," 2023, https://wearablesensing.com, [Retrieved: March, 2023].
26. B. Daniel, "The Emergence of Brain-Computer Interfaces: Unlocking the Potential of Direct Neural Controls," 2023, https://www.linkedin.com/pulse/emergence-brain-computer-interfaces-unlocking-potential-daniel-bron/, [Retrieved: March, 2023].
27. A. Gonfalonieri, "What Brain-Computer Interfaces Could Mean for the Future of Work," 2020, https://hbr.org/2020/10/what-brain-computer-interfaces-could-mean-for-the-future-of-work, [Retrieved: March, 2023].
28. L. Drew, "The Ethics of Brain-Computer Interfaces," Nature, vol. 571, pp. S19–S21, 2019. https://doi.org/10.1038/d41586-019-02214-2.
29. S. Burwell, M. Sample, and E. Racine, "Ethical Aspects of Brain-Computer Interfaces: A Scoping Review," BMC Medical Ethics, vol. 18, article no. 60, 2017, https://doi.org/10.1186/s12910-017-0220-y.
30. M. Ienca, and P. Haselager, "Hacking the Brain: Brain-Computer Interfacing Technology and the Ethics of Neurosecurity," Ethics and Information Technology, vol. 18, pp. 117–129, 2016.
31. G. Schalk and E. C. Leuthardt, "Brain-Computer Interfaces Using Electrocorticographic Signals," in IEEE Reviews in Biomedical Engineering, vol. 4, pp. 140–154, 2011, https://doi.org/10.1109/RBME.2011.2172408.
32. F. Gilbert, and T. O'Brien, "Neuroenhancement and Neuroprosthetics should we be Permitted to Enhance our Brains," Current Opinion in Psychiatry, vol. 24, pp. 562–567, 2011.
33. A. Coin, M. Mulder, and V. Dubljević, "Ethical Aspects of BCI Technology: What is the State of the Art?," Philosophies, vol. 5, issue 4, p. 31, 2020. https://doi.org/10.3390/philosophies5040031.
34. N. Bostrom, and A. Sandberg, "Cognitive Enhancement: Methods, Ethics, Regulatory Challenges," Science and Engineering Ethics, vol. 15, pp. 311–341, 2009, https://doi.org/10.1007/s11948-009-9142-5.
35. A. Koene, and E. Hildt, "An Ethical Perspective on Cyberpsychology and Brain–Computer Interfaces," Journal of Cognitive Enhancement, vol. 2, pp. 1–12, 2018.
36. J. J. Shih, D. J. Krusienski, J. R. Wolpaw, "Brain-Computer Interfaces in Medicine," in Proc. the Mayo Clinic. Elsevier, 2012, Vol. 87, Issue 3, pp. 268–279. https://doi.org/10.1016/j.mayocp.2011.12.008
37. M. Poulos, T. Felekis, A. Evangelou, "Is it Possible to Extract a Fingerprint for Early Breast Cancer via EEG Analysis?," Medical Hypotheses, vol. 78, issue 6, pp. 711–716, 2012, https://doi.org/10.1016/j.mehy.2012.02.016.
38. Z. Eksi, A. Akgül and M. R. Bozkurt, "The Classification of EEG Signals Recorded in Drunk and Non-Drunk People," International Journal of Computer Applications, vol. 68, issue 10, 2013.
39. M. Sharanreddy, and P. Kulkarni, "Automated EEG Signal Analysis for Identification of Epilepsy Seizures and Brain Tumour," Journal of Medical Engineering & Technology, vol. 37, pp. 511–519, 2013.
40. S-F. Liang, F-Z. Shaw, C-P. Young D-WChang, Y-C Liao, "A Closed-Loop Brain Computer Interface for Real-Time Seizure Detection and Control," in Proc. the Annual International Conference of the IEEE Engineering in Medicine and Biology Society. IEEE, 2010, pp. 4950–53. https://doi.org/10.1109/IEMBS.2010.5627243.
41. K. K. Ang, C. Guan, K. S. Geok, K. S. Chua, B. T. Ang, C. Kuah, C. Wang, K. S. Phua, Z. Y. Chin, and H. Zhang, "Clinical Study of Neurorehabilitation in Stroke Using EEG-Based Motor Imagery Brain Computer-Interface with Robotic Feedback," in Proc. the Annual International Conference of the IEEE Engineering in Medicine and Biology Society. IEEE, 2010, pp. 5549–52. https://doi.org/10.1109/IEMBS.2010.5626782.

42. D. Plass-Oude Bos, B. Reuderink, B. van de Laar, H. Gürkök, C. Mühl, M. Poel, A. Nijholt, et al. (2010). Brain-computer interfacing and games. In D. S. Tan & A. Nijholt (Eds.), Brain-Computer Interfaces. Applying our Minds to Human-Computer Interaction (pp. 149-178). London: Springer London. https://doi.org/10.1007/978-1-84996-272-8_10
43. A. S. Royer, A. J. Doud, M. L. Rose and B. He, "EEG Control of a Virtual Helicopter in 3-Dimensional Space Using Intelligent Control Strategies," IEEE Transactions on Neural Systems and Rehabilitation Engineering, vol. 18, no. 6, pp. 581–589, Dec. 2010, https://doi.org/10.1109/TNSRE.2010.2077654.
44. M. Spüler, T. Krumpe, C. Walter, C. Scharinger, W. Rosenstiel and P. Gerjets, "Brain-Computer Interfaces for Educational Applications," In: Buder, J., Hesse, F. (eds) Informational Environments. Springer, Cham, 2017, https://doi.org/10.1007/978-3-319-64274-1_8.
45. C. Wegemer, "Brain-computer interfaces and education: the state of technology and imperatives for the future," International Journal of Learning Technology, Vol. 14. Issue 2, pp. 141–161, 2019, https://doi.org/10.1504/IJLT.2019.101848
46. N. Birbaumer, S. Ruiz, and R. Sitaram, "Learned Regulation of Brain Metabolism," Trends in Cognitive Sciences, vol. 17, issue 6, pp. 295–302, 2013.
47. S. Soman, and B. K. Murthy, "Using Brain Computer Interface for Synthesized Speech Communication for the Physically Disabled," Procedia Computer Science, vol 46, pp. 292–298, 2015, https://doi.org/10.1016/j.procs.2015.02.023.
48. C. Lin, B. Lin, F. Lin, and C. Chang, "Brain Computer Interface-Based Smart Living Environmental Auto-Adjustment Control System in UPnP Home Networking." *IEEE Systems Journal, vol. 8,* issue 2, pp. 363–370. 2012, https://doi.org/10.1109/JSYST.2012.2192756.
49. A. Czech, "Brain-Computer Interface Use to Control Military Weapons and Tools," In: Paszkiel, S. (eds) Control, Computer Engineering and Neuroscience. ICBCI 2021. Advances in Intelligent Systems and Computing, vol. 1362. Springer, Cham. 2021, https://doi.org/10.1007/978-3-030-72254-8_20.
50. R. T. Marler, and A. L. Binnendijk, and E. M. Bartels, "Brain-Computer Interfaces: U.S. Tactical Military Applications and Implications," Technical Report, 2020, https://www.researchgate.net/publication/342184562_Brain-Computer_Interfaces_US_Tactical_Military_Applications_and_Implications/citations.
51. A. Bonci, S. Fiori, H. Higashi, T. Tanaka, and F. Verdini, "An Introductory Tutorial on Brain–Computer Interfaces and Their Applications," Electronics, vol. 10, issue 5, 560, 2021, https://doi.org/10.3390/electronics10050560.
52. G. Vecchiato, L. Astolfi, F. De Vico Fallani, S. Salinari, F. Cincotti, F. Aloise, D. Mattia, M. G. Marciani, L. Bianchi, R. Soranzo, and F. Babiloni, "The Study of Brain Activity During the Observation of Commercial Advertising by Using High Resolution EEG Techniques," in Proc. the Annual International Conference of the IEEE Engineering in Medicine and Biology Society. IEEE, 2009, pp. 57–60. https://doi.org/10.1109/IEMBS.2009.5335045.
53. J. N. Mak and J. R. Wolpaw, "Clinical Applications of Brain-Computer Interfaces: Current State and Future Prospects," IEEE Reviews in Biomedical Engineering, vol. 2, pp. 187–199, 2009, https://doi.org/10.1109/RBME.2009.2035356.
54. D. L. Thomas. "Brain-Computer Interface: Huge Potential Benefits and Formidable Challenges," 2019, https://www.news-medical.net/news/20190911/Brain-computer-interface-huge-potential-benefits-and-formidable-challenges.aspx, [Retrieved: February, 2023].
55. M. Sharanreddy, and P. Kulkarni, "Detection of Primary Brain Tumor Present in EEG Signal Using Wavelet Transform and Neural Network," International Journal of Biological & Medical Research, vol. 4, issue 1, 2013.
56. N. R. P. Rao, A. Stocco, M. Bryan, D. Sarma, T. M. Youngquist, J. Wu, and C. S. Prat, "A Direct Brain-To-Brain Interface in Humans," PLOS ONE, vol. 9, issue 11, e111332, 2014. https://doi.org/10.1371/journal.pone.0111332.
57. Brainproducts, "Solutions for Neurophysiological Research," 2023, https://www.brainproducts.com, [Retrieved: March, 2023].
58. X. Gu, Z. Cao, A. Jolfaei, P. Xu, D. Wu, T.-P. Jung, and C.-T. Lin, "EEG-Based Brain-Computer Interfaces (BCIs): A Survey of Recent Studies on Signal Sensing Technologies and

Computational Intelligence Approaches and Their Applications," IEEE/ACM Transactions on Computational Biology and Bioinformatics, vol. 18, pp. 1645–1666, 2021.

Using Artificial Neural Networks to Predict Critical Displacement and Stress Values in the Proximal Femur for Distinct Geometries and Load Cases

Ana Pais, Jorge Lino Alves, and Jorge Belinha

Abstract In computational mechanics, the finite element method (FEM) is a very common discretization numerical technique. The complexity of numerical applications, however, is rising today. As a result, classic solution methods typically require more processing power and exhibit higher computational costs. To lower the computing cost associated with the numerical analysis, machine learning approaches can be coupled with the FEM and used as surrogate solvers or as a prediction tool. This alternative was examined in order to demonstrate the possibilities of fusing artificial NN with FEM for a biomechanical application. The proximal femur was used as a numerical example. Thus, distinct geometries were generated and to each discretized model different load cases were applied. Then, all the discrete models were analyzed with the FEM, and the initial conditions (geometry and load cases) and the obtained results (displacements and stresses) were organized as input and output data, respectively. The ANN was trained and then its accuracy was verified. It was observed that artificial NN can accurately forecast displacements and stresses while also saving a significant amount of computing time.

Keywords Artificial neural networks · Machine learning · Finite element method · Biomechanics · Elasticity

A. Pais
Institute of Science and Innovation in Mechanical and Industrial Engineering (INEGI), Porto, Portugal

J. L. Alves
Department of Mechanical Engineering, Faculty of Engineering, University of Porto (FEUP), Porto, Portugal
e-mail: falves@fe.up.pt

J. Belinha (✉)
Department of Mechanical Engineering, School of Engineering, Polytechnic University of Porto (ISEP), Porto, Portugal
e-mail: job@isep.ipp.pt

K. Daimi et al. (eds.), *Cutting Edge Applications of Computational Intelligence Tools and Techniques*, Studies in Computational Intelligence 1118,
https://doi.org/10.1007/978-3-031-44127-1_2

1 Introduction

The field of computer science known as artificial intelligence (AI), which aims to create software and robots with intelligence similar to that of a human, is proving to be an effective replacement for traditional modeling discretization methods. Artificial intelligence (AI), utilizing artificial neural networks (NN), the primary deep learning (DL) technique, has already been put to the test and used in important computational mechanics domains [1]. According to their architectural design, artificial neural networks can be categorized as either "feed-forward neural networks" or "mutually connected neural networks". In machine learning (ML), training data is used to create models that can then be used to forecast trends. In addition, when compared to traditional computational approaches, artificial NN are ideal to provide a solution for multi-variable problems in a fraction of the time since they are capable of nonlinear mapping.

Recently, there has been growing interest in the application of machine learning or deep learning techniques in the field of computational mechanics [1], such as modeling of composites [2]; homogenization of heterogeneous materials [3] and in non-linear structural analysis [4]. In the field of biomechanics some examples are the prediction of the mechanical response and load in long bones [5, 6], body posture, [7–9], prediction of scaffold properties [10].

This work uses artificial NN to predict the normal stress at the middle of the femoral shaft and the displacement at the top of the femoral head. Using artificial NN to predict the structural response of regular domains (such as rectangular, circular plates) submitted to uniform loads is a straightforward application, since such solid mechanics problems possess an analytical solution easily recognized by shallow artificial NN. However, for more demanding problems, such as irregular bio-structures under complex load cases, it is necessary to explore and investigate the performance and efficiency of such machine learning frameworks.

2 Materials and Methods

2.1 *Neural Networks*

Any hidden or output node applies a non-linear transformation to the weighted sum of the values of the nodes in the previous layer (input layer or hidden layer) passed through an activation function g, and is shown in Eq. (1)

$$z = g\left(b + \mathbf{x}^T \cdot \mathbf{w}\right) = g\left(b + \sum_{i=1}^{m} x_i \cdot w_i\right) \tag{1}$$

where x_i is the value from the node i in the previous layer, w_i is the weight corresponding to that node and b is the bias value. The weights and bias are obtained during the neural network training. The activation function used in this work is the hyperbolic tangent:

$$f(\Sigma) = \frac{e^{+\Sigma} - e^{-\Sigma}}{e^{+\Sigma} + e^{-\Sigma}} \quad (2)$$

where Σ is, according to Eq. (1), the weighted sum of the nodes from the previous layer [11].

The final network is therefore able to perform some non-linear transformation and create decision boundaries capable of accurately classifying complex sets of data. To evaluate the accuracy of the network predictions, the mean squared error (MSE) calculates the difference between the target vector and the output vector according to Eq. (3),

$$MSE = \frac{1}{m}\sum_{i=1}^{m}(t_i - y_i)^2 \quad (3)$$

where t_i is the element of the target vector and y_i is the prediction made by the neural network for the same inputs that originate the target t_i. The differences are squared in order to make sure that negative differences are not subtracted form positive ones [11].

During training data undergoes forward and backwards propagation. During back-propagation the cost function is minimized by adjustment of the weights and bias. The weights of the neurons in a neural network are closely tied to the cost/loss function. By using gradient descent, the weights can be modified in a way that moves them towards the steepest descent to reach the minimum of the cost function, thereby leading to a set of network weights that will result in the global minimum of the cost function. The most effective method for achieving this in feed-forward neural networks, such as multi-layer perceptrons, is referred to as the backpropagation of error. The precise weight-adjustment formulas for each neuron can be determined by computing the gradient of its transformation function. Since each neuron contributes differently to the overall error, some neurons may cause higher errors than others, and therefore the weight adjustment of those neurons should be greater. If the activation function used is the sigmoid function, the contribution of a neuron to the overall error can be calculated using the following method: For a neuron in the output layer

$$\delta_i^{(1)} = y_i(1 - y_i)(t_i - y_i) \quad (4)$$

where $(t_i - y_i)$ is the difference between the output i and the correct target which then is multiplied by $y_i\ (1 - y_i)$, a term brought to a minimum when $y_i = 0$ or $y_i = 1$ and to a maximum when $y_i = 0.5$. Likewise, if this term is minimized the neuron has a high error contribution and if the term is maximized, the neuron is neutral. If the neuron is in the hidden layer the contribution is the following:

$$\delta_i^{(2)} = h_j(1 - h_j) \sum_i \delta_i^{(1)} w_{ji} \tag{5}$$

The contribution of the weight of a neuron in the hidden layer is calculated by backpropagating the error contribution of the neuron's in the output layer through the term $\sum_i \delta_i^{(1)} w_{ji}$. Each $\delta_i^{(1)}$ is multiplied by the weight of the link connecting the output neuron i to the hidden neuron j. From the values calculated before of δ_i, it is possible to calculate the new weight. The weight update of a neuron in the output layer is:

$$w_{ji}^{(1)} := w_{ji}^{(1)} + \mu \delta_i^{(1)} h_j \tag{6}$$

and the weight update for a neuron in the hidden layer is:

$$w_{kj}^{(2)} := w_{kj}^{(2)} + \mu \delta_j^{(1)} x_k \tag{7}$$

learning rate is μ, which theoretically should be a value between 0 and 1. Every time data passes through the network is called an epoch. Depending on whether the batch size matches the quantity of the training data, an epoch may include one or more iterations.

Various algorithms exist for adjusting the weights in artificial neural networks, and the selection of these algorithms should be based on the specific problem being analyzed. The type of problem, whether it is a regression or classification problem, determines the appropriate algorithm. The primary distinction between the two tasks is that a classification problem involves predicting a label, while a regression problem involves predicting a quantity.

2.2 *Problem Summary*

Differences in the anatomy of this bone occur naturally in the population and will lead to very different structural responses. Taking for example the angle of inclination, shown in Fig. 1, the conditions of *coxa vara* and *coxa valga* both alter hip biomechanics.

The aim of this problem is to predict stresses and displacements at the proximal femur. The variable meaning is shown in Fig. 2. Therefore, P_1 is any random point located in the greater trochanter and P_2 is any random point approximately in the center of the femur head. In summation, in this problem, the objective is to predict the normal stresses σ_{xx} and σ_{yy} at the proximal diaphysis, and displacements d_x and d_y in the femoral head. The variables h, $P_1P_2(l)$ quantify size differences in the model, height, and width respectively, and θ accounts for the model distortions.

The magnitude of F_2 was defined as being approximately $0.3 \times F_1$, and F_1 is enough to lead the material to its elastic limit. The model used to gather the training and test data is discretized into 3300 nodes and 3150 quadrilateral elements. The

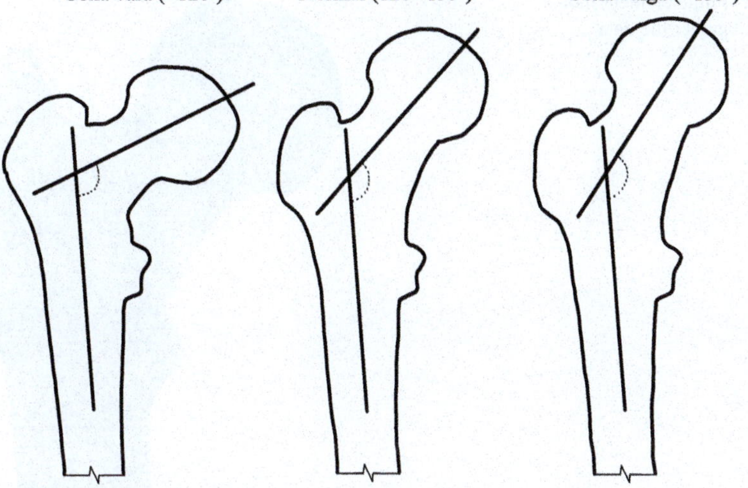

Fig. 1 Differences in the anatomy of the proximal femur

Fig. 2 Variable summary of the studied problem

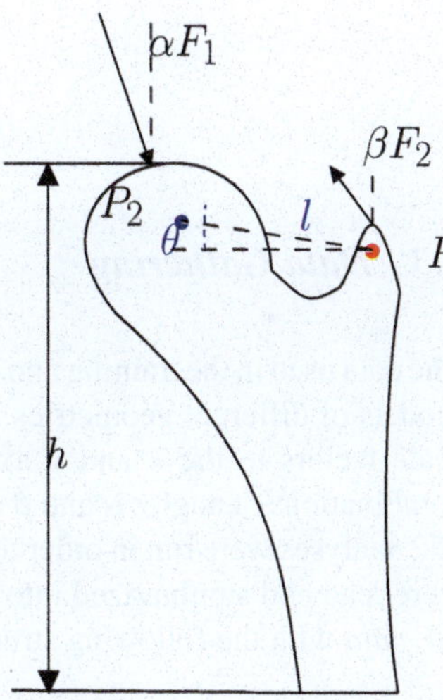

elastic modulus considered for the model is 33 GPa, meaning that the whole model is considered to be cortical bone (for simplicity's sake). The finite element model used to gather the data is shown in Fig. 3.

Fig. 3 Finite element model used to gather the data

2.3 *Data Gathering*

The data used in the training and testing stages was obtained through FEM analysis on models of different geometries. Each geometry was obtained by applying different scale factors to the x and y axes. The different load cases consisted of different combinations of angles α and β at which the loads are applied. A total of $n_{instances} = 152$ analyses were run in order to obtain the necessary data. First, the simulation files were read and summarized into a table containing all the inputs and outputs named *all_data* with the following structure:

h	$P_1P_2(l)$	θ	α	β	d_x	d_y	σ_{xx}	σ_{yy}
...	...	...	...	...	...	...	...	...

so the inputs correspond to the first 5 columns of the table and the targets correspond to the last four columns of the table.

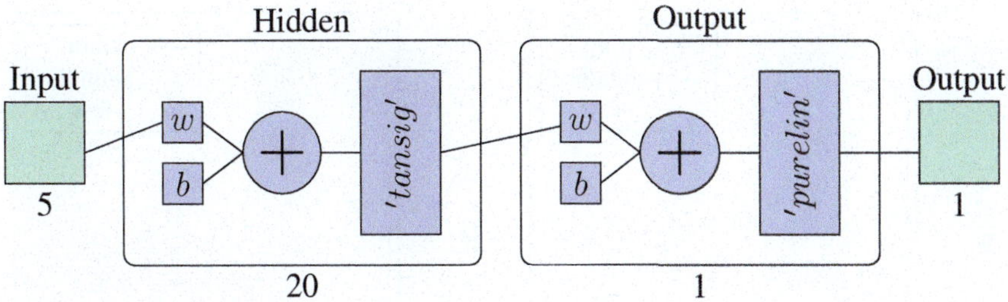

Fig. 4 Network architecture used to obtain the displacements d_x and d_y and the normal stresses σ_{xx} and σ_{yy}

2.4 *Neural Network Architecture*

For this problem, the network architecture consists of one hidden network with 20 nodes and the output layer with one node for the output. Thus, a network for each output variable was trained. Figure 4 shows the network for all variables. The activation function used in the hidden layer was the symmetric sigmoid (*'tansig'*) and the output layer also consisted of a linear transformation (*'purelin'*), where the input equals the output. The training function defined by default in the MATLAB *feedforwardnet* function is used, which is the *trainlm* function which uses the Levenberg-Marquardt algorithm. This training algorithm converges faster than for example gradient descent. From the gathered data 106 samples are used in training, 23 samples are used in validation, and 23 samples are used in testing. The samples used in the training were chosen randomly from the training data. Additionally, 12 samples (that had never been used before to train the model) were used to test the accuracy of the network. These 12 samples all present the same distortion but vary between themselves the α and β parameters.

Finally, because the weights are initialized randomly and the network's final performance is dependent on the initial weights, the weights were initialized 1000 times except for the network which outputs the σ_{xx}, which was initialized 5000 times. Out of all trained networks, the selected network is the network that leads to the lowest percentage error within the 12 samples that are not used in training (Fig. 5).

3 Results and Discussion

The error results of the trained networks for each of the displacements and stresses are shown in Fig. 7. Further, in order to evaluate the network capacity to output values that correspond to the target values, regression plots are shown in Fig. 6. It can be seen that for all ranges of output values, the network is able to output values that are close to their target value, as also can be seen from the MSE plots in Fig. 7.

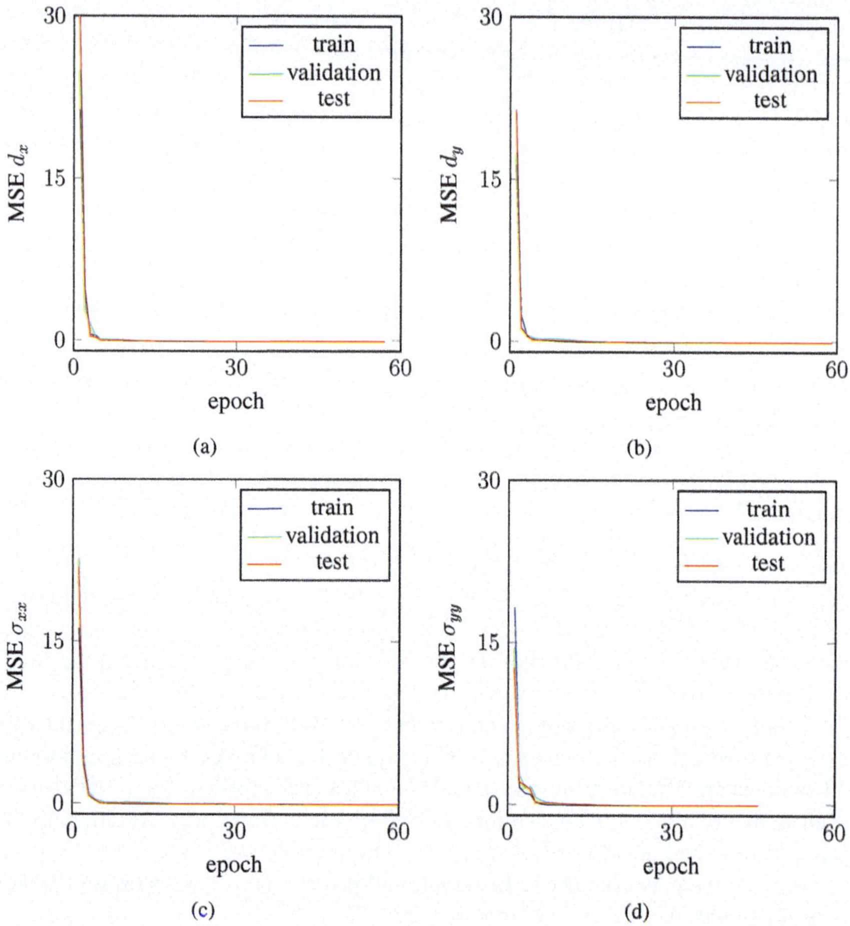

Fig. 5 MSE plots for all the tested variables on the network with the best performance

The relative error on the 12 samples (which were never used in the training of the network) is shown in Figure for the four variables.

It is possible to visualize that with enough tries, the weight initialization provided errors that are acceptably below 10%. More, from Table 1, which shows how the artificial neural network compares to the FEM performance, it is possible to visualize how a well-trained neural network outperforms the FEM in terms of computational time with satisfactory accuracy.

Due to the relative small size of the training data, the training of one network took, in some cases, a lower time than the prediction. The limited amount of data combined with the weight initialization algorithm's randomization required many attempts. This quantity was necessary to obtain a neural network with adequate performance. As a consequence of the necessary number of tries, the total training

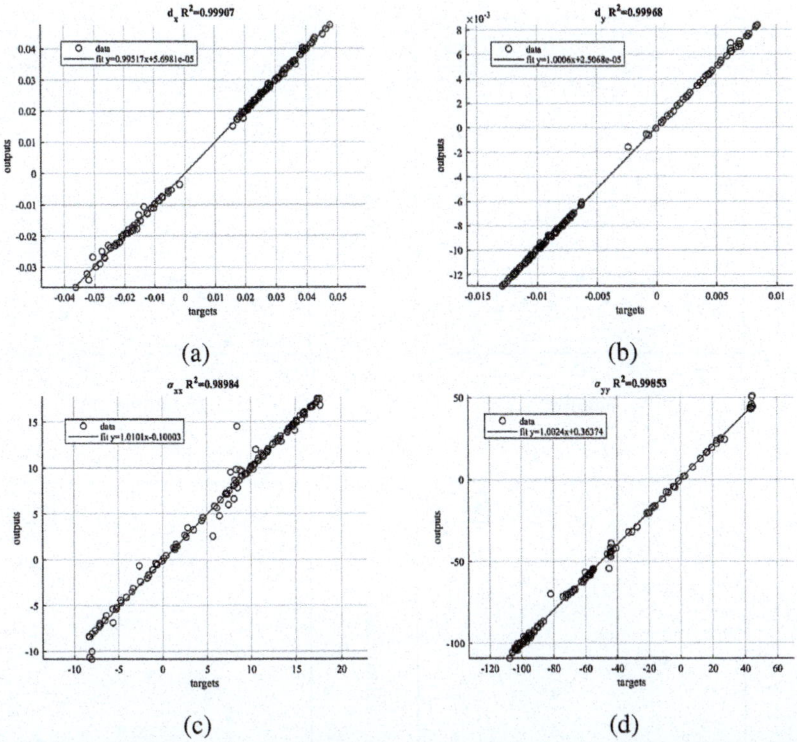

Fig. 6 Regression plots for the best network for each variable, indicating a good correlation between the target and the output

time increased drastically. It is visible in the work of Mouloodi et al. [5] the amount of data that is required in order to train the artificial neural network so it presents a low error [5].

However, the network could be trained to have an acceptable amount of inaccuracy. Despite the fact that the stresses and displacements are only needed at two sites in the model, FEM analysis is necessary to get those variables due to the complicated structure of the femur. Since there are so few parameters to calculate, the trained artificial neural network offers accurate enough data to skip the FEM phase.

Using the method proposed by Olden et al. [12], it is also possible to evaluate the significance of each variable for the computation of each output. The matrix of connections weights between the input and hidden layer W_I with dimensions $[N_{in} \times M]$ is multiplied by the matrix of connection weights between the hidden and output layer W_O with dimensions $[M \times N_o]$ to achieve the relative importance. N_in, N_out and M are the number of input, output, and hidden neurons respectively The importance result is given by

$$I = W_I \cdot W_O \tag{8}$$

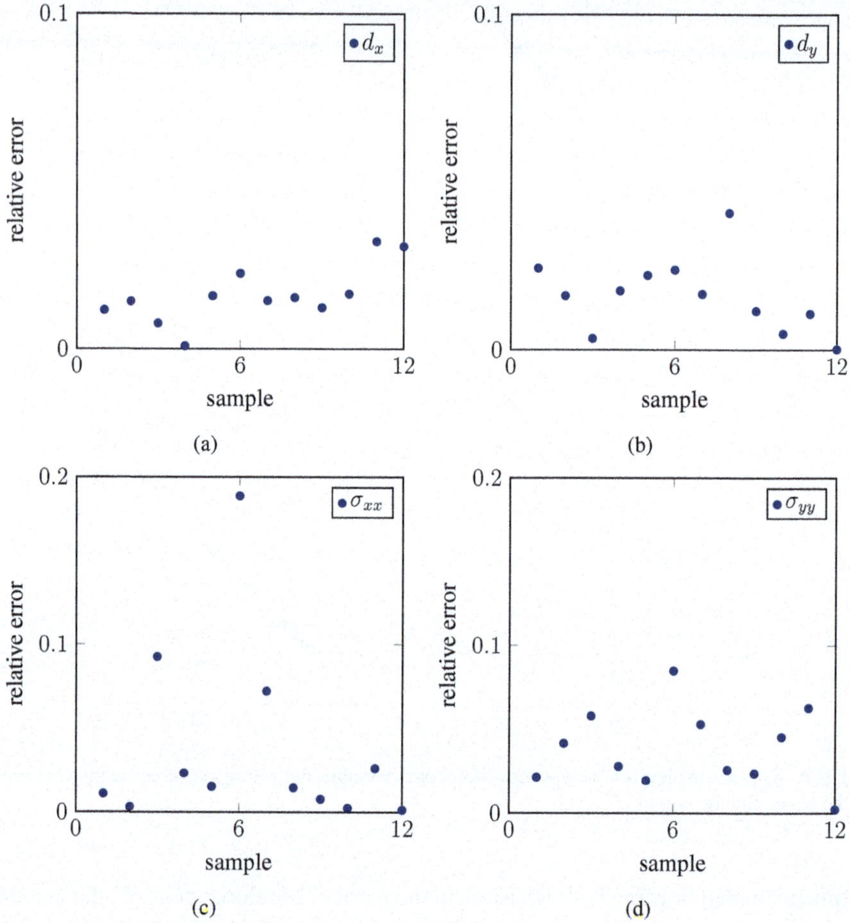

Fig. 7 Relative error of the 12 samples not seen by the neural network

Table 1 Performance comparison between the artificial neural networks and the reference results, obtained through the FEM

	Best net (s)	Total training (s)	Prediction time (s)	FEM time (s)	MAPE (%)	Necessary runs
u_x	1.199	1110.135	1.374	30	1.63	1000
u_y	1.807	1635.129	1.374	30	1.62	1000
σ_{xx}	1.627	9269.034	1.374	30	5.20	5000
σ_{yy}	2.578	2452.89	1.374	30	3.90	1000

Table 2 Importance of each variable using approach by Olden et al. [12]

	d_x		d_y		σ_{xx}		σ_{yy}	
	Importance	Rank	Importance	Rank	Importance	Rank	Importance	Rank
α	−1.433177231	2	−6.012677872	1	0.273148366	4	−8.961103865	1
β	−1.195840853	3	−3.15426349	2	−0.08160777	5	−3.403068005	2
θ	2.847490417	1	−0.072197728	5	2.162784543	1	−0.732870909	3
h	0.062400554	5	−1.415369025	3	−0.489007928	3	0.553022086	5
l	0.837122148	4	0.08092881	4	0.893937726	2	−0.657534966	4

where I is a $[N_{in} \times N_o]$ matrix and consists in the importance that each input variable has for the output result. If this value is negative, it implies a negative correlation between the input variable with the output. Table 2 ranks the variables. Olden's approach has certain advantages over other feature selection methods because it additionally discloses information about a variable's excitation or inhibition effect.

Each neural network relies differently on distinct features since each challenge is unique. Given the physical explanation for the problem, it is expected that the angle of the force applied to the femoral head α would consistently rank higher than the angle of the force applied to the greater trochanter β since the magnitude of the first load is higher than the magnitude of the second load. Rerunning the training process with each network omitting the variables with the least or very low relevance would be interesting. The dataset utilized for the investigation could also be the cause of other discrepancies.

4 Conclusion

In summary, this work demonstrates how artificial neural networks can significantly reduce computation time, especially for models with a lot of nodes, when calculation time rises sharply. To get a good approximation from the neural network, however, a lot of data must be collected. The training time must also be considered. Despite this issue, the training time for one network was lower than the FEM analysis. Furthermore, even though artificial intelligence is frequently a "black box" approach, the neural network was able to capture the physical meaning of the problem because the most crucial variables for each case agree with the expectations derived from the mechanics of the studied problem. This led to the conclusion that the neural network was capable of capturing the physical meaning of the problem despite this fact. The main contribution of the proposed technique, beside the reduction of the overall computational cost of the structural analysis, is its capability to provide sufficiently accurate solutions in a fraction of the time required by traditional hard coding procedures. Such advantage will allow, in the near future, its inclusion in nonlinear frameworks capable of predicting complex structural behaviours with higher efficiency.

Acknowledgements The author truly acknowledges the funding provided by LAETA, under project UIDB/50022/2020 and the doctoral grant SFRH/BD/151362/2021 financed by the Portuguese Foundation for Science and Technology (FCT), Ministério da Ciência, Tecnologia e Ensino Superior (MCTES), with funds from the State Budget (OE), European Social Fund (ESF) and PorNorte, under the MIT Portugal Program.

References

1. Atsuya Oishi and Genki Yagawa. Computational mechanics enhanced by deep learning. *Computer Methods in Applied Mechanics and Engineering*, 327:327–351, 2017. ISSN 0045-7825.
2. Fei Tao, Xin Liu, Haodong Du, and Wenbin Yu. Finite element coupled positive definite deep neural networks mechanics system for constitutive modeling of composites. *Computer Methods in Applied Mechanics and Engineering*, 391:114548, 2022. ISSN 00457825. https://doi.org/10.1016/j.cma.2021.114548.
3. Florent Pled, Christophe Desceliers, and Tianyu Zhang. A robust solution of a statistical inverse problem in multiscale computational mechanics using an artificial neural network. *Computer Methods in Applied Mechanics and Engineering*, 373:113540, 2021. ISSN 00457825. https://doi.org/10.1016/j.cma.2020.113540.
4. Marcus Stoffel, Franz Bamer, and Bernd Markert. Artificial neural networks and intelligent finite elements in non-linear structural mechanics. *Thin-Walled Structures*, 131(April):102–106, 2018. ISSN 02638231. https://doi.org/10.1016/j.tws.2018.06.035.
5. Saeed Mouloodi, Hadi Rahmanpanah, Soheil Gohari, Colin Burvill, and Helen M.S. Davies. Feedforward backpropagation artificial neural networks for predicting mechanical responses in complex nonlinear structures: A study on a long bone. *Journal of the Mechanical Behavior of Biomedical Materials*, 128(January):105079, 2022. ISSN 18780180. https://doi.org/10.1016/j.jmbbm.2022.105079.
6. Saeed Mouloodi, Hadi Rahmanpanah, Colin Burvill, and Helen MS Davies. Prediction of displacement in the equine third metacarpal bone using a neural network prediction algorithm. *Biocybernetics and Biomedical Engineering*, 40(2):849–863, 2020. ISSN 02085216. https://doi.org/10.1016/j.bbe.2019.09.001.
7. A. Gholipour and N. Arjmand. Artificial neural networks to predict 3D spinal posture in reaching and lifting activities; Applications in biomechanical models. *Journal of Biomechanics*, 49(13):2946–2952, 2016. ISSN 18732380. https://doi.org/10.1016/j.jbiomech.2016.07.008.
8. Mahdi Mohseni, Farzad Aghazadeh, and Navid Arjmand. Improved artificial neural networks for 3D body posture and lumbosacral moment predictions during manual material handling activities. *Journal of Biomechanics*, 131(December 2021):110921, 2022. ISSN 18732380. https://doi.org/10.1016/j.jbiomech.2021.110921.
9. F. Aghazadeh, N. Arjmand, and A. M. Nasrabadi. Coupled artificial neural networks to estimate 3D whole-body posture, lumbosacral moments, and spinal loads during load-handling activities. *Journal of Biomechanics*, 102:109332, 2020. ISSN 18732380. https://doi.org/10.1016/j.jbiomech.2019.109332.
10. B. S. Reddy, Kim Hong In, Bharat B. Panigrahi, Uma Maheswera Reddy Paturi, K. K. Cho, and N. S. Reddy. Modeling tensile strength and suture retention of polycaprolactone electrospun nanofibrous scaffolds by artificial neural networks. *Materials Today Communications*, 26(February):102115, 2021. ISSN 23524928. https://doi.org/10.1016/j.mtcomm.2021.102115.
11. Miroslav Kubat. *An Introduction to Machine Learning*. Springer, 2017. ISBN 9783319639130. https://doi.org/10.1007/978-3-319-63913-0.
12. Julian D. Olden, Michael K. Joy, and Russell G. Death. An accurate comparison of methods for quantifying variable importance in artificial neural networks using simulated data. *Ecological Modelling*, 178(3–4):389–397, 2004. ISSN 03043800. https://doi.org/10.1016/j.ecolmodel.2004.03.013.

An Integrated Model for Automated Identification and Learning of Conversational Gestures in Human–Robot Interaction

Aditi Singh and Arvind K. Bansal

Abstract Conversational gestures are an important part of communication and interaction to augment intention, emphasis, and emotion in speech, signal attention, express mental states, describe entities and actions in dialogs, and provide object localization. Cognitive psychologists have proposed many classifications of human gestures, based upon discourse analysis, head-motion, and hand-motion and shapes. Many machine and deep learning techniques have been used to detect limited head or hand-based gestures during multi-agent interactions using a combination of image and motion analysis. However, research on a general purpose integrated model to detect, train and automated learning and adaptation of conversational gestures by social robots is limited. Current models do not consider spatiotemporal synchronization and integration of dialog and motion analysis. This book chapter reviews major psycholinguistic subclasses of conversational gestures, and the application of machine learning and deep learning techniques for conversational gesture recognition. It describes a Synchronous Colored Petri Net (SCPN) model for automated training of social robots to detect and learn major subclasses of conversational gestures based upon synchronization of organ-motions and speech. The model automatically learns and adapts to conversational gestures by observation, mimicking and statistical analysis. The model integrates vision and machine learning techniques on real-time videos to extract spatiotemporal relationships between organ-movements and spoken phrases. Using synchronization reduces perceptual distortion in gesture detection and labeling. The approach also integrates contour analysis of hand-motion with conceptual dependency model for spoken phrases to identify iconic gestures. The model also applies decision trees for ambiguity resolution during gesture labeling.

A. Singh (✉) · A. K. Bansal
Department of Computer Science, Kent State University, Kent, OH, USA
e-mail: asingh37@kent.edu

A. K. Bansal
e-mail: akbansal@kent.edu

K. Daimi et al. (eds.), *Cutting Edge Applications of Computational Intelligence Tools and Techniques*, Studies in Computational Intelligence 1118,
https://doi.org/10.1007/978-3-031-44127-1_3

Keywords Application · Conversational gestures · Human–robot interaction · Deep learning · Machine learning · Petri net · Synchronization · Vision

1 Introduction

Social robots have grown in significance in recent years for much-needed human–robot interaction, particularly in industries like healthcare, education, disaster management, exploration, and entertainment [1]. Social robots would fulfill various roles, including providing help with either physical tasks or emotional support. By combining their capabilities into various setting, social robots have the potential to revolutionize human–machine interactions, enabling social robots to become valuable companions in our daily lives [1–5].

Speech, dialogs, facial expressions, emotions, and gestures are vital elements in everyday human interactions, and they serve multiple roles to enhance understanding human intentions and responsiveness. Interpreting and generating these modalities is important for realistic human–robot interactions. Social robots, including intelligent machines, need to be trained to recognize, learn automatically by observation, adapt automatically, and generate voice, gestures, and facial expressions to interact meaningfully with humans [1, 2, 6–9]. Ultimately, this could lead to a future in which humans and robots can communicate seamlessly, enhancing our ability to work and interact with machines in a wide range of contexts in meaningful ways [10].

Conversational gestures are a major class of gestures for meaningful interaction between two agents. It involves the synchronized integration of speech, dialogs, upper-body motions such as head-motion, hand-motion including palm and finger movements, and eye-motion [11]. The significance of conversational gestures in social settings cannot be overstated, as they hold a pivotal role in facilitating and enriching social interactions by communicating intentions and behavioral modes such as agreement, disagreement, emphasis, attentiveness, mental states, reference to entities and symbolic description of objects and actions [12, 13].

As human–robot interfaces become ubiquitous in modern society, the study of conversational gestures and their subclasses, such as deictic and iconic gestures, will contribute significantly to the interface for improved communication and comprehension. By accurately detecting and interpreting various subclasses of human gestures, gesture recognition holds the potential to bridge the gap between humans and machines, fostering a more natural and intuitive mode of communication for enhanced collaboration in various domains.

Despite the significance of conversational gestures in human–robot interaction, there are limited general-purpose techniques for their recognition and interpretation. Current approaches lack modeling of the spatiotemporal relationship between motions and speech in conversational gestures and lack complete subclassification due to previous focus on either textual discourse by cognitive psychologists [7, 14] or hand-shape based posture analysis by computational scientists [15]. Gesture classification based solely on discourse analysis of simple sentences lack gestures based upon mental, haptics, and sensory perception, eye-motion, and vision-based analysis of motion and synchronization [16–21]. Computational approaches solely based on vision analysis of postures and hand-shapes lack the information present in dialogs, motion and spatiotemporal synchronization of motion [22, 23]. A comprehensive conversational gesture recognition and comprehension requires an extended classification of conversational gestures that integrates analysis based upon vision analysis, speech analysis of dialogs and spatiotemporal synchronization of motions of various organs and speech [24–27].

This chapter describes gesture classification and provides an enhanced conversational gesture classification based on the addition of haptics, mental states, enumeration, and emphasis [24]. Conceptual dependency analysis acts as a bridge between discourse-based comprehension and vision-based analysis of motion [20, 21, 28]. The chapter also summarizes the research and limitations of machine learning and deep learning gesture recognition techniques [20, 29–55]. It proposes a Synchronous Colored Petri Net (SCPN) model as a general purpose integrated computational model to incorporate spatiotemporal synchronization for gesture modeling and recognition [20, 21, 24, 54, 55]. This chapter also describes a computational methodology for detecting and accurately labeling major subclasses of conversational gestures in real-time, which has been used for training and automated learning by social robots [54, 55]. The SCPN-based model uses frame-based video analysis and the notion of motion-vectors and silence vectors [24]. The model also exploits a machine learning technique (decision-trees) for the ambiguity resolution during gesture recognition [24]. The extended version of the SCPN model integrates hand-based motion analysis, conceptual dependency, and contour pattern analysis to detect and label complex hand-motion based deictic and iconic gestures [54, 55].

The roadmap of this chapter is as follows. Section 2 describes the background and definitions related to gestures, Petri net, synchronization, conceptual dependency, and deep learning. Section 3 describes conversational gesture subclassification by cognitive psychologists, their limitations and the extended classification that integrates subclasses of deictic and iconic gestures based on vision and synchronization based analysis. Section 4 summarizes machine learning and deep learning approaches for gesture recognition and automated learning of gestures by mimicking. Section 5 describes SCPN, and its extension to model and detect different subclasses of gestures, including composite motions to detect contours in iconic gestures. Section 6 describes the techniques used to recognize conversational head-gestures, deictic gestures, and iconic gestures. Section 7 describes the limitations and future work, and Section 8 concludes the book chapter.

2 Background and Fundamentals

2.1 Gesture

Gesture is a non-verbal language exchanged between interacting agents that combines postures, actions, or body movement to communicate intentions, reference, behavior modes, mental states, and attributes of entities and actions augmenting verbal communication to reduce ambiguity in cognitive comprehension [7, 22]. Many times, human–robot interactions require comprehension of integrated gestures involving speech (including silence) and motions, and their integration with a combination of facial expression, dialogs and modulated speech [11, 23]. Cognitive psychologists have classified conversational gestures (co-speech gestures), based upon dialog and discourse analysis, head-motion, and hand-shape and hand-motion analysis [7, 14]. The major subclasses are *deictic gestures, iconic* (*kinetographic* and *pictographic*), *metaphoric*, and *beat gestures* [7, 14, 56].

Deictic gestures refer to an entity or group of entities present in a scene, including places, objects, concepts, people, and even the self in the current spatiotemporal context. Figure 1 illustrates instances of deictic gestures. *Iconic gestures* (see Fig. 2) sketch attributes (such as shape, size or perimeter) of an entity or action, or physical proximity between two entities, including the interlocutor, in a spoken phrase using a combination of posture and synchronized motions of palms or fingers.

Metamorphic gestures pictorially model abstract concepts. They differ from iconic gestures because iconic gestures model tangible entities such as room, mountains, and valleys. Examples of metaphorical gestures is modeling abstract concepts such as speed, brightness using hand-motions. *Beat* gestures are rhythmic hand-motions used during knowledge retrieval. However, they do not associate with any entity or action. *Kinetographic* gestures are a subclass of iconic gestures which are

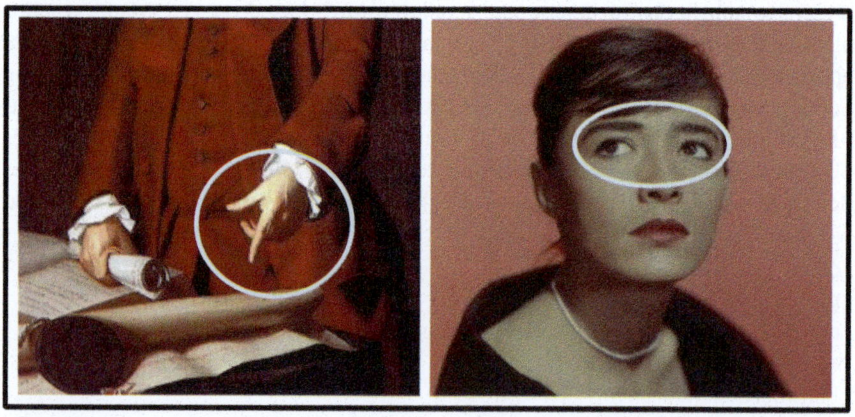

Fig. 1 Instances of deictic gestures adapted from wikimedia commons public domain [57, 58]

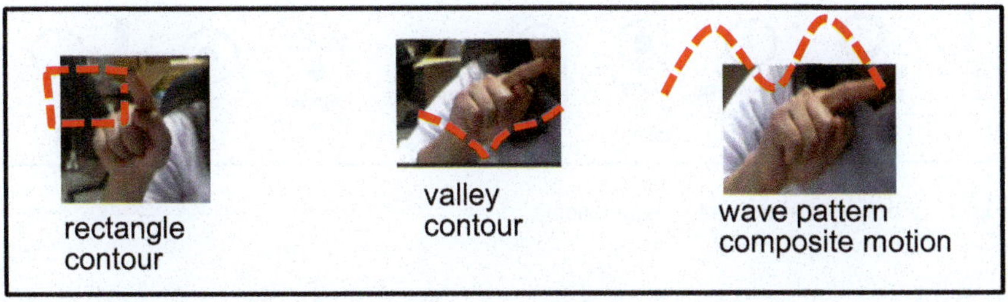

Fig. 2 An illustration of iconic gesture using finger motion

depicted using hand-motions [56]. *Pictographic* gestures are another subclass of iconic gestures which are depicted using hand-shapes [56]. *Emblem* gestures are culturally defined gestures such as hand waving, thump up. *Illustrators* illustrate words uttered in spoken phrases. *Adapters* model internal states such scratching head to show confusion or indecisiveness. *Regulators* control the flow of conversation such as nodding, raising hand. These definitions are based on classification made by cognitive psychologists, and many of them overlap.

2.2 Petri Net

Petri net is a directed weighted graph-based abstraction to represent concurrent events [59]. It uses two types of nodes: *place-nodes*, and *transition-nodes*. The flow of tokens models firing and execution of concurrent tasks. Directed edges connect *place-nodes* to *transition-nodes* and vice versa. Tokens move from one place-node to another place-node through transition-nodes during a transition. A transition occurs when the number of tokens in a place-node is greater than or equal to the weight of the corresponding outgoing edge. If a transition-node has two incoming edges, tokens combine to create a single output token. If a place-node is connected to multiple transition-nodes, tokens flow through only one transition-node exhibiting nondeterministic behavior. Pictorial representation of a Petri net traditionally uses circles to denote places, bars to denote transitions, and black dots to denote tokens.

Petri Nets are formally represented as a 5-tuple $\{P, Tr, A, I, O\}$. where P is a finite set of places; Tr is a finite set of transitions; A is a finite set of arcs connecting places to transitions or transitions to places; I is a set of input functions that map from a place to a transition; O is a set of output functions that map from a transition to a place. The distribution of tokens in place-nodes is described by a marking, which is a mapping from P to the set of natural numbers. Hierarchical modeling is possible by embedding a Petri net into one node. Figure 3 illustrates the components of a simple Petri net and the Petri net. The five sets in the Petri net are: $P = \{p_1, p_2\}$; $Tr = \{tr_1\}$; $A = \{(p_1, tr_1), (tr_1, p_2)\}$; $I = \{p_1 \rightarrow tr_1\}$; $O = \{tr_1 \rightarrow p_2\}$. Before firing, a token is in place p_1. After firing, the token moves to place p_2.

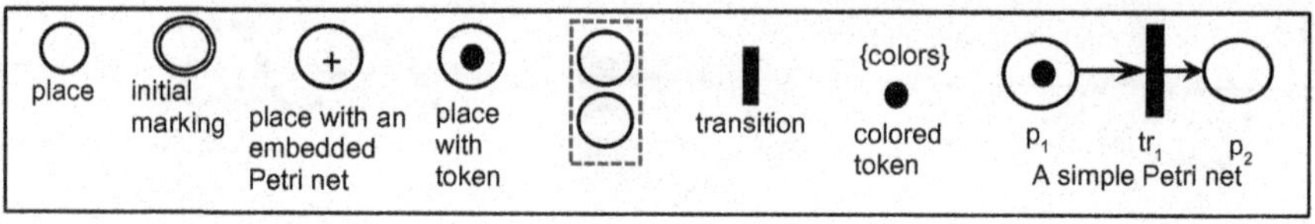

Fig. 3 A simple petri net and its components

Petri net can be further classified as *colored Petri net*, *timed Petri net* and *synchronized Petri net*. *Colored Petri nets* allow simplification of Petri nets by representing similar parts using distinct color-tokens, which can be integers or label sets [60]. Colored Petri nets are useful in modeling multidimensional events, such as head and hand motions, where each dimension is modeled using a distinct color.

Timed Petri nets are used to model discrete events and synchronization of concurrent events, using *guards* and *delays* [61]. These nets have two types of delays—*enabling-delay* and *firing-delay*, which are associated with transitions. A transition is enabled to fire after an *enabling-delay* and then fires after a *firing-delay*.

2.3 Synchronization in Gesture Motions and Speech

Synchronization refers to coordinating tasks to maintain specific spatial or temporal constraints. *Temporal synchronization* involves temporal constraints between two or more events. *Spatial synchronization* involves spatial constraints between two entities. Gestures based on motion require both temporal and spatial constraints. Both types of synchronization are employed to ensure the accuracy and comprehension of events by minimizing perceptual distortion. Allen introduced 13 temporal synchronization constraints to determine the temporal relationship between the start and end of two events [62]. A subset of six synchronization relation have been used to model conversational gestures [20, 55].

The subset of synchronization relations used in conversational gesture analysis is {*sequential, start-synchronization, end-synchronization, strict synchronization, during synchronization, overlap synchronization*}. Given two tasks T_1 and T_2, the corresponding start-time $S(T_1)$ and $S(T_2)$, and the end-time $E(T_1)$ and $E(T_2)$, the temporal constraints are described in Table 1 where ε is the perceptional threshold.

2.4 Models for Deep Learning

In the last decade, deep learning models have been developed and applied to vision and natural language understanding. There are many models of deep learning such as Convolution Neural Network (CNN) and its variants for image classification [37, 47–49], Recurrent Neural Network (RNN) and its variants (Gated RNN and LSTM)

Table 1 Temporal constraints in conversational gestures

Tasks	Temporal constraints
Sequential	$start(T_1) < end(T_1) < start(T_2) < end(T_2)$
Start-synchronization	$\lvert start(T_1) - start(T_2)\rvert < \varepsilon$
End-synchronization	$\lvert end(T_1) - end(T_2)\rvert < \varepsilon$
Strict-synchronization	$(\lvert start(T_1) - start(T_2)\rvert < \varepsilon) \wedge (\lvert end(T_1) - end(T_2)\rvert < \varepsilon)$
During-synchronization	$(start(T_1) < start(T_2)) \wedge (end(T_2) < end(T_1)$
Overlap-synchronization	$(start(T_1) < start(T_2)) \wedge (end(T_1) < end(T_2)$

for natural language translation and motion analysis [37, 39–51], and transformers for both natural language translation and image classification [52, 53].

CNN uses a cascade of convolution layers (convolution filters + RELU + pooling layer) followed by a classifier such as feed-forward neural network. Convolution filters collect the local features of an image; RELU filters out the noise below a threshold; and pooling layer aggregates the features reducing the computational complexity and combining the local features for better comprehension.

RNN combines the past output from the neural network with the current input. The feedback of the past output provides context sensitivity and improves the accuracy in natural language translation. However, along with the output, error is also fed back, which can be further amplified. To reduce the ill-effect of introduced error, Long Short Term Memory (LSTM) models introduce *forgetfulness* at the individual cell level. Still, context in RNN is limited by the depth of the past output included with the current input.

Transformers use statistical analysis and attention-based encoders and decoders to predict the probability of the next translated word in the sequence [52]. This concept has revolutionized natural language translation and has been extended to vision analysis by projecting 2D image macroblocks to a sequence of macroblocks and analyzing the sequence using transformers [53].

2.5 *Conceptual Dependency Analysis*

Conceptual dependency is a linguistic graphical model to express the semantic connectivity of actors, actions, objects, and their proximity to other objects [28, 63]. Gestures model the interlocutor as an actor directly expressing actions, attributes of physical entities, or proximity between entities in the shared world of a dialog to the listeners; the direction of motion is expressed with respect to the body of the interlocutor. The formalism is based on semantic networks and infers cognitive models from stories or dialogs, expressed as a sequence of phrases [63, 64]. For example, consider the sentence “Aditi is writing a book”. Here Aditi is an agent; action is ‘write’. The object is ‘book’. Figure 4 illustrates the conceptual; dependency graph for the sentence.

Fig. 4 An illustration of conceptual dependency

3 Conversational Gestures Classifications

3.1 Discourse Based Gesture Classification by Cognitive Psychologists

Gesture classifications, as proposed by cognitive psychologists Kendon [7], McNeil [14], and Ekman and Friesen [56], serve as a fundamental framework for understanding gestures. Kendon classification includes *emblem gestures*, *illustrator gestures*, *adapter gestures*, *regulator gestures,* and *beat* gestures. McNeil's classification encompasses *iconic gestures*, *metaphoric gestures*, *deictic gestures*, and *beat* gestures [8]. Ekman and Friesen present a slightly different scheme with *kinetographic gestures*, *pictographic gestures*, *spatial gestures*, *ideographic gestures*, *deictic gestures*, and *baton gestures* [56]. These classifications provide valuable insights about the applications of gestures to model objects, abstract concepts, spatial relationships, and emphasize speech. McNeil's iconic gestures subsume Ekman's *kinetographic gestures*, *pictographic gestures*, *spatial gestures*, and *ideographic* gestures.

Cognitive psychologists have also focused on analyzing behavior patterns based upon head-motions. They have identified multiple conversational head-gestures. Conversational head-gestures use head-postures and a combination of head-nod, head-shake and head-tilt along with spoken phrases to convey behavior modes. There are 36 major head-gestures: *accept, admire, acknowledge, affirmation, argument, avoidance, intent, backchannel, confidence, defensive, discourage, encourage, greeting, interest, denial, disagreement, agreement, appreciation, arrogance, dominance, confusion, expectation, inclusion, interject, permission, persuade, plead, question, defiance, reject, request, ridicule, encouragement, interrogation, frustration, unsure* [9, 65–68].

However, these classifications lack comprehensiveness as they do not consider gestures that involve haptics, spatial and temporal coordination of relevant body parts, and the expression of mental states. Deictic gestures combine with haptics (touch) to create composite *touching gestures.* Deictic gestures also combine with metaphoric gestures to create composite *exhibit gestures*. For instance, a mother may point towards an egg and use an exhibit gesture to encourage her child to eat breakfast. Another major limitation is that these classifications are based solely on discourse analysis. However, gestures require a combination of visual, textual and synchronized motion analysis, which requires further subclassifications as described in the following subsections.

3.2 *Extending Deictic Gestures Subclassification*

Deixis is characterized by referential spoken words such as *here, there, this*, and *that*. In the deictic gestures, these referential words are accompanied by a combination of head, hand, and finger motions, or eye-pointing towards the entity [69, 70]. However, spatial and temporal use of these referent words can be complex; the same word can be used either for spatial or temporal annotations. The vision-based analysis of deictic gesture requires analyzing different scenarios for different organs. The classification has to integrate the discourse analysis along with a combination of synchronized organ-motions in a particular setting.

Table 2 describes different scenarios that require different combinations. Deictic gestures differ from conversational head-gestures due to the acyclic nature of head-motion and the use of index-finger or palm to point at an object in addition to the use of referential words.

We identify three major subclasses of deictic gestures: *pointing*, *presenting*, and *grouping*. *Pointing gestures* have only one object to point and require synchronization of head, hand (including index-finger or palm), and eye-motion. *Presenting-gestures* have entities distributed in the interaction space in different field-of-vision such as a classroom setting. This requires a synchronized repeated combination of head, hand (including index-finger or palm) and eye-movement. Grouping gestures have entities in the same field-of-vision such as close meeting. This requires head-movement only once to face towards the entities followed by repeated synchronized combination of hand-movements (including index-finger or palm) and eye-movements.

These subclasses have further subclasses. For example, pointing is done either by a combination of *head and hand motion* or *head and eye-motion*. Table 3 describes three subclasses of *pointing-gestures*: (1) DG_1–synchronization of hand, finger, and index-finger movements; (2) DG_2–eye movements only; (3) DG_3–synchronous head and eye movements followed by speech [58]. Table 4 outlines characterizations of *presenting-gestures* (DG_4) and *grouping-gestures* (DG_5).

Table 2 Examples of spatial deictic gestures

	Sentence	Referential words	Motions
1	The book is here	Here	Synchronous hand, finger, and index-finger movements to point to the book
2	Look there	There	Only eye movements to refer to the object
3	Listen to me	N/A	Synchronous head and eye movements followed by speech to seek attention
4	Remove that mug	That	Synchronous head, hand, finger, and speech

Table 3 Pointing-gesture subclasses

Subclass	Description
Deictic gesture 1 (DG_1)	Synchronous hand, finger, and index-finger movements to point to the referred entity
Deictic gesture 2 (DG_2)	Only eye movements to refer to the object
Deictic gesture 3 (DG_3)	Synchronous head and eye movements followed by speech to seek attention

Table 4 Presenting-gesture (DG_4) and grouping gesture (DG_5) characterization

Subclass	Description
DG_4	Multiple objects, synchronous head, hand, finger, and speech for first object, then repeated head, arm, eye movements and speech for remaining object
DG_5	Multiple objects in same field-of-vision, synchronous head, hand, finger, and speech for the first object, then repeated synchronous arm, eye movements and speech

3.3 Extending Iconic Gestures Subclassification

Although pictographic iconic gestures are well researched, kinematic iconic based on finger or palm generated contours are not well researched in the gesture analysis domain. Kinematic iconic gesture involve organ-movements synchronized with the word (in the spoken phrase or dialog), which is being depicted. Organ movements derive a contour which needs to be refined and recognized as a regular form despite noise in the hand-motion.

The classification has to integrate the *contour segment pattern analysis* with the speech. We identify four kinds of iconic gestures: *closed contour (CC), composite closed contour (CCC), regular open contour (ROC), irregular form open contour (IOC)*. To express iconic gestures, fingers can be folded; palm can be stretched and moved; fingers or palm can draw a contour [14]. Table 5 illustrates examples of iconic gesture using finger motions.

Table 5 Examples of iconic gestures imagery using finger motions

Action/attribute	Initial posture	Final posture	Contour
Eating a sandwich	Hand holding sandwich	Hand moving to mouth	Semi-circle (ROC)
Pouring liquid	Hand holding pitcher	Hand pouring	Downward arc (ROC)
Writing a letter	Hand holding pen	Hand moving across paper	Zigzag or curved line (CCC)
Turning a key in a lock	Hand holding key	Hand twisting key	Rotation (CC)

3.4 Extending Conversational Classification for Integrated Computational Analysis

Figure 5 describes an extended classification of conversational gestures using new proposed extensions. The figure has two types of arrows: *single-parent subclass* and *multiple-parents subclass*. In addition, all the enclosed boxes within a rectangle are subclasses in the class represented by the enclosing rectangle. For example, *agreement, disagreement, appreciation* are subclasses of conversational head-gestures. Similarly, *action, haptics,* and *emphasis* are subclasses of *meaningful hand-gestures*. An arrow from a rectangle containing multiple subclasses also includes all the enclosed subclasses. An arrow to specific subclass in a rectangle is specific to that subclass only. For example, *mental-states* is a subclass of *haptic gestures*.

Conversational gestures have three major subclasses: *conversational head-gestures, conversational hand-gestures,* and *eye-motion gestures*. Conversational hand-gestures have two major subclasses: *meaningful gestures* and *nondeliberate gestures*. *Meaningful gestures* are further classified as *iconic gestures, deictic gestures, metaphoric gestures, enumerative gestures, emphasis gestures, mental states, haptic gestures, kinetic-haptic gestures, spatial gestures,* and *attributional*

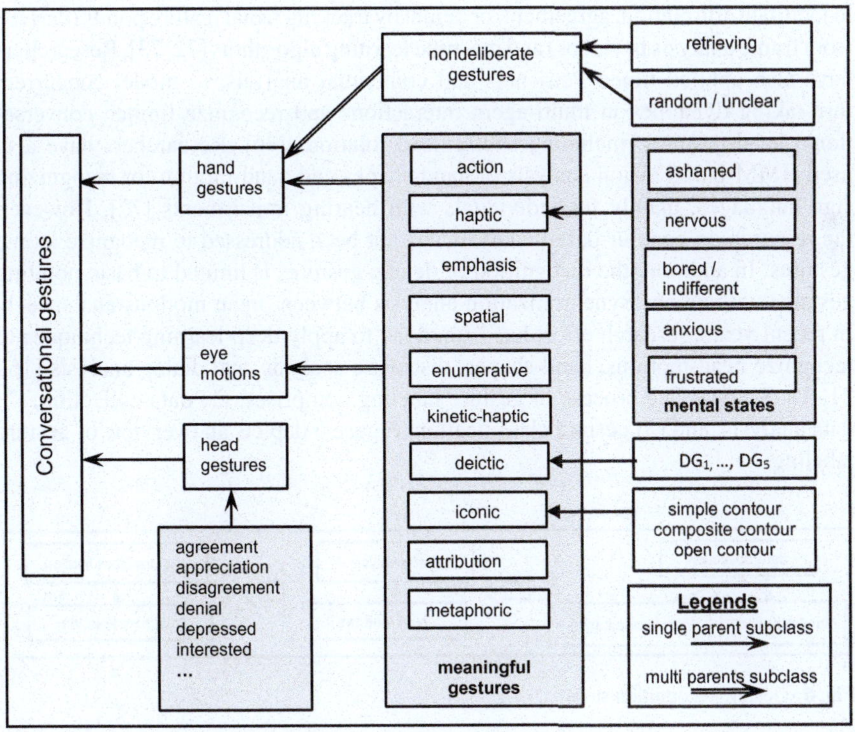

Fig. 5 An extended conversational gesture classification

gestures. Mental states can be *anxiety, frustration, ashamed, nervousness*, and *bored/indifferent*. Eye-motions are mainly used for deictic gestures and some head-gestures for focusing and tracking.

Emphasis gestures, enumeration gestures, and *beat gestures* use rhythmic motion. However, they are separated by their semantics and the motion type. *Emphasis gestures* are used to emphasize certain parts-of-speech and is characterized by simple cyclic motion of finger going up and down. *Enumeration gestures* are associated with spoken phrase related to counting entities and is characterized by a composite motion of index-finger that includes the finger going up and down and translating in another direction. *Beat gestures* use hand going from near the body to away from the body symbolizing extraction of knowledge. *Attributional gestures* describe the attributes using contours. However, they differ from iconic gestures because iconic gestures also describe entities by their attributes.

4 Gesture Recognition Approaches

Previous studies have contributed to different aspects of gesture recognition [71–75]. Researchers have explored recognition of head-nod, head-shake, and head-tilt, to facilitate affirmation, agreement, or denial by tracking facial feature points (derived from frame analysis in videos) and machine learning algorithms [72, 73]. Researchers have also applied timed Petri nets and contextual analysis, to model concurrent turn-taking dynamics in multi-agent interactions and recognize limited conversational head-gestures, including subtle head rotations [76]. Researchers have also used HMM-based vision analysis of hand-shapes and hand-motion for recognizing sign languages, mainly for individuals with hearing impairments [77]. However, the research on contour-based analysis has not been addressed to recognize iconic gestures. In addition, the recognition of deictic gestures is limited to basic pointing gestures without any synchronization analysis between organ motions and speech. In recent years, research effort has been done to apply deep learning techniques to recognize head-motions, hand-shapes, discourse analysis, and dialog analysis [16–21, 39–51, 64]. The process of gesture labeling comprises: (1) data collection, (2) data analysis, and (3) gesture classification. Figure 6 depicts an overview of gesture labeling.

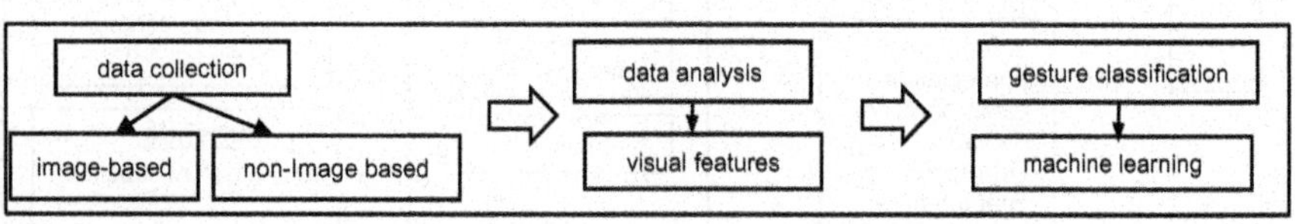

Fig. 6 Gesture recognition steps

Table 6 Data collection methods

Type	Sensor	Description	Application
Image-based	Camera	Capture image and video	Human–robot interaction, video surveillance, home automation
	Depth-sensor camera	Depth data in 3d objects	Gaming, virtual reality
	Motion capture	Marker-based tracking	Healthcare, sports analysis
Non-image based	Wearable sensors	Sensors attached to body or clothing	Fitness, healthcare, sports analysis
	Pressure sensors	Sensors embedded in the floor	Human–robot interaction, home automation
	IMUs	Sensors to capture movement	Healthcare, sports, robotics

4.1 Data Collection and Analysis

Data collection involves various forms, including images, videos, and sensor readings, to extract feature-vectors and analyze gesture patterns [29, 30]. Data analysis involves denoising and feature-extraction from the collected data. Gesture classification uses machine and deep learning techniques as described in section IV(B). Table 6 summarizes data collection steps, including image and non-image based methods. Image-based methods capture images or video using a 2D camera, depth information using depth sensors, and motion information using video frames and feature-points or marker-based systems. Non-image based sensors such as IMU are also used to provide motion information.

During data analysis, visual features such as feature-points of face, eyes, index-finger, wrist, and shoulder are recognized. The centroids of feature-points are derived. The changes in centroid-coordinates are used to derive motion-vectors.

4.2 Machine and Deep Learning Based Gesture Classification

Gesture recognition is done using popular machine and deep learning techniques such as K-nearest neighbors (KNN), Hidden Markov Model (HMM), Artificial Neural Network (ANN), decision trees, Long Short-Term Memory (LSTM), Convolutional Neural Network (CNN) and their combinations. Table 7 summarizes some popular machine learning techniques used for gesture recognition, along with their advantages and limitations.

KNN (K-Nearest Neighbors) based approach classifies data based on its proximity to labeled examples. It works by calculating the distance between a given data

Table 7 Machine and deep learning approaches

Technique	Advantages	Limitations
K-nearest neighbors	Simple and easy to implement	Performance depends inversely on the database-size and dimensions
HMM	Effective for recognizing complex gestures	Sensitive to lighting conditions during segmentation
RNN	Able to recognize dynamic gestures	Information loss over long time-lags
CNN	Effective in extracting depth features	Limited in temporal patterns
LSTM	Effective in processing time-series data	Requires extensive training data
Decision tree + SCPN	General model for upper body gestures. Exploits synchronization. Identifies major subclasses. Automated learning by mimicking	Not integrated with context analysis, complex dialogs and deep learning

point and its K nearest neighbors, where K is a predefined parameter. The algorithm assigns the majority label among the nearest neighbors to the data point [15]. The performance of KNN classification depends on the size of the database (large amount of feature-vector) and the feature-vector dimensions. A feature-vector with a large number of dimensions is computationally prohibitive. KNN is not suitable for conversational gesture recognition due to its lack of temporal modeling, sensitivity to feature representation, and computational inefficiency with large datasets.

HMM-based models have been used for gesture-based control of robot [16] and hand-gesture recognition [17, 38]. HMMs have limitations in conversational gesture recognition due to their simplistic temporal and statistical structure, which fails to capture the intricate dynamics and complex variations involved in co-articulation, synchronization and overlapping phases between motions and speech, which cannot be effectively represented by HMMs.

RNN has been used along with skeleton-based model to recognize a dynamic hand-gesture [36, 37]. The skeleton-based model uses angles of the bones to derive finger-movements. The movements of a hand are described using combinations of rotations and translations. Finger-motion features along with the global features act as an input to the RNN. However, RNN based systems have significant time-delay in learning. In addition, the predicted finger-motion error could be amplified, leading to an inaccurate gesture labeling.

RNN has been coupled with CNN to augment hand-shape and hand-motion based gesture recognition [37]. The system recognizes dynamic gestures using both depth-feature and temporal features. CNN uses depth-features along with 2D images to classify 3D gestures. RNN extracts temporal patterns from the movement of skeleton-points within a sequence. However, this architecture suffers from information loss when considering a dynamic gesture over a longer period due to vanishing gradient.

LSTM has been used in speech recognition [40–42], handwriting recognition [43], and image captioning [44]. Variants of LSTM have been used in dynamic hand-gesture recognition [45–49]. In LSTM based dynamic gestures recognition, hand-shapes are modeled as feature-vectors in a frame. Gesture recognition requires processing of time-series of frames. The LSTM-cell takes input as the current frame, previous LSTM cell output, and content of the memory-cells from previous time-step. The current frame comprises vectors for each joint (*positions, speed*). The output of the LSTM-cell at each time-step is fed into a logistic regression model that classifies the frame as either inside or outside a gesture.

CNN is coupled with LSTM to form a convolutional long short-term memory network (ConvLSTM), which has been widely used for recognizing dynamic hand-gestures. Dynamic gestures require processing of a sequence of images, as in video analysis, along with other dynamically changing data such as depth measurements. LSTM has been used to learn behavior based upon multiple sequences of joints-movement data from both arms of a humanoid robot Nao [50]. The joints include shoulder pitch, shoulder roll, elbow roll, elbow yaw, and wrist yaw for both left and right arms. The joints are extracted as vectors with positions.

SCPN based model integrates synchronization between motions and speech, conceptual dependency, and analyzes for multiple subclasses of conversational as described in Section V. It also uses decision-tree for gesture disambiguation. However, it is currently not integrated with deep learning techniques which can provide context resolution and better integration with dialog and discourse analysis.

4.3 Automated Learning by Mimicking

Robot learning by imitation is a type of machine learning technique where robots learn to perform a task by observing and imitating a human or another robot performing the same task [51, 55]. The process of robot learning by imitation involves multiple steps. First, the robot observes a demonstration of the task being performed by a human or another robot. This demonstration is provided as video recordings or direct demonstrations. Next, the robot analyzes the demonstration to identify the key features and actions necessary to perform the task. Finally, the robot uses this analysis to generate its own plan for performing the task.

There are several advantages to learning by imitation. It allows robots to learn new skills and behaviors quickly and efficiently with no explicit programming or complex algorithms. It enables robots to adapt to new situations and environments easily, as they can learn from different demonstrations and adjust their behavior using pattern and statistical analysis. It also improves the safety, acceptability, and reliability of robots, as they can learn from human experts and avoid errors or dangerous behaviors.

5 Synchronous Colored Petri Net (SCPN) Model

SCPN combines *synchronization* with *colored Petri nets* using delays [78] to model a combination of synchronized motions and speech and analyze conversational gestures. Using colors allows the modeling of motions in multiple dimensions. Each place-node models the start or end-point of a motion, or the relaxed organ for a longer duration marking the gesture-boundary. Synchronization between two or more motions or between speech and motions is achieved by introducing delays in the place-nodes, transition nodes, and edges.

Colored tokens are used to model 3D motions of the same organ—one color for each dimension. *Place* is defined as a point of *stillness* when motion-type changes, or *total stillness,* as in the relaxed-state. A *transition-node* denotes the continued motion of an organ or speech. The labels associated with colored tokens are placed as a subset within curly brackets.

Start-synchronization occurs when two tasks (motion or speech) start within an imperceptible duration. *End- synchronization* occurs when two tasks end within an imperceptible duration. *Strict-synchronization* occurs when two or more tasks exhibit both start-synchronization and end-synchronization. *During-synchronization* occurs when the second task starts after the first task and ends before the first task. *Overlap-synchronization* occurs when the second task starts after the first task and ends after the first task.

A *trigger-node* is used to spawn two or more synchronized tasks. *Start-synchronization* is modeled by assigning a delay δ as a weight of the edge connecting the trigger-node to the next place-node. *End-synchronization* is modeled using multiple transition-nodes connecting to the same place-node, with the delay δ placed on the connecting edges. The modeling of synchronization in SCPN is described in Fig. 7.

The delay before a movement from a place is modeled as an enabling delay δ^{E}. This corresponds to the activation delays in muscles or delays in the deactivation of moving muscle during direction reversal or movement-stop. The delay during transition between two places is modeled as a *firing-delay*, denoted by δ^{F}.

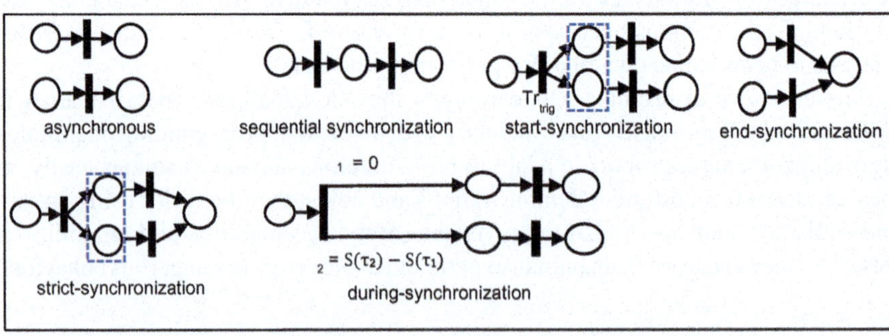

Fig. 7 Overall synchronization construction

There are two kinds of transitions: *regular transition* and *trigger-transition*, denoted by tr^{trig}, for triggering two or more synchronous tasks. *Trigger-transition* has two or more outgoing edges from the trigger-transition-node showing the firing of two or more tasks (such as speech and head-motion) with timing constraints. Each outgoing edge of a trigger-transition is associated with a *firing-delay*, denoted by δ^{trF}. The firing-delay δ^{trF} for the ith outgoing-edge is equal to relative start-time $S(\tau_i)$ for the i_{th} task τ_i. This modeling is useful for *during synchronization* and *start-synchronization*. The *firing-delay* δ^{F} is associated with the duration of the *regular motion* of an organ from the current place to the next place.

SCPN-graph is modeled as a matrix that is dynamically built as the places are recognized. After reaching the gesture-boundary—a relaxed state with a delay longer than the gesture-boundary threshold, the matrix is analyzed for cyclic motions, and meta-attributes, which are fused to form the signature of the derived gesture.

5.1 Modeling Composite Synchronized Motions

Composite motions like waveforms are handled using a new concept that splits motion-dimensions at the beginning of *composite motions* such as waveform and joins the motion-dimensions at the end of composite motions. The splitting of motion-dimensions is called *color-splitting*, and the joining of motion-dimension when a hand (or both hands) starts making simple motion again is called *color-joining*, as illustrated in Fig. 8 [54].

Example 1 Consider the waveform motion. It has two synchronized simple motions: (1) cyclic motion in x direction with index-finger going up and down; (2) translational motion in y direction. At the start of wave-motion, color {x, y} is split into colors {x} and {y}. At the end of wave motion both split colors are joined again as shown in Fig. 8.

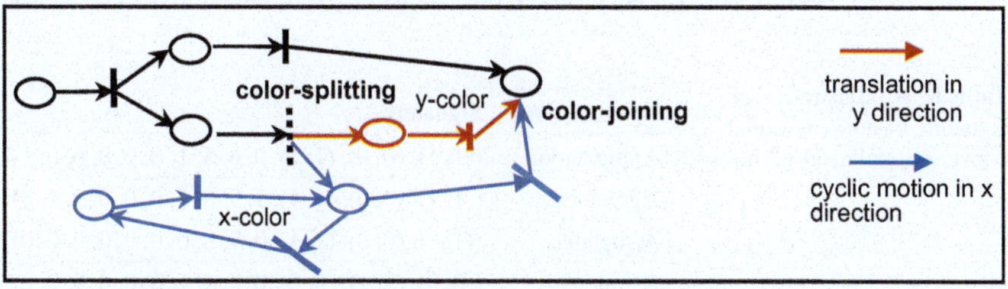

Fig. 8 Color-splitting and color-joining to model composite motions

5.2 *Signature of a Gesture*

Every gesture is modeled using a unique signature—an n–tuple comprising *n* meta-attributes derived from the corresponding SCPN. These meta-attributes include the information about *head-nods, head-shakes, head-tilts, eye-focus, the number of places, transitions, start-synchronization, end-synchronization, strict-synchronization, during-synchronization, concurrent asynchronous actions, cycles,* and *speech*. Head-nod, head-shake, and head-tilt are pairs comprising binary values for motion and motion-direction; eye-focus, and speech are modeled using binary values; All other parameters are modeled using natural numbers.

Table 8 describes signatures of a subset of conversational head-gestures. Each signature is a 13-tuple of the form (*head-node, head-shake, head-tilt, eye-focus, number of places, number of transitions, number of start-synchronization, number of end-synchronization, number of strict-synchronization, number of during-synchronization, cycles, speech)*. The first three elements are pairs of the form (*head-motion, motion-direction*).

The absence of head-nod, head-shake, or head-tilt is denoted by '0'. Head-nod starting in upward direction is denoted by '1', while the downward direction is denoted by '0'. Head-shake in right rotation is denoted as '1', and head-shake in left rotation is denoted as '0'. A head-tilt in the right direction is denoted by '1' while the left direction is denoted by '0'. Non-specific head-movements starting in either direction are denoted by '*', which can match either '0' or '1'. An unfocused eye is denoted as '0'; a focused eye is denoted as '1'. The absence of a specific synchronization is denoted by '0'.

6 Conversational Gesture Recognition Using SCPN

Gesture recognition comprises five modules, as shown in Fig. 9. The first module uses video analysis and speech intensity analysis to derive a *motion-vector* and a *speech-vector*. These vectors are analyzed to derive synchronization and cycles. The second

Table 8 An illustration of signature of non-emotional conversational head-gestures

Head-gestures	Signature
Appreciation	((1, 1), (0, 0), (1, 1), 0, 6, 5, 1, 0, 0, 0, 0, 0, *)
Agreement	((1, 1), (0, 0), (1, 1), 1, 9, 8, 1, 0, 0, 0, 0, 1, *)
Arrogance	((1, 1), (0, 0), (1, 1), 0, 9, 8, 0, 0, 0, 0, 0, 0, 0)
Avoid	((0, 0), (1, 1), (0, 0), 0, 5, 4, 1, 0, 0, 0, 0, 0, 0)
Backchannel	((1, 1), (0, 0), (0, 0), 1, 7, 6, 1, 0, 0, 0, 0, 1, *)
Confusion	((0, 0), (1, 1), (1, 1), 0, 7, 6, 1, 0, 0, 0, 0, 0, *)
Defensive	((0, 0), (1, 1), (1, 1), 1, 9, 8, 1, 0, 0, 0, 0, 0, *)
Denial	((0, 0), (1, 1), (0, 0), 1, 5, 4, 0, 0, 0, 1, 0, 1, *)

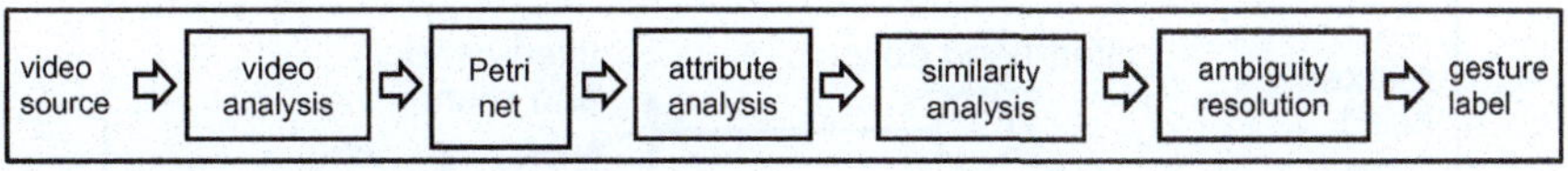

Fig. 9 Conversational gesture recognition steps

module generates matrices dynamically to derive places, transitions, and delays for each gesture based on the corresponding SCPN graph. The third module derives the signature of each SCPN-graph. The fourth module matches derived signatures with archived labeled-signatures using similarity analysis to label the derived gestures. The fifth module disambiguates derived gestures using decision-trees and resolves errors introduced in identifying places and transitions due to the choice of thresholds.

6.1 Recognizing Conversational Head-Gestures

Conversational head-gestures are based on modeling and analyzing synchronization of head-motions and spoken phrases using video analysis. The synchronization is modeled using start-synchronization, end-synchronization, strict-synchronization and during-synchronization.

Conversational head-gestures comprise a sequence of head-nods, head-shake, or head-postures, including the relaxed-state. Two adjacent head-gestures are separated by a gesture boundary. Video analysis extracts speech-amplitude and motion of feature-points. Random motions less than spatial thresholds are treated as *stillness*. Head-postures are classified in finite fuzzy states such as *head-up, head-down, head-left, head-right, head-tilted-up, head-tilted-down,* and *relaxed head*.

A head-gesture starts with a relaxed-head and silence and ends in the next relaxed-head and silence. Motion is modeled as a motion-vector—a binary vector which denotes motion by '1' and stillness or localized random motion below a threshold by '0'. Similarly, speech and silence are modeled using a speech-vector—a binary vector which denotes the speech by '1' if the sound intensity is greater than or equal to 35 dB and '0' if the sound intensity is less than 35 dB. Both temporal vectors depend upon the frame-sampling rate. These two vectors implicitly carry the information about temporal synchronization needed to model transitions in SCPN.

Example 2 Figure 10 illustrates an SCPN-graph to model conversational head-gesture for appreciation [21]. The gesture is modeled as a head-tilt in either direction, followed by speaking a short phrase such as "good" while moving the tilted head down and returning the head back to the relaxed position after a small finite delay. First, transition is single-colored (rotation around Z-axis for head-tilt); second and third transitions are two-colored (rotation around Y-axis for head-nod and existing Z-axis rotation). The top path models head-motions; the bottom path models speech. Both head-motions and speech are synchronized using start-synchronization. The

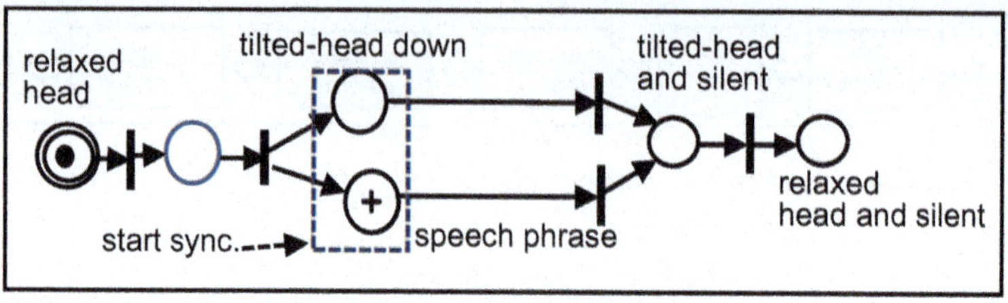

Fig. 10 SCPN model of a conversational head-gesture 'appreciate'

transition tr_2 is a *trigger-node* that starts synchronously speech and motion. The '+' symbol in the speech phrase represents an embedded Petri net.

6.2 *Recognizing Deictic Gestures*

Deictic gestures require the coordinated movement of the head, eyes, hand, index finger, or palm [55]. The start of a spoken phrase can be delayed because *localization* precedes *utterance*. Speech can also end later such as in a presentation gesture due to longer explanation related to the referred object after the pointing. Additionally, speech can also continue after the initial pointing gesture, particularly in situations like a presentation, where there may be a longer explanation or discussion related to the object being referred to.

To model deictic gestures effectively, it is necessary to analyze the synchronized movements of the hand (raising the arm), non-repetitive motions of the head, movements of the index-finger. This synchronization occurs because the process of locating or pointing to something with the hand motion comes before speaking. Once the speech is finished, the arm and index-finger maintain their final position to attract attention. Each organ motion and speech is modeled as one path between the current relaxed-state and the next relaxed-state. These paths have various synchronizations as described in Tables 2 and 3.

Signatures for deictic gestures are modeled as 18-tuple of the form *((head-nod, direction), (head-shake, direction), (head-tilt, direction), (hand-motion, direction), (index-finger/palm, direction), eye-tracking, eye-focus, number of places, number of transitions, number of start-synchronization events, number of end-synchronization events, number of strict-synchronization events, number of during-synchronization events, number of overlap-synchronization events, number of concurrent asynchronous actions, head-motion cycle, hand-motion cycle, Petri net cycle, speech))*.

6.3 Recognizing Iconic Gestures—Contour Segment Pattern (CSP) Analysis

Iconic gestures are classified as: (1) finger/palm postures; (2) closed contour (CC); (3) composite closed contour (CCC); (4) regular open contour (ROC); (5) irregular open contour (IOC) as described in section III. *Iconic gestures* are modeled as a triple (*initial posture, motion* of the *hand-components* → *contour, final posture*). Postures and motions are modeled using image-coordinates, orientation, and their changes with time.

Contours are characterized by a pattern of coordinates and orientation-changes [54]. Contour signature analysis [54] is based on: (1) sampling the changes in x, y coordinates and the embedded orientations in the motion trajectory; (2) denoising using threshold analysis to account for slight variations in hand-motions; (3) merging the changes in adjacent coordinate-values $f_{\dagger}^{\Delta x}$ (or $f_{\dagger}^{\Delta y}$), if the orientation changes are below a threshold, to form bigger segments; (4) mapping merged segments to a normalized value $\in \{+1, -1, 0\}$, after identifying significant orientation change, depending upon the increase in value above the threshold, the change in value below the threshold, or absolute magnitude of the decrease in value above the threshold; (5) removing low frequency magnitude and orientation changes to remove noise caused by jitters in hand-motion during drawing; (6) pattern matching (using similarity analysis) for known contour-signature pattern(s) in the knowledgebase.

Magnitudes vary and are specific to individuals [55]. Hence, coordinate changes, after significant orientation changes occur, are mapped to a value $\in \{-1, 0, +1\}$. The coordinate-change $f_{\dagger}^{\Delta x} \geq \varepsilon^{X} \rightarrow +1$ (or $f_{\dagger}^{\Delta y} \geq \varepsilon^{Y} \rightarrow +1$); $f_{\dagger}^{\Delta x} < -\varepsilon^{X} \rightarrow -1$ (or $f_{\dagger}^{\Delta y} < -\varepsilon^{Y} \rightarrow -1$) where ε^{X} and ε^{Y} are threshold-values. Coordinate-changes map to 0 if absolute value $|f_{\dagger}^{\Delta x}| < \varepsilon^{X}$ (or $|f_{\dagger}^{\Delta y}| < \varepsilon^{Y}$) to remove noise. We introduce the concept of *contour segment pattern* (*CSP*) to model and recognize various contours. CSP is defined as a sequence of pair (*x and y coordinates, changes from the previous-value*). Changes from the previous value is triple of the form (*change in x-coordinate, change in y-coordinate, change in orientation*).

The second level of noise occurs due to the approximate motion of fingers/palms to draw a regular shape. This irregular motion causes spurious minor changes in orientation. As a result, contour-drawing using fingers or palm have minor jitters. To filter such jitters, a frequency-analysis is done, and motions with smaller frequency below a threshold are filtered out. This frequency-based filtering also helps in the recognition of smooth curves such as circles, which are approximated as N-polygons with large N. To reduce this noise, the sequence is scanned using a parametric window of size *w*, and the pairs having a frequency of movements with an orientation change below a threshold are pruned.

Invariance of contours due to the reversal of direction of traversal or due to starting from any vertex in a closed contour is ascertained using the application of two mathematical functions *inversion* and *rotation* on CSP. A closed contour can be drawn clockwise or counterclockwise. It can also be drawn starting from any vertex in a polygon. Clockwise and counterclockwise contour-drawing are invariant under the *inversion* of the CSP. The set of closed contours drawn from any vertex randomly form an equivalence set under the *rotation* of a CSP.

Under *inversion*, the signs of the patterns are inverted from +1 to −1 and vice versa. In addition, the sequence is traversed from right to left. Thus, a CSP for <(x_1, y_1), (x_2, y_2), …, (x_N, y_N)> for clockwise traversal becomes <($-x_N$, $-y_N$), ($-x_{N-1}$, $-y_{N-1}$), …, ($-x_1$, $-y_1$)>. For example, the CSP <(0, +1), (+1, 0), (0, −1), (−1, 0)>, drawn for clockwise contours drawing, would look as <(+1, 0), (0, +1), (−1, 0), (0, −1)>) under inversion, which is also the CSP for the same contour when drawn counter-clockwise starting from the same vertex.

Rotation of a CSP <(x_1, y_1), (x_2, y_2), …, (x_N, y_N)> gives N equivalent CSPs, one each for starting a contour from a different vertex. The CSPs can be rotated either in the left or right direction. For example, the CSP <(x_1, y_1), (x_2, y_2), …, (x_N, y_N)> when rotated left by one position gives <(x_2, y_2), (x_3, y_3), …, (x_N, y_N)(x_1, y_1)>.

$$
\begin{aligned}
\theta &= \tan^{-1}(f_{\dagger}{}^{\Delta x}/f_{\dagger}{}^{\Delta y}) - \text{orientation}\\
\text{sign}(\text{X} - \text{coordinate}) &= -1 \text{ if} \cos(\theta) \le -\varepsilon^{\theta};\, 0 \text{ if } |\cos(\theta)| < \varepsilon^{\theta};\; +1 \text{ if} \cos(\theta) \ge +\varepsilon^{\theta}\\
\text{sign}(\text{Y} - \text{coordinate}) &= -1 \text{ if} \sin(\theta) \le -\varepsilon^{\theta};\, 0 \text{ if } |\sin(\theta)| < \varepsilon^{\theta};\; +1 \text{ if} \sin(\theta) \ge +\varepsilon^{\theta}
\end{aligned}
\quad (1)
$$

The invariance of contours drawn from any vertex randomly is modeled using two steps: generating CSP using Eq. (1) for sign change when orientation changes, followed by the *rotation* of the CSP. There are only eight combinations of individual elements in CSP: (0, +1), (0, −1), (+1, −1), (+1, 0), (−1, 0), (−1, +1), (+1, +1), (−1, −1). The pair (0, 0) is not possible because it shows no movement. Each of these eight combinations are modeled as one byte. Thus, each CSP is a sequence of bytes, which is converted as a key for knowledge-based lookup.

After identifying a noise-free CSP, its equivalent combinations using *sign-version* and *rotation* are derived and coded into a byte-sequence. This set of byte-sequences becomes keys for looking up in the knowledge-base to label the contour and derive its properties and associations.

Example 3 Consider a sequence of coordinate-pairs of a rectangle S_1 = <(0, 0), (20, 0), (20, 10), (0, 10)> describing the positions of the vertices *A*, *B*, *C*, and *D* of a rectangle. The sequence expressed as ($f_{\dagger}{}^{\Delta x}$, $f_{\dagger}{}^{\Delta y}$) pairs is <(+10, 0), (0, +10), (−10, 0), (0, −10)>. The corresponding CSP is <(+1, 0), (0, +1), (−1, 0), (0, −1)>. After the inversion, the corresponding CSP is <(0, +1), (+1, 0), (0, −1), (−1, 0)>. Each of these two CSPs are rotated four times to give a set of eight equivalent CSPs: {<(+1, 0), (0, +1), (−1, 0), (0, −1)>, <(0, +1), (+1, 0), (0, −1), (−1, 0)>, <(0, +1), (−1, 0), (0, −1)(+1, 0)>, <(−1, 0), (0, −1), (+1, 0), (0, +1)>, <0, −1), (+1, 0), (0, +1), (−1, 0)>, <(+1, 0), (0, −1), (−1, 0), (0, +1)>, <(0, −1), (−1, 0), (0, +1), (+1, 0)>, <(−1, 0), (0, +1), (+1, 0), (0, −1)>}.

Let us assume that (0, +1) →0, (0, −1) →1, (+1, −1) → 2, (+1, 0) →3, (−1, 0) → 4, (−1, +1) → 5, (+1, +1) → 6, (−1, −1) → 7. Thus, the set of equivalent keys is {3041, 0314, 0413, 4130, 1304, 3140, 1403, 4031}. The properties of the rectangle and associated entities or actions, associated with these keys, are looked up.

6.4 Ambiguity Resolution Using Decision Trees

Sampling-rate, missing frames and feature-point-extraction errors can result in a missed place during video analysis. Since the motion-analysis is based upon the detection of place-nodes, this error may alter the original signature of the gesture, resulting in inaccurate labeling. Similarly, a cycle may be missed if the location of the moving organ is not currently identified due to fast sampling-rate or missed-frame. The resulting signature may mislabel an actual gesture as a different gesture.

This error has been reduced using decision-trees. The gestures are grouped into various subclasses based upon their semantics and mutual exclusiveness of the motion. For example, head-nod and head-shake are mutually exclusive: head-nod shows positive assertion, while head-shake is associated with negative intent such as denial or disagreement. Similarly, gestures based upon the eye-focus and the unfocussed-eye are mutually exclusive.

Within the same class, gestures are separated further using *motion direction, synchronization information, place information* and *transition information* for further resolution. A decision-tree is formed to separate different subclasses based upon the signature of actual gestures. Using the decision-tree, errors and ambiguities in labeling are reduced.

7 Limitations and Future Work

The recall rate of the investigated gestures varies from 82 to 92%, depending on the gesture. Gesture mislabeling is caused by (1) missed feature-points and frames in video analysis, resulting in missed places and the corresponding transitions; (2) missing small undetectable motions in gestures; (3) imprecise spatial and temporal threshold values. Unfortunately, despite statistical analysis to find an optimum threshold, mislabeling occurs due to thresholds being dynamic, context dependent, and individual specific. This will require transformer based deep learning analysis in the future.

Signatures also lack the result of speech analysis, dialog-context, motion-speed, facial-expression analysis, and modeling of composite gestures. The mislabeling can be reduced by improving the confidence factor using dialog and context analysis [63, 79].

8 Conclusion

The chapter extends the classification of conversational gestures, described earlier by cognitive psychologists, based upon haptics, mental states, visual and synchronized motion based analysis. Applications of machine learning and deep learning techniques for gesture recognition have been described. New subclassifications and gesture recognition techniques for iconic and deictic gestures have been proposed based upon the integration of video analysis, dialog analysis, temporal synchronization of speech and motion, temporal synchronization of motions of various organs, and conceptual dependency for improved comprehension. The scheme is general enough for automated learning by mimicking, and automated adaptation [55].

A promising direction in gesture recognition and classification is the integration of multiple sources such as speech (or dialog) and context to improve the overall performance [61, 64]. By leveraging multiple modalities, researchers can develop robust systems to interpret human intentions and behavioral responses.

Another promising area of research is the use of deep learning in posture analysis, motion analysis, and gesture classification. While the use of deep learning in gesture recognition is still in its early stages, it has the potential to significantly improve gesture recognition.

References

1. C. Breazeal, A. Brooks, J. Gray, G. Hoffman, C. Kidd, , H. Lee, et el., "Humanoid Robots as Cooperative Partners for People," International Journal of Humanoid Robots, vol. 1, no. 2, pp. 1–34, 2004.
2. M.A. Diftler, J.S. Mehling, M.E. Abdallah, N.A. Radford, L.B. Bridgewater, A.M. Sanders, et el., "Robonaut 2 – The First Humanoid Robot in Space," in Proc. the IEEE International Conference on Robotics and Automation, Shanghai, China, 2011, pp. 2178–2183, https://doi.org/10.1109/ICRA.2011.5979830.
3. R. M. Agrigoroaie, and A. Tapus, "Developing a Healthcare Robot with Personalized Behaviors and Social Skills for the Elderly," in Proc. 11th ACM/IEEE International Conference on Human-Robot Interaction (HRI), Christchurch, New Zealand, 2016, pp. 589–590, https://doi.org/10.1109/HRI.2016.7451870.
4. D. H. García, P. G. Esteban, H. R. Lee, M. Romeo, E. Senft, and E. Billing, "Social Robots in Therapy and Care," in Proc. the 14th ACM/IEEE International Conference on Human-Robot Interaction (HRI), Daegu, Korea, 2019, pp. 669–670, https://doi.org/10.1109/HRI.2019.8673243.
5. R. Rosenberg-Kima, Y. Koren, M. Yachini, and G. Gordon, "Human-Robot Collaboration (HRC): Social Robots as Teaching Assistants for Training Activities in Small Groups," in Proc. the 14th ACM/IEEE International Conference on Human-Robot Interaction (HRI), Daegu, South Korea, 2019, pp. 522–523, https://doi.org/10.1109/HRI.2019.8673103.
6. J. Wainer, D. J. Feil-seifer, D. A. Shell, and M. J. Mataric, "The Role of Physical Embodiment in Human-Robot Interaction," in Proc. the 15th IEEE International Symposium on Robot and Human Interactive Communication (ROMAN), Hatfield, UK, 2006, pp. 117–122, https://doi.org/10.1109/ROMAN.2006.314404.
7. A. Kendon, "Gesture: Visible Actions as Utterance," Cambridge University Press: Cambridge, UK, 2004.

8. J. P. De Ruiter, "The Production of Gesture and Speech," In: D. McNeill (editor), Language and Gesture, pp. 248–311, Cambridge University Press: Cambridge, UK, 2000.
9. A. Singh, and A. Bansal, "Declarative Modeling and Implementation of Robotic Head-based Gestures for Human-Robot Interaction," International Journal of Computers and Their Application, vol. 26, no. 2, pp. 49–66, 2019.
10. S. W. Cook, and M. K. Tanenhaus, "Embodied Communication: Speakers' Gestures Affect Listeners' Actions," Cognition, vol. 113, no.1, pp. 98–104, 2009, https://doi.org/10.1016/j.cognition.2009.06.006.
11. A. Csapo, E. Gilmartin, J. Grizou, J. Han, R. Meena, D. Anastasiou, et el., "Multimodal Conversational Interaction With a Humanoid Robot," in Proc. the 3rd IEEE International Conference on Cognitive Infocommunications (CogInfoCom), Kosice, Slovakia, 2012, pp. 667–672, https://doi.org/10.1109/CogInfoCom.2012.6421935.
12. Z. Shen, A. Elibol, and N. Y. Chong, "Inferring Human Personality Traits in Human-Robot Social Interaction," in Proc. the 14th ACM/IEEE International Conference on Human-Robot Interaction (HRI), Daegu, South Korea, 2019, pp. 578–579, https://doi.org/10.1109/HRI.2019.8673124.
13. L. P. Morency, C. Sidner, C. Lee, and T. Darrell, "Contextual Recognition of Head Gestures," in Proc. the International Conference on Multimodal Interfaces (ICMI), Trento, Italy, 2005, pp. 18–24, https://doi.org/10.1145/1088463.1088470.
14. D. McNeill, "Hand and Mind: What Gestures Reveal about Thought," University of Chicago Press: Chicago, IL, USA, 1992.
15. C. Li, K. Bredies, A. Lund, V. Nierstrasz, P. Hemeren, and D. Högberg, "K-Nearest-Neighbor Based Numerical Hand Posture Recognition Using a Smart Textile Glove," in Proc. the Fifth International Conference on Ambient Computing, Application Services and Technologies (AMBIENT), Nice, France, 2015, pp. 36–41.
16. H. Liu, and L. Wang, "Gesture Recognition for Human-Robot Collaboration: A Review," International Journal of Industrial Ergonomics , vol. 68, pp. 355–367, 2018, https://doi.org/10.1016/j.ergon.2017.02.004.
17. H. S. Park, E. Y. Kim, S. S. Jang, S. H. Park, M. H. Park, and H. J. Kim, "HMM-Based Gesture Recognition for Robot Control," in Proc. the Second Iberian Conference on Pattern Recognition and Image Analysis (IbPRIA), Estoril, Portugal, 2005, LNIP, vol. 3522, pp. 607–614, Springer: Berlin / Heidelberg, Germany, 2005, https://doi.org/10.1007/11492429_73.
18. M. A. Moni, and A. B. M. S. Ali, "HMM Based Hand Gesture Recognition: A Review on Techniques and Approaches," in Proc. the 2nd IEEE International Conference on Computer Science and Information Technology, Beijing, China, 2009, pp. 433–437, https://doi.org/10.1109/ICCSIT.2009.5234536.
19. X. Cucurull, and A. Garrell, "Continual Learning of Hand Gestures for Human-Robot Interaction," 2023, arXiv:2304.06319, https://arxiv.org/pdf/2304.06319.pdf, [Retrieved: April, 2023].
20. A. Singh, A. Bansal, and C.C. Lu, "Synchronous Colored Petri Net Based Modeling and Video Analysis of Conversational Head-Gestures for Training Social Robots," in Proc. the Future Technologies Conference (FTC), LNNS, vol 359, pp. 476–495, Springer: Cham, Switzerland, https://doi.org/10.1007/978-3-030-89880-9_36.
21. A. Singh, and A. Bansal, "Towards a Synchronous Model of Non-emotional Conversational Gesture Generation in Humanoids," in Proc. the Intelligent Computing Conference, 2022, LNNS, vol. 283, pp. 737–756, 2022, Springer: Cham, Switzerland, https://doi.org/10.1007/978-3-030-80119-9_47
22. J. M. Iverson, and S. Goldin-Meadow, "Why Do People Gesture as They Speak," Nature, vol. 396, pp. 228, 1998.
23. D. Efron, "Gesture and Environment." King's Crown Press: Morningside Heights, New York, USA, 1941.
24. A. Singh, and A. Bansal, "Automated Real-Time Recognition of Non-emotional Conversational Head-Gestures for Social Robots," in Proc. the Future Technologies Conference (FTC), vol. 3, 2022, LNNS, vol 561, pp. 432–450, 2022, Springer: Cham, Switzerland, https://doi.org/10.1007/978-3-031-18344-7_29

25. P. Wagner, Z. Malisz, and Z. S. Kopp, "Gesture and Speech in Interaction - An Overview," Speech Communication, vol. 57, pp. 209–232, 2014, https://doi.org/10.1016/j.specom.2013.09.008.
26. S. Goldin-Meadow, "The Role of Gesture in Communication and Thinking," Trends in Cognitive Sciences, vol. 3, no. 11, pp. 419–429, 1999, https://doi.org/10.1016/S1364-6613(99)01397-2.
27. S. D. Kelly, C. Kravitz, and M. Hopkins, "Neural Correlates of Bimodal Speech and Gesture Comprehension," Brain and Language, vol. 89, no. 1, pp. 253–260, 2004, https://doi.org/10.1016/S0093-934X(03) 00335-3.
28. R. C. Schank, "Conceptual Dependency: A Theory of Natural Language Understanding," Cognitive Psychology, vol. 3, no. 4, pp. 552–631, 1972, https://doi.org/10.1016/0010-0285(72)90022-9.
29. S. Mitra, and T. Acharya, "Gesture Recognition: A Survey," IEEE Transactions on Systems, Man, and Cybernetics, Part C (Applications and Reviews), vol. 37, no. 3, pp. 311–324, 2007, https://doi.org/10.1109/TSMCC.2007. 893280.
30. R. Zhao, Y. Wang, P. Jia, C. Li, Y. Ma, and Z. Zhang, "Review of Human Gesture Recognition Based on Computer Vision Technology," in Proc. the IEEE 5th Advanced Information Technology, Electronic and Automation Control Conference (IAEAC), Chongqing, China, 2021, pp. 1599–1603, https://doi.org/10.1109/IAEAC50856. 2021.9390889.
31. P. K. Pisharady, and M. Saerbeck, "Recent Methods in Vision-based Hand-gesture Recognition: A Review," Computer Vision and Image Understanding, vol. 141, pp. 152–165, 2015, https://doi.org/10.1016/j.cviu.2015.08.004.
32. "Gesture Recognition Market Size, Share & Trends Analysis Report by Technology (Touch-based, Touchless), By Industry (Automotive, Consumer Electronics, Healthcare), By Region, and Segment Forecasts, 2022 – 2030," https://www.grandviewresearch.com/industry-analysis/gesture-recognition-market, [Retrieved: April, 2023].
33. M. J. Cheok, Z. B. Omar, and M. H. Jaward, "A Review of Hand Gesture and Sign Language Recognition Techniques," International Journal of Machine Learning and Cybernetics, vol. 10, pp.131–153, 2019, https://doi.org/10.1007/s13042-017-0705-5.
34. Z. Černeková, N. Nikolaidis, and I. Pitas, "Single Camera Pointing Gesture Recognition Using Spatial Features and Support Vector Machines," in Proc. the 15th European Signal Processing Conference, Poznan, Poland, 2007, pp. 130–134.
35. K. V. Eshitha, and S. Jose, "Hand Gesture Recognition Using Artificial Neural Network," in Proc. the International Conference on Circuits and Systems in Digital Enterprise Technology (ICCSDET), Kottayam, India, 2018, pp. 1–5, https://doi.org/10.1109/ICCSDET.2018.8821076.
36. X. Chen, G. Wang, H. Guo, C. Zhang, H. Wang, and L. Zhang, "Motion Feature Augmented Recurrent Neural Network for Skeleton-Based Dynamic Hand Gesture Recognition," in Proc. the IEEE International Conference on Image Processing (ICIP), Beijing, China, 2017, pp. 2881–2885, https://doi.org/10.1109/ICIP.2017.8296809.
37. K. Lai, and S. N. Yanushkevich, "CNN + RNN Depth and Skeleton based Dynamic Hand Gesture Recognition," in Proc. the 24th International Conference on Pattern Recognition (ICPR), Beijing, China, 2018, pp. 3451–3456, https://doi.org/10.1109/ICPR.2018.8545718.
38. Z. Yang, Y. Li, W. Chen, and Y. Zheng, "Dynamic Hand Gesture Using Hidden Markov Model," in Proc. the 7th International Conference on Computer Science & Education (ICCSE), Melbourne, Australia, 2012, pp. 360–365, https://doi.org/10.1109/ICCSE20062.2012.
39. S. Shin, and W. Sung, "Dynamic Hand Gesture Recognition for Wearable Devices with Low Complexity Recurrent Neural Networks," in Proc. the IEEE International Symposium on Circuits and Systems (ISCAS), Montreal, QC, Canada, 2016, pp. 2274–2277, https://doi.org/10.1109/ISCAS.2016.7539037.
40. J. Jo, S. Hwang, S. Lee, and Y. Lee, "Multi-Mode LSTM Network for Energy-Efficient Speech Recognition," in Proc. the International SoC Design Conference (ISOCC), Daegu, South Korea, 2018, pp. 133–134, https://doi.org/10.1109/ISOCC.2018.8649913.

41. J. Billa, "Dropout Approaches for LSTM Based Speech Recognition Systems," in Proc. the IEEE International Conference on Acoustics, Speech and Signal Processing (ICASSP), Calgary, AB, Canada, 2018, pp. 5879–5883, https://doi.org/10.1109/ICASSP.2018.8462544.
42. A. Graves, N. Jaitley, and A.-R. Mohamed, "Hybrid Speech Recognition with Deep Bidirectional LSTM," in Proc. the IEEE Workshop on Automatic Speech Recognition and Understanding, Olomouc, Czech Republic, 2013, pp. 273–278, https://doi.org/10.1109/ASRU.2013.6707742.
43. P. P. Sahu, V. Singh, I. Kiran, V. Veera, T. Abhinav, A. Vijay, and S. M. Venkatesan, "Personalized Handwriting Recognition Using Continued LSTM Training," in Proc. the 14th IAPR International Conference on Document Analysis and Recognition (ICDAR), Kyoto, Japan, 2017, pp. 218–223, https://doi.org/10.1109/ICDAR.2017.44.
44. M. Wang, L. Song, X. Yang, and C. Luo, "A parallel-fusion RNN-LSTM architecture for image caption generation," in Proc. the IEEE International Conference on Image Processing (ICIP), Phoenix, AZ, USA, 2016, pp. 4448–4452, https://doi.org/10.1109/ICIP.2016.7533201.
45. T.-M. Tai, Y.-J. Jhang, Z.-W. Liao, K.-C. Teng, and W.-J. Hwang, "Sensor-Based Continuous Hand Gesture Recognition by Long Short-Term Memory," IEEE Sensors Letters, vol. 2, no. 3, Article id. 6000704, 2018, https://doi.org/10.1109/LSENS.2018. 2864963.
46. G. Zhu, L. Zhang, P. Shen, and J. Song, "Multimodal Gesture Recognition Using 3-D Convolution and Convolutional LSTM," IEEE Access, vol. 5, pp. 4517–4524, 2017, https://doi.org/10.1109/ACCESS.2017.2684186.
47. L. Zhang, G. Zhu, P. Shen, J. Song, S. A. Shah, and M. Bennamoun, "Learning Spatiotemporal Features Using 3d CNN and Convolutional LSTM for Gesture Recognition," in Proc. the IEEE International Conference on Computer Vision (ICCV), Venice, Italy, 2017, pp. 3120–3128, https://doi.org/10.1109/ICCVW.2017.369.
48. C. R. Naguri, and R. C. Bunescu, "Recognition of Dynamic Hand Gestures from 3D Motion Data Using LSTM and CNN Architectures," in Proc. the 16th IEEE International Conference on Machine Learning and Applications (ICMLA), Cancun, Mexico, 2017, pp. 1130–1133, https://doi.org/10.1109/ICMLA.2017.00013.
49. Y. Wu, B. Zheng, and Y. Zhao, "Dynamic Gesture Recognition Based on LSTM-CNN," in Proc. the Chinese Automation Congress (CAC), Xi'an, China, 2018, pp. 2446–2450, https://doi.org/10.1109/CAC.2018.8623035.
50. D. N. T. How, K. S. M. Sahari, H. Yuhuang, and L. C. Kiong, "Multiple Sequence Behavior Recognition on Humanoid Robot using Long Short-term Memory (LSTM)," in Proc. the IEEE International Symposium on Robotics and Manufacturing Automation (ROMA), Kuala Lumpur, Malaysia, 2014, pp. 109–114, https://doi.org/10.1109/ROMA.2014.7295871.
51. S. Calinon, and A. Billard, "Learning of Gestures by Imitation in a Humanoid Robot," In C. Nehaniv & K. Dautenhahn (Eds.), Imitation and Social Learning in Robots, Humans and Animals: Behavioral, Social and Communicative Dimensions, pp. 153–178, Cambridge University Press: Cambridge, UK, https://doi.org/10.1017/CBO9780511489808.012.
52. A. Vaswani, N. Shazeer, N. Parmar, J. Uszkoreit, L. Jones, A. N. Gomez, et el., "Attention Is All You Need" In Proc. the 31st Conference on Neural Information Processing Systems (NIPS), Long Beach, CA, USA, 2017, Article 30.
53. A. Dosovitskiy, L. Beyer, A. Kolesnikov, D. Weissenborn, X. Zhai, T. Unterthiner, et el., "An Image is Worth 16x16 Words: Transformers for Image Recognition at Scale", In Proc. The Ninth International Conference on Learning Representations (ICLR), 2021, available at https://openreview.net/pdf?id=YicbFdNTTy, [Retrieved: May, 2023].
54. A. Singh, and A. Bansal, "An Integrated Analysis for Identifying Iconic Gestures in Human-Robot Interactions," in Proc. the IntelliSys Conference, Amsterdam, The Netherlands, 2023, in press.
55. A. Singh, and A. Bansal, "Synchronized Colored Petri Net based Multimodal Modeling and Real-time Recognition of Conversational Spatial Deictic Gestures," in Proc. the Computing Conference, London, United Kingdom, 2023, in press.
56. P. Ekman, and W. V. Frisen, "The Repertoire of Nonverbal Behavior: Categories, Origins, Usage and Coding," Semotica, vol. 1, 49–98, 1969.

57. J. S. Copley, Wikimedia Commons, SamuelAdamsLarge - Category:Samuel Adams - Wikimedia Commons [Retrieved: April, 2023].
58. P. Pellicer, Wikimedia Commons, https://commons.wikimedia.org/wiki/File:Pina_Pellicer_publicity_photos_for_One-Eyed_Jacks_(1961)_(cropped).jpg, [Retrieved: April, 2023].
59. W. Reising, "Understanding Petri Nets: Modeling Techniques, Analysis Methods, Case Studies," Springer-Verlag: Berlin, 2013, https://doi.org/10.1007/978-3-642-33278-4.
60. K. Jensen, "A Brief Introduction to Colored Petri Nets: Tools and Algorithms for the Construction and Analysis of Systems," in Proc. the International Workshop on Tools and Algorithms for the Construction and Analysis of Systems, LNCS, vol. 1217, pp. 203–208. Springer: Heidelberg, Germany, 1997, https://doi.org/10.1007/BFb0035389.
61. J. Wang, "Timed Petri Net: Theory and Applications," Springer Science + Business Media: New York. 1998, https://doi.org/10.1007/978-1-4615-5537-7.
62. J. F. Allen, "Maintaining Knowledge about Temporal Intervals," Communications of the ACM, vol. 26, no. 11, pp. 832–843, 1983, https://doi.org/10.1145/182.358434.
63. M. Chein, and M. L. Mugnier, "Conceptual Graphs: Fundamental Notions," Revue d'Inteligence Artificielle, vol. 6, no. 4, pp. 365–406, 1992.
64. L-P. Morency, I. Kok, and J. Gratch. "Context-based Recognition during Human Interactions: Automatic Feature Selection and Encoding Dictionary," in Proc. the Tenth International ACM Conference on Multimedia Interfaces (ICMI), Chania, Crete, Greece, 2008, pp. 181–188, https://doi.org/10.1145/1452392.1452426.
65. G. Ball, and J. Breese, "Relating Personality and Behavior: Posture and Gestures," In Proc. the International Workshop on Affective Interactions (IWAI), Siena, Italy, 1999, Springer: Heidelberg, Germany, LNCS 1814, pp. 196–203, 2000.
66. P. Bremner, A. Pipe, C. Melhuish, M. Fraser, and S. Subramanian, "Conversational Gestures in Human-Robot Interaction," In Proc. the IEEE International Conference on Systems, Man, and Cybernetics, San Antonio, TX, USA, 2009, pp 1645–1649.
67. M. Salem, S. Kopp, I. Wachsmuth, and F. Joublin, "Towards Meaningful Robot Gesture," Human Centered Robot Systems: Cognitive Systems Monographs, H. Ritter, G. Sagerer, R. Dillmann, and M. Buss (eds.), Springer: Berlin, Germany, vol. 6, pp. 173–182, 2009.
68. J. Stolzenwald, and P. Bremner, "Gesture Mimicry in Social Human-Robot Interaction," In Proc. the 26th IEEE International Symposium on Robot and Human Interactive Communication (RO-MAN), Lisbon, Portugal, 2017, pp 430–436.
69. C. J. Fillmore, "Towards a Descriptive Framework for Spatial Deixis," Speech Place and Action, Studies in Deixis and Related Topics, pp. 31–59, 1982.
70. A. Stukenbrock, "Deixis, Meta-perceptive Gaze Practices and the Interactional Achievement of Joint Attention," Frontiers in Psychology, vol. 11, Article 1779, 2020, https://doi.org/10.3389/fpsyg.2020.01779.
71. C.T. Ishi, C. Liu, H. Ishiguro, and N. Hagita, "Head Motion during Dialogue Speech and Nod Timing Control in Humanoid Robots," in Proc. the 5th ACM/IEEE International Conference on Human-Robot Interaction (HRI), Osaka, Japan, 2010, pp. 293–300, https://doi.org/10.1109/HRI.2010.5453183.
72. A. Kapoor, and R. W. Picard, "A Real-time Head nod and Shake Detector," in Proc. the Workshop on Perceptive User Interfaces (ICMI-PUI), Orlando, Florida, USA, 2001, pp. 1–5, https://doi.org/10.1145/971478.971509.
73. W. Tan, and G. Rong, "A Real-time Head Nod and Shake Detector using HMMs," Expert Systems with Applications, vol. 25, no. 3, pp. 461–466, 2003, https://doi.org/10.1016/S0957-4174(03)00088-5.
74. J. Saunders, D. S. Syrdal, K. L. Koay, N. Burke, and K. Dautenhahn, "Teach Me–Show Me—End-User Personalization of a Smart Home and Companion Robot," IEEE Transactions on Human-Machine Systems, vol. 46, no. 1, pp. 27–40, 2016, https://doi.org/10.1109/THMS.2015.2445105.
75. L. Dong, Y. Jin, L. Tao, and G. Xu, "Recognition of Multi-Pose Head Gestures in Human Conversations," in Proc. the Fourth International Conference on Image and Graphics (ICIG), Chengdu, China, 2007, pp. 650–654, https://doi.org/10.1109/ICIG.2007.176.

76. C. Chao, and A. L. Thomaz, "Timing in Multimodal Turn-taking Interactions: Control and Analysis using Timed Petri Nets," Journal of Human-Robot Interaction, vol. 1, no. 1, pp. 4–25, 2012, https://doi.org/10.5898/JHRI.1.1.Chao.
77. L. Zheng, B. Liang, and A. Jiang "Recent Advances of Deep Learning for Sign Language Recognition," In Proc. the International Conference on Digital Image Computing: Techniques and Applications (DICTA), Sydney, Australia, 2017, pp. 1–7, https://doi.org/10.1109/DICTA.2017.8227483.
78. W. Liu, and Y. Du, "Modeling Multimedia Synchronization using Petri Nets," Journal of Information Technology, vol. 8, no. 7, pp. 1054–1058, 2009, https://doi.org/10.3923/itj.2009.1054.1058.
79. C. C. Chiu, L.-P. Morency, and S. Marsella, "Predicting Co-verbal Gestures: A Deep and Temporal Modeling Approach," In Proc. the International Conference on Intelligent Virtual Agents (IVA), Los Angeles, CA, USA, 2015. LNCS, vol 9238, pp. 152–166, Springer: Cham, Switzerland, https://doi.org/10.1007/978-3-319-21996-7_17.

Computational Intelligence Methods for User Matching

Yongjun Li, Shuai Yang, and Wenqiang He

Abstract User identification refers to the process of user matching accounts across various social media platforms, which has numerous real-world applications. However, there are still many issues here, mainly in efficiency and effectiveness. As the time complexity of direct one-to-one user matching is $O(mn)$ (Suppose there are m users on one platform and n users on another platform), the computation time increases exponentially as the number of users grows. Therefore, we explored methods to reduce the number of matching pairs. Before beginning formal computation, we propose method to filter users by record data, thereby eliminating the vast majority of unlikely candidate pairs and retaining as many real candidate pairs as possible. This approach can significantly reduce computation time. Besides, current user trajectory-based methods tend to focus separately on spatial and temporal data and fail to fully leverage the interdependence between them. In contrast, our approach integrates spatial-temporal information to enhance user identification accuracy, through a three-step process. First, we use kernel density estimation to measure the similarity of users' trajectories, taking both spatial and temporal information into account. Second, we assign weights to each check-in record to prioritize discriminative ones. Finally, we utilize inconsistencies among check-in records to compute penalties for trajectory similarity. By identifying account pairs with similarity scores above a predefined threshold, we can determine whether they belong to the same user. We evaluated our method on three ground-truth datasets, demonstrating its competitive performance.

Keywords User matching · Efficiency · Effectiveness · Spatial-temporal data · Locality-sensitive hashing

---

Y. Li (✉) · S. Yang
Northwestern Polytechnical University, Xian, China
e-mail: lyj@nwpu.edu.cn

W. He
University of Science and Technology of China, Hefei, China

K. Daimi et al. (eds.), *Cutting Edge Applications of Computational Intelligence Tools and Techniques*, Studies in Computational Intelligence 1118,
https://doi.org/10.1007/978-3-031-44127-1_4

1 Introduction

As social media continues to grow in popularity, more and more users are registering accounts on multiple platforms. Cross-site user identification (CSUI) has emerged as a popular method for leveraging user data across social networks, enabling the linking of different social sites, enhancing user profiles, and facilitating research on information diffusion [1]. The benefits of CSUI extend to a variety of applications, including recommendation systems [2] and marketing efforts.

Current approaches mainly focus on measuring the similarity between two user accounts, and then converting the matching task into a classification problem. These approaches can be broadly divided into three categories. The first category is based on user profile information such as the display name, gender, and age to calculate similarity. However, these attributes can be hidden or falsified for various purposes. The second category measures similarity based on the friendship network, but the availability of this information is limited due to privacy concerns [3, 4], and even if available, the connections may be sparse [5]. The third category measures similarity based on user-generated content (UGC), such as posting time, location, writing style, and semantics. However, extracting and representing UGC in an appropriate manner remains a challenging research question. As previously mentioned, current methods face difficulty in obtaining enough genuine user data or informative features for user identification, which may restrict the robustness of the current approaches [6].

Recent developments have emphasized the utilization of spatial-temporal information to match user accounts across various social sites [7–11] due to the widespread use of GPS-enabled mobile devices. This allows users to share their statuses that are timestamped and tagged with location information. These tagged statuses offer numerous advantages: (1) they are publicly available with fewer privacy limitations; (2) they are more consistent across social sites; and (3) they are relatively simple to process when represented as longitude and latitude. So Riederer et al. [8] divided time and location into bins and calculated the similarity between two accounts based on the co-occurrences in the bins. However, this discretization approach may result in information loss. Chen et al. [12] addressed this issue by using continuous variables to represent time and location and handling them separately. However, spatial and temporal information in check-in records are often closely related, and their work did not capture this coupling relation well.

Apart from the aforementioned issues, current methods primarily concentrate on extracting behavioral patterns of users, devising similarity metrics for accurate identification, and neglect the challenge of extensive computations required for pairwise comparison. This bottleneck undoubtedly hampers the applicability of CSUI techniques on large-scale social platforms [13]. We will discuss some methods to address these issues in the next section.

2 Efficiency and Effectiveness of User Matching

In this section, we mainly studied user matching from two aspects: efficiency and effectiveness. In terms of efficiency, we preprocessed candidate pairs using Locality-Sensitive Hashing (LSH) to reduce the number of candidate pairs. In terms of effectiveness, we investigated the kernel density function that considers the coupled information of spatial and temporal.

2.1 Efficiency

As shown in Fig. 1, assuming there are m users on platform A and n users on platform B. If we directly match the users between the two platforms, the time complexity would be $O(mn)$. However, the maximum number of true matches (user u_1 on platform A and user u_2 on platform B are the same person) is less than $min\{m, n\}$. Therefore, most candidate pairs are not true matches, so before calculating the similarity of the candidate pairs ($\{u_1, u_2\}$ represent a candidate pair), it is necessary to first filter out a large number of impossible matches using a low time complexity approach, while preserving as many true candidate pairs as possible becomes an essential task [13].

In our previous study [14], we proposed a modulo hashing function for Euclidean distance to tackle the issue of identical spatial data but different timestamps, which ignored the temporal data. Based on this idea, statuses with the same location but different timestamps were treated as the same.

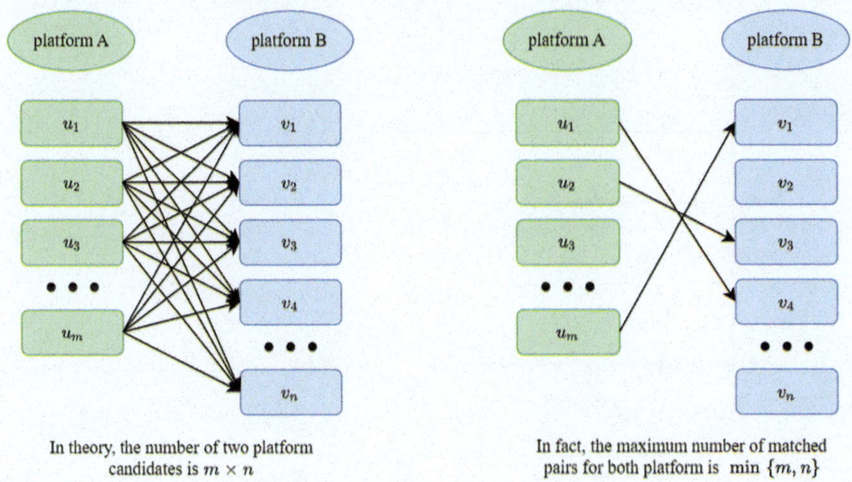

Fig. 1 An instance of two platform's matching

We propose a new method called Binary-search-based Locality-Sensitive Hashing (BLSH) [13] to address the issue of identifying similar spatial-temporal records. Our approach is trajectory-oriented and uses a BLSH encoding function to convert each record into a binary hash code for easier processing. This function is inspired by the technology of locality-sensitive hashing (LSH) [15–17] and binary search. We also incorporate the idea of nearest neighbors searching [18, 19] to narrow down each user's candidates. By searching BLSH buckets where similar user statuses are clustered and dissimilar ones are separated, we can efficiently build a cluster of k_1-nearest neighbors for each user, called *Neighborhood*, thereby avoiding full-scale pairwise comparison and significantly reducing computation costs.

Moreover, we have developed a hierarchical discrete attention mechanism in our approach, which carefully takes into account the spatial arrangement of BLSH buckets and also harmonizes the spatial-temporal attributes of each record [7]. This results in improved discriminability. Thus, our BLSH method has now been enhanced into a hierarchically attentioned binary-search-based LSH (HA-BLSH).

2.2 Effectiveness

In order to enhance the accuracy of matching, considering time and space together is necessary. This is specifically demonstrated in Fig. 2, which can help address a portion of the false positive problem. Suppose there are two users, u_1 and u_2, from different social sites, each with three check-in records. User u_1 visited three different locations, l_1, l_2, and l_3, at t_1, t_2, and t_3, respectively. User u_2 visited l_2, l_3, and l_1 at t_1, t_2, and t_3, respectively. In other words, the two users visited different locations at the same time. In this scenario, it is evident that u_1 and u_2 are not the same user.

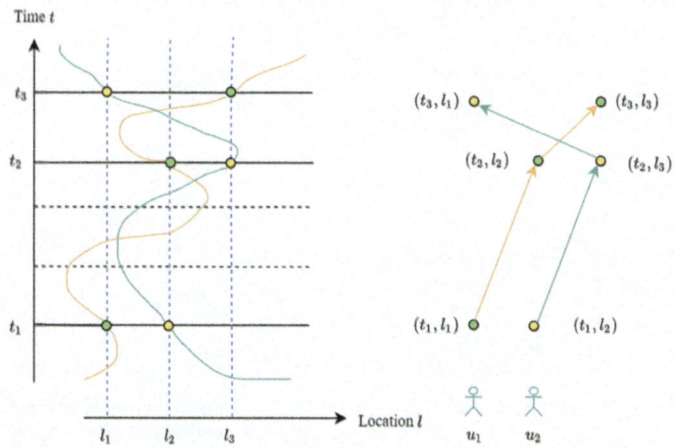

Fig. 2 An instance of two users' check-in records

However, according to [8, 12], u_1 and u_2 are considered to be the same user since they have the same records of visited locations or visit times.

To tackle the aforementioned problem, we propose a novel user identification method with spatiotemporal awareness(UIDwST) [7]. Firstly, we use kernel density estimation (KDE) to measure the proximity between two users' check-in records, which considers the relationship between spatial and temporal data in a check-in record. We assign different weights to the check-in records to improve the effectiveness, giving more weight to the more informative ones. We then introduce a penalty term to the proximity measurement if two check-in record sets have conflicting records. Finally, we calculate the similarity of the two accounts. If the similarity exceeds a certain threshold, we conclude that the two accounts belong to the same user.

3 The Process of User Matching

We elaborate on the pre-filtering process in user matching and the similarity calculation method that combines spatiotemporal coupling information in detail in this chapter. In the last section, we present some practical examples of user matching.

3.1 *Pre-filtering*

Figure 3 illustrates an overview of our BLSH, whose input is a set of user records and output is the matched pairs. Our method includes four components illuminated as follows.

(1) *BLSH Encoding:* We encode each record into a hash code with the binary-search-based LSH function.
(2) *Buckets Filtering:* We project all records into the hash buckets established on the BLSH codes, which would efficiently cluster similar or identical records on the condition of a certain precision.

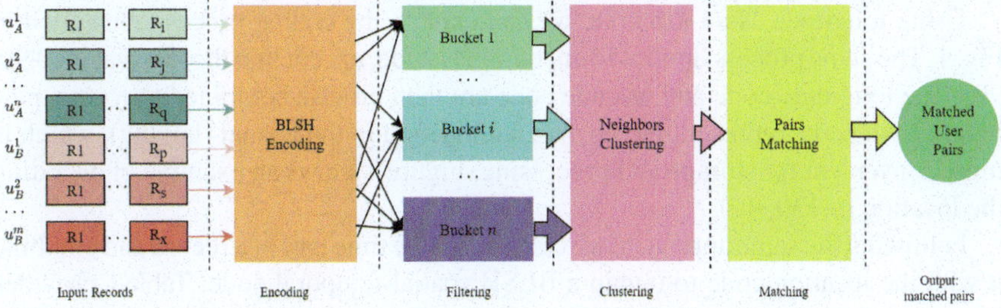

Fig. 3 Overview of the BLSH method

(3) *Neighbors Clustering:* By traversing the BLSH buckets, we construct a cluster of k_1-nearest neighbors for each user as its *Neighborhood*.
(4) *Pairs Matching:* In a user's *Neighborhood*, we compute the KDE-based trajectory similarity between it and each of its neighbors with a hierarchy of attention and then pick out top $k_2 (k_2 \leq k_1)$ similar users as its matched pairs.

3.1.1 BLSH Encoding

A data record consists of two components: temporal and spatial information, which makes similarity measurement computationally challenging. Although the spatial information can be converted into longitude and latitude, the computational complexity remains significant for large datasets. Thus, it is essential to simplify the representation of a record.

LSH technology [15, 16] is a promising solution to address the challenge of the (R, c)-nearest neighbor problem and reduce the computational cost. In this chapter, we propose a binary search-based LSH function [17] to encode each record into a unique binary hash code. To illustrate the encoding process, we first use a 1-D interval, such as longitude, as an example, and then explain the encoding process of the record. Given an interval (l,r) to be encoded with an encoding error of α, the encoding process is as follows.

Step I: Bisect interval (l,r) into two subintervals, left and right, with 0 being the first bit of binary hash code on the left and 1 being the first bit on the right.

Step II: In the ith ($i \geq 2$) iteration, bisect each of subintervals with 0 being the ith bit on the left and 1 being the ith bit on the right. That is, adding a bit to the binary hash code bisects a subinterval, effectively zooming in to a more detailed interval.

Step III: Repeat the Step II until $|r' - l'| \leq \alpha$(suppose that (l', r') is an interval after the ith iteration). At this point, each subinterval is encoded as a binary hash code of length i.

Suppose that number $n (n \in (l, r))$ needs to be encoded. After the above i iterations, number n must fall into one of the subintervals, denoted by ϑ. The binary hash code of the subinterval ϑ is the code of the number n. It is obvious that, $\forall m \in \vartheta$, m and n have identical code. Furthermore, the closer the two numbers are, the more bits their codes have in common. In other words, α is the coding resolution. The smaller α, the more precise the location and the more bits in code.

Using longitude as an example, we can explain the coding process illustrated in Fig. 4. The same process applies to encoding latitude, except that the interval is (−90, 90). The longitude code and latitude code are then bitwise combined to form a new binary hash code, representing the original location in the format (lat, lng), which is more convenient for storage and processing. Figure 5 shows an example of encoding the location (−130, 40).

Following the same approach, we can encode the time part of a record and combine it with the location code to obtain a BLSH spatial-temporal code. Table 1 provides an example of a record and its BLSH code.

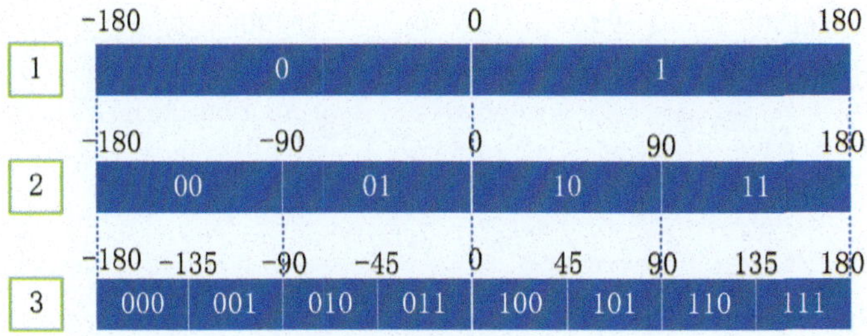

Fig. 4 Procedure of the BLSH encoding longitude

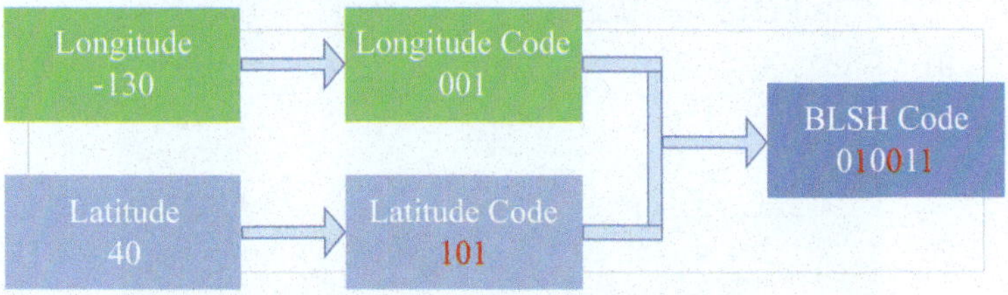

Fig. 5 Procedure of the BLSH encoding location

Table 1 Example of BLSH encoding where α for location is 0.01 and α for time is 1

Record	Time:20150104 17:19, Location:(−5.99420320851,37.3972548371)
Time code	11000000001000100010010010100 0x18044494
Location code	01101111100110111000111011101 0x1BE6E3BD
Spatial-temporal	0x1BE6E3BD1917FFFD

3.1.2 Buckets Filtering

From BLSH's encoding rules, it is clear that records with similar locations will share more common bits in their codes from the beginning. This property of BLSH encoding makes it an effective method for preparing data for BLSH buckets filtering, which can efficiently cluster similar records and separate dissimilar ones.

To construct BLSH buckets, we use the first n (where $n = 2k, k \in N$) bits of location codes as their ID numbers and map them to spatial areas that are evenly partitioned by bisecting longitude and latitude. In other words, during the BLSH encoding process, we divide the space into 2^n cells, and each cell acts as a bucket containing records that share the same first n bits in their location codes. Figure 6 provides a visual example of what BLSH buckets might look like in 2-D space.

The partitioning and distribution of BLSH buckets in space depend on the combination mode of longitude and latitude codes. While a bucket can take any shape, such

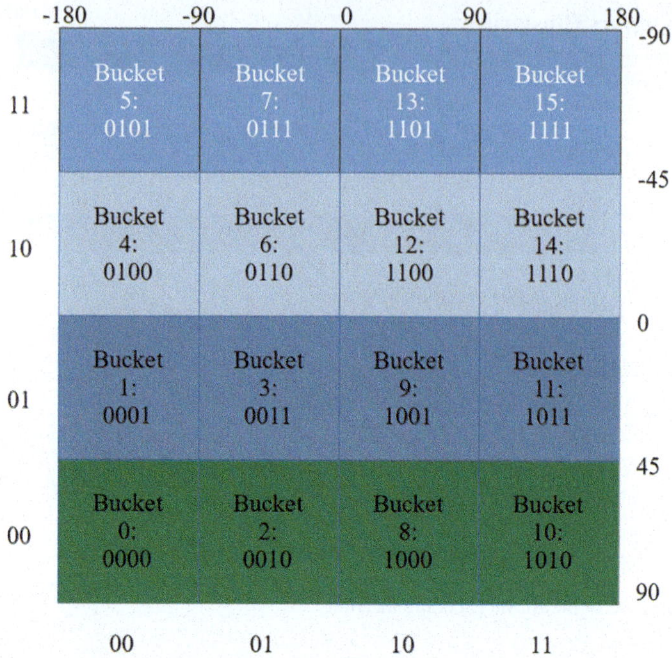

Fig. 6 Example of BLSH buckets where $\alpha = 90$ for longitude and $\alpha = 45$ for latitude

as a circle, square, segment, or sphere, it is reasonable to assume that more buckets in space result in smaller partitioned spaces allocated to each bucket, thereby increasing the precision that each bucket represents.

Next, we project all records into the buckets according to their first n bits of location codes. By doing this, we successfully make the similar records cluster together in high efficiency. Figure 7 visualizes the separating and clustering process as follows.

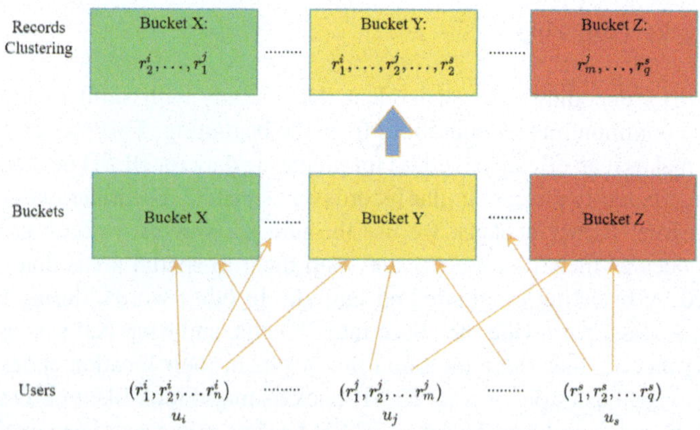

Fig. 7 Procedure of bucket's classificaiton

3.1.3 Neighbors Clustering

After clustering similar records, we create a *Neighborhood* for each user by searching for neighbors.

If we simply traverse each BLSH bucket and consider the neighbors of the target user as its *Neighborhood*, we may encounter a large number of false positives due to the existence of popular records. For instance, many people may visit the same restaurant at dinner time on weekends. To tackle this issue, we calculate a simple statistical similarity, referred to as Sim_s, between the target user and each of its neighbors to identify the nearest k_1 ones. Finally, we add the k_1-nearest neighbors to the target user's *Neighborhood*.

The statistical similarity is defined based on the Term Frequency-Inverse Document Frequency(TF-IDF) technique [20, 21], which examines the frequency patterns. The concept behind the statistical similarity is quite straightforward: the closer the frequency patterns of two users are, the more likely they are to be genuine pairs. Therefore, to measure the distance between their frequency patterns, we use the geometric mean, which better reconciles the TF-IDF values of any two users.

Assume that there are two users, u and v, and they share n BLSH buckets. The statistical similarity between u and v is defined as Eq. 1:

$$Sim_s(u, v) = \sum_{i=1}^{n} \left(idf_i \times \sqrt{tf_i^u \times tf_i^v} \right) \tag{1}$$

In the formula, idf_i represents the IDF value of the ith bucket, which indicates how effective the bucket is at distinguishing dissimilar users. tf_i^u represents the TF value of user u in the ith bucket, which reflects the importance of u's records in that bucket to u's trajectory.

After calculating a simple similarity score, we output the top k_1 users with the highest similarity score for each user, which form a set of candidate matches. We can then use other methods to further refine the user matching process.

3.2 User's Similarity with Spatiotemporal Awareness

Numerous methods for user matching already exist, and some of them use check-in records for processing. However, the main challenge lies in measuring the proximity between the check-in records of two users, u_1 and u_2. Our work introduces a novel approach UIDwST [7] that takes into account the correlation between the spatial and temporal information in a check-in record. Although our problem deals with identifying users on two social sites, it can be generalized to identifying users across multiple social sites by measuring the pairwise similarity.

The idea behind UIDwST is to identify whether two user accounts u_1 and u_2 from two different social sites belong to the same individual by considering the proximity and non-conflict of their check-in records. UIDwST consists of three components: (1) measuring the proximity between two sets of check-in records, (2) assigning weights to each check-in record, and (3) penalizing the similarity measure when conflicting check-in records occur. The key is to identify if some of the check-in records on both sites are close to each other, and if they do not conflict, which would indicate that the accounts are likely to belong to the same individual. We explain these three components in more detail below.

3.2.1 Promimity of Check-In Record Sets

Typically, the more similar online activities that exist in the user trajectories of u_1 and u_2, the more likely it is that the two accounts belong to the same user. However, it can be challenging to align check-in records generated by the same user for the same offline activity on different social sites. This is because: (1) check-in records from users tend to be sparse, [10, 22] (2) users may not publicly share all their behaviors on all registered sites, and (3) the location and time of the same user activity may differ on different social sites. These factors make it difficult to measure the similarity of user check-in records. To address these challenges, we propose a basic kernel-density-estimation-based method to calculate the proximity of check-in records.

To illustrate the approach, we provide an example in Fig. 8. Consider the case of user u_1 with his check-in records $R^1 = \{r_j^1 | 1 \leq j \leq n\}$ and user u_2 with $R^2 = \{r_i^2 | 1 \leq i \leq m\}$. Generally, users may not always publish their behaviors on different sites at the exact same time but rather with minor deviations. Li et al. [5] found that this deviation on different sites may vary up to 5 d, which we denote as δ. Namely, the check-in record r_i^2 may correspond to the same behavior of a record in the set of check-in records $R_{t_i}^1$, where $R_{t_i}^1 = \{r_k^1 | t_i - \delta \leq r_k^1.t \leq t_i + \delta\}$. Inspired by the idea

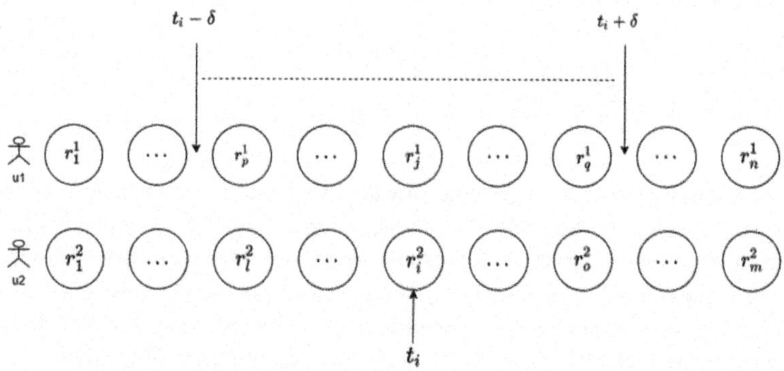

Fig. 8 An illustraion of UIDwST. Each check-in record at t_i of u_2 is compared with the ones of u_1 within a time window $[t_i - \delta, t_i + \delta]$,where δ reflects our observation that there is usually a minor deviation in the check-in time

of [10, 23], the density of r_i^2 over $R_{t_i}^1$ can be defined as $f\left(r_i^2 | R_{t_i}^1, h\right)$. Which be used to measure the proximity between r_i^2 and $R_{t_i}^1$. The larger the value of $f\left(r_i^2 | R_{t_i}^1, h\right)$ is, the more similar r_i^2 is to the check-in records in $R_{t_i}^1$. Based on the above definition, we introduce the definition of the proximity between check-in records of u_1 and u_2 as follows.

$$Sim\left(R^1, R^2\right) = \frac{1}{|R^2|} \sum_{r_i^2 \in R^2} f\left(r_i^2 | R_{t_i}^1, h\right) \tag{2}$$

We want to point out that Eq. 2 is not symmetric, meaning that $Sim\left(R^1, R^2\right)$ is not always equal to $Sim\left(R^2, R^1\right)$. This lack of symmetry can potentially result in different identification outcomes, especially when user trajectories are frequently tracked, such as when using GPS for taxi tracking. However, in this chapter, we focus on the check-in activities of users. Since the number of check-in records for a user is usually not large, the proximity between $Sim\left(R^1, R^2\right)$ and $Sim\left(R^2, R^1\right)$ is typically small. Therefore, the asymmetry of Eq. 3 does not have a significant impact on the proposed method.

3.2.2 Weighting of Check-In Records

People's daily activities exhibit regular patterns. For instance, students usually have lunch around noon and leave school for home in the late afternoon. These typical activities are less informative for user identification as they are commonly shared by a certain group of people. On the other hand, if a check-in record indicates that a student had lunch at a location outside the school area, it can be strong evidence for identifying that student from others. Therefore, check-in records for different activities may convey different volumes of meaningful information, and we need to prioritize them accordingly. We propose a TF-IDF-based weighting scheme to achieve this.

The time required for different individuals to carry out the same activity can vary, such as different students having lunch in different time slots. Furthermore, the same individual may share the same behavior on Facebook at 10 am and wait to post it on Twitter at 11 am. Hence, the time and/or location of check-in records for the same activity may differ. To capture these deviations in identifying the same user behavior, we introduce the definition of *identical check-in records* with two relaxed conditions.

$$K_h(r_i^2, r_k^1) = \frac{1}{\sqrt{2\pi}h} exp\left(-\frac{1}{2}\left(\frac{||r_i^2.l - r_k^1.l||_2}{h}\right)^2\right) \tag{3}$$

$$f_w(r_i^2 | R_{t_i}^1, h) = \frac{1}{|R_{t_i}^1|} \sum_{r_k^1 \in R_{t_i}^1} (w(r_k^1) \cdot K_h(r_i^2, r_k^1)) \tag{4}$$

$$Sim_w(R^1, R^2) = \frac{1}{|R^2|} \cdot \sum_{r_i^2 \in R^2} \left(w(r_i^2) \cdot f_w(r_i^2 | R_{t_i}^1, h)\right) \tag{5}$$

The specific formulas are shown above. Where h is the variance parameter and the $K_h(\cdot)$ is the kernel function. $w(r_k^1)$ represents the weight of r_k^1 in $R_{t_i}^1$. $Sim_w(R^1, R^2)$ represents the simiarity of R^1 and R^2.

3.2.3 Penalty for Similarity

Generally, it is very unlikely that a user checks in at different locations in the same time slot. In the context of user identification, for any two check-in records, (l_1, t_1) and (l_2, t_2) , of the same user, if the check-in time is the same, i.e., $t_1 = t_2$, the difference between the check-in locations $||l_1 - l_2||_2$ should be small enough such that (l_1, t_1) and (l_2, t_2) can be identified as the same place geographically. In the cases where $||l_1 - l_2||_2$ are large enough to be identified as two separate locations, then the two check-in records are less likely to be of the same user. We refer to these check-in records with the same temporal but different spatial information as conflicting check-in records.

$d(R^1, R^2)$ represents the number of conflicting check-in records between R^1 and R^2. We define the penalty term based on the variant of the *sigmoid* function as follows, which falls within the range [0,1], The term gives a larger value when the number of conflicting check-in records is larger.

$$P(R^1, R^2) = \frac{2}{1 + exp(-d(R^1, R^2))} - 1 \tag{6}$$

Based on Eqs. 5 and 6,we define the similarith between u_1 and u_2, as follows.

$$Sim(u_1, u_2) = S_w(R^1, R^2) - P(R^1, R^2) \tag{7}$$

3.3 *User Matching*

Given a set of cross-site users, we calculate the similarities of the user. We treat the processing results differently according to different scenarios. For example, for the recommendation, we don't need an extremely precise one-to-one matching. Specifically, given a user u_i, the top n similar candidates are the matching users, as shown in Eq. 8.

$$C(u_i) = Top_n\{Sim(u_i, u_k), \forall u_k \in O_2\} \tag{8}$$

where O_2 represents other platform and $C(u_i)$ is the set of candidates of user u_i.

Another type of application scenario is the accurate matching. Given a user u_i, the similarities between u_i and the candidate users are calculated, and then ranked. We select the most similar user u_k as the matching user, as shown in Eq. 9.

$$C(u_i) = max\{Sim(u_i, u_k), \forall u_k \in O_2\} \tag{9}$$

If user u_i and user u_k are the correct matching, we match them correctly, otherwise we fail.

The last type of application scenario is to try to find all users that match with u_i as much as possible. We select all user u_k as the matching users, as shown in Eq. 10.

$$C(u_i) = \{Sim(u_i, u_k) \geq Sim_\theta, \forall u_k \in O_2\} \quad (10)$$

Sim_θ is the threshold.If the ground-truth matched user u_k is one element of $C(u_i)$, we match them correctly, otherwise we fail.

4 Other Models for User Matching

Basically, the idea of existing works is to measure the similarity of cross-site users based on their profiles [24, 25], their friendship networks [26, 27], or their user-generated content [28], and then utilize the similarity to determine whether two cross-site users are the same person, which focus on the effectiveness of matching a pair of users.

4.1 Based on Username and Display Name

Previous studies primarily rely on the extensive online profiles or activities of users. However, the availability, completeness, and reliability of this online data can be limited due to privacy settings or specific reasons, causing existing methods to not function properly. Nevertheless, users often publicly display their usernames or screen names on various social networks. These names often contain abundant redundant information belonging to the same user, presenting an opportunity to address the matching problem[29].

We are addressing the challenge of user matching accounts on social networks using only usernames and display names. This task involves two main objectives: (1) identifying the information redundancies present in usernames and display names; and (2) using these redundancies to match user accounts. To achieve this, we propose a solution called User Identification across Social Network based on Username and Display name (UISN-UD), which comprises three key components: 1) extracting features that leverage the information redundancies among names based on user naming patterns; (2) training a two-stage classification framework to tackle user identification based on the extracted features; and (3) employing the Gale-Shapley algorithm to eliminate one-to-many or many-to-many relationships in the identification results.

The UISN-UD implementation framework, illustrated in Fig. 9, comprises two major components: (1) user identification based on username and display name; (2) user identification refinement through a one-to-one constraint. To be specific, we refer to part (1) as UISN-UD, and the combination of part (1) and part (2) as UISN-

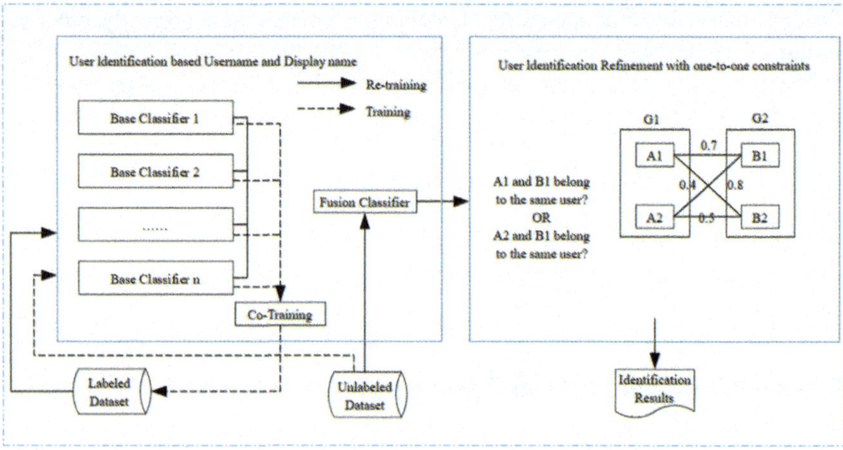

Fig. 9 Framework of UISN-UD

UD with one-to-one constraint. The main objective of UISN-UD is to determine whether a pair of accounts belong to the same user or not. In contrast, UISN-UD with one-to-one constraint is designed to identify the optimal matching scheme for two groups of accounts. More specifically, the framework for UISN-UD with one-to-one constraint includes four key aspects: feature extraction, basic classifier construction, fusion classifier construction, and one-to-one constraint implementation.

4.2 Based on User Friendship

Identifying users has been a topic of interest in academic research. Friendship-based methods have been proposed to enhance the identification accuracy, due to the challenge of replicating friendship networks. However, the impact of information redundancies in k-hop ($k > 1$) neighbors on user identification has not been thoroughly studied in existing literature. Analyzing these issues would aid in comprehending the problem of friendship-based user identification and developing more efficient solutions [30].

To fully characterize the information redundancies in the friendship network for user identification, we begin by obtaining ground-truth friendship networks from three popular social sites. We then analyze the similarities of k-hop neighbors and apply these redundancies in several classifiers to determine their contributions to user identification. Additionally, we combine the friendship-based and display-name-based redundancies to improve the performance and universality of the identification method. Our experiments reveal that the similarities of 1-hop neighbors are most effective for user identification, but the information redundancies of k-hop neighbors ($k > 1$) are also highly useful. Furthermore, jointly applying display-name-based information redundancies can lead to improved performance.

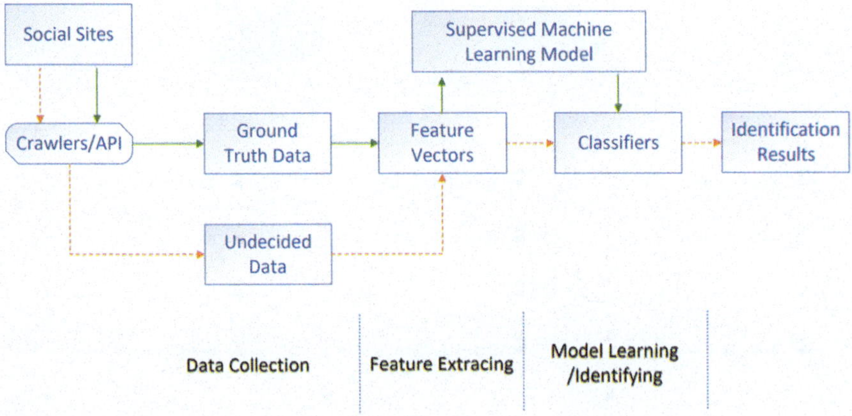

Fig. 10 User friendship-based framework

The friendship network-based user identification model is presented in Fig. 10. The model has a "learning route" (represented by a solid green line) and an "identifying route" (represented by a dotted red line). The ground-truth data is obtained from social media sites using specialized crawlers or API in the learning route. For every pair of user identities, their friendship networks are converted into feature vectors that capture their similarities. These feature vectors serve as the input for a supervised machine learning algorithm. By deploying different learning algorithms, several classifiers can be obtained through this process. In the identifying route, we analyze the similarities of two accounts and their corresponding friendship networks, and encode these similarities as a feature vector. The identification result is established by applying the learned classifier to this feature vector.

4.3 Based on User Generated Content

If two users share multiple similar user-generated contents (UGC), such as having similar content, posting time, and posting location, then it is highly likely that these two users belong to the same offline individual [5].

We introduce a supervised machine learning approach with three main steps. First, we use various algorithms to measure the spatial, temporal, and content similarity between two user-generated contents (UGCs). Second, we extract the corresponding features from these similarities. Finally, we utilize machine learning techniques to match user accounts.

The framework of U-UIM is illustrated in Fig. 11. Each UGC set (TW_i^1, TW_k^2) is represented as a bag of feature vectors $X_{ik} = \{X_{ik}^l, X_{ik}^t, X_{ik}^W\}$. Assuming a set of identified users $\{X_{ik}, y_{ik}\}$ is available for training, a cascaded three-level classifier is proposed, based on the labeled data and the U-UIM identification model, to match user accounts across different OSNs.

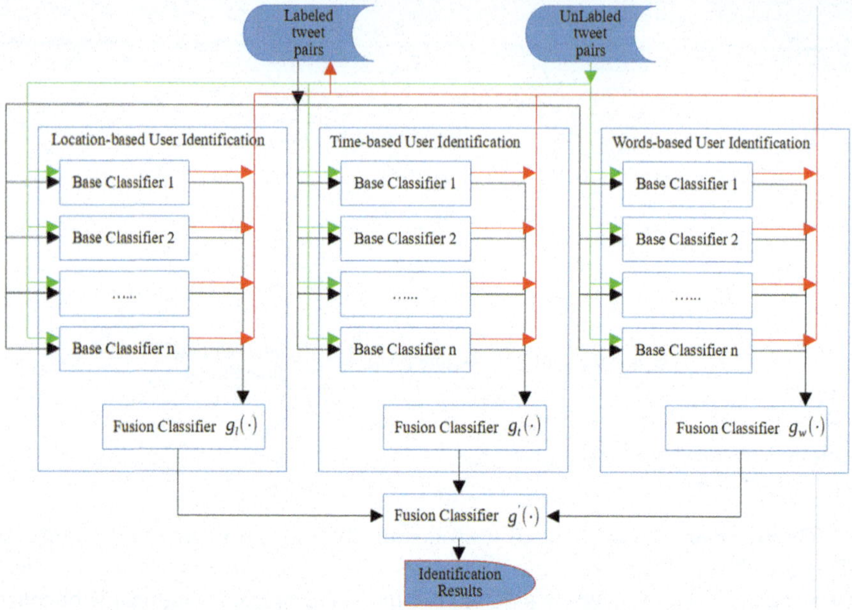

Fig. 11 Framework of U-UIM

5 Challenges and Future of User Matching

There are still several challenges in cross-platform user matching. One major challenge is the issue of data heterogeneity, as different social networks may have different data formats, structures, and levels of data availability. Another challenge is the issue of privacy, as users may have different privacy preferences across different platforms, which can impact the amount and quality of data available for matching. Additionally, the problem of data sparsity, where users may not have a significant presence or activity across all platforms, can also pose a challenge. Finally, the issue of scalability, where the matching process needs to handle a large number of users and platforms, can also be a significant challenge.

Regarding the first data heterogeneity issue, we have investigated username, user relationship, user-generated content, and time-space information of check-in records up to now. However, there are still new challenges, such as the varying granularity of geographic locations for check-ins on different platforms. We use grid partitioning to solve the problem of location data granularity, but grid data can also lead to new issues. If two locations are located in the border area of the grid, they may be assigned to different grids, causing us to consider them as belonging to different people and resulting in some data loss.

Regarding the second issue of user privacy, users typically exhibit different characteristics on different platforms due to the varying social attributes of each platform. Using only user-generated content to determine if two users are the same person is difficult. We still use check-in records to address this issue because even if the content posted is different, if it is posted at a similar time and location, it is still possible that it belongs to the same person. However, there are also challenges because many users may not post check-in records at similar times on multiple platforms, leading to information loss. In fact, when we are able to identify two users as the same person, it is usually because they are active users. It is difficult to identify silent users.

As for the third issue of data sparsity, we also use the grid partitioning method to solve it. The grid can alleviate the problem of data sparsity, and by adjusting the size of the grid, some sparse points can be put into the same grid. We also use the TF-IDF method to address the issue of hotspots. We assign smaller weights to the places where most people go, and assign higher weights to the places where few people go, which are considered as unique places for a user.

For the last scalability issue, we adopted the idea of Locality Sensitive Hashing (LSH) to encode user location data and put them into hash buckets. We first form candidate pairs of users within each hash bucket, then calculate a simple similarity metric for each candidate pair, and finally filter the candidate pairs using a threshold, only keeping those pairs with a similarity score above the threshold., greatly reducing the computational complexity. However, the worst-case time complexity of this method is still $O(mn)$. In theory, the best-case time complexity is $O(\frac{mn}{g})$, where g is the number of grids with users in them. Actually, currently the best case scenario is the ability to filter out 90% of the candidate pairs.

In addition, we are currently researching the application of space-filling curves in candidate pair filtering. Space-filling curves are a method of reducing high-dimensional data to one dimension, and the adjacent points in high dimensions still retain their proximity characteristics when reduced to one dimension, making them suitable for handling location data. Apart from that, we are also researching one-to-many matching and many-to-many matching, which, if successful, could further reduce computation time.

6 Conclusion

With the development of social networks, more and more users are active on multiple platforms. User matching can help better depict user profiles (as individual platform behaviors are often one-sided), thereby providing users with more accurate recommendations for content they are interested in, and also assisting businesses in pushing more accurate advertisements. This has great potential, but there are still many problems with cross-platform user matching. Therefore, we propose a method for filtering candidate pairs, which effectively reduces the time required for user matching. Additionally, we introduce a kernel density method that utilizes both time and space, enhancing the effectiveness of user matching. In addition, we also present some models that utilize other information for user identification, hoping to provide some inspiration to the researchers in the field.

References

1. K. Shu, S. Wang, J. Tang, R. Zafarani, and H. Liu, "User identity linkage across online social networks: A review," *ACM SIGKDD Explor.Newslett.* , vol. 18, no. 2, pp. 5–17, 2017.
2. A. Sapountzi, K.E. Psannis, Social networking data analysis tools and challenges, Future Generation Computer Systems 86 (2018) 893–913.
3. C. Stergiou, K.E. Psannis, T. Xifilidis, A.P. Plageras, B.B. Gupta, Security and privacy of big data for social networking services in cloud, in: Proceedings of 2018 IEEE Conference on Computer Communications Workshops (INFOCOM WKSHPS), Honolulu, HI, United states, 2018, pp. 438–443.
4. Z. Zhang, B.B. Gupta, Social media security and trustworthiness: overview and new direction, Future Generation Computer Systems 86 (2018) 914–925.
5. Y. Li, Z. Zhang, Y. Peng, et al, Matching user accounts based on user generated content across social networks, Future Generation Computer Systems 83 (2018) 104–115.
6. X. Zhou, X. Liang, H. Zhang, et al, Cross-platform identification of anonymous identical users in multiple social media networks, IEEE Transactions on Knowledge and Data Engineering 28 (2) (2016) 411–424.
7. Y. Li, W. Ji, X. Gao, Y. Deng, W. Dong, and D. Li, "Matching user accounts with spatio-temporal awareness across social networks," Inf. Sci., vol. 570, pp. 1–15, Sep. 2021.
8. C. Riederer, Y. Kim, A. Chaintreau, et al, Linking users across domains with location data: theory and validation, in: Proceedings of the 25th International Conference on World Wide Web, 2016, pp. 707–719.
9. X. Han, L. Wang, L. Xu, and S. Zhang, "Social media account linkage using user-generated geo-location data," in Proc. IEEE Conf. Intell. Secur. Informat. (ISI), 2016, pp. 157–162.
10. W. Chen, H. Yin, W. Wang, L. Zhao, and X. Zhou, "Effective and efficient user account linkage across location based social networks," in Proc. IEEE 34th Int. Conf. Data Eng. (ICDE), Los Alamitos, CA, USA, 2018, pp. 1085–1096.
11. J. Feng et al., "DPLink: User identity linkage via deep neural network from heterogeneous mobility data," in Proc. World Wide Web Conf. (WWW), New York, NY, USA, 2019, pp. 459–469.
12. W. Chen, H. Yin, W. Wang, et al, Exploiting spatio-temporal user behaviors for user linkage, in: Proceedings of the 2017 ACM on Conference on Information and Knowledge Management, 2017, pp. 517–526.
13. W. He, Y. Li, Y. Zhang, X. Li,A Binary-Search-Based Locality-Sensitive Hashing Method for Cross-Site User Identification, IEEE Transactions on Computational Social Systems, vol. 10, no. 2, April 2023.
14. Y. Li, X. Li, J. Yang, and C. Gao, "Matching user accounts across large-scale social networks based on locality-sensitive hashing," in Proc. IEEE Int. Conf. Parallel Distrib. Process. Appl., Big Data Cloud Comput., Sustain. Comput. Commun., Social Comput. Netw. (ISPA/BDCloud/SocialCom/SustainCom), 2020, pp. 802–809.
15. A. Gionis, P. Indyk, and R. Motwani, "Similarity search in high dimensions via hashing," in Proc. 25th Int. Conf. Very Large Data Bases, San Francisco, CA, USA, 1999, pp. 518–529.
16. M. Datar, N. Immorlica, P. Indyk, and V. S. Mirrokni, "Locality-sensitive hashing scheme based on p-stable distributions," in Proc. 20th Annu. Symp. Comput. Geometry (SCG), New York, NY, USA, 2004, pp. 253–262.
17. Y. Wang, H. Shen, J. Gao, and X. Cheng, "Learning binary hash codes for fast anchor link retrieval across networks," in Proc. World Wide Web Conf., New York, NY, USA, May 2019, pp. 3335–3341.
18. V. Verroios and H. Garcia-Molina, "Top-K entity resolution with adaptive locality-sensitive hashing," in Proc. IEEE 35th Int. Conf. Data Eng. (ICDE), 2019, pp. 1718–1721.
19. S. Har-Peled, P. Indyk, and R. Motwani, "Approximate nearest neighbor: Towards removing the curse of dimensionality," Theory Comput., vol. 8, no. 1, pp. 321–350, Jul. 2012.
20. G. Salton and C. T. Yu, "On the construction of effective vocabularies for information retrieval," in Proc. Meeting Program. Lang. Inf. Retr. (SIGPLAN), New York, NY, USA, 1973, pp. 48–60.

21. Y. Zhang, J. Tang, Z. Yang, J. Pei, and P. S. Yu, "COSNET: Connecting heterogeneous social networks with local and global consistency," in Proc. 21th ACM SIGKDD Int. Conf. Knowl. Discovery Data Mining, New York, NY, USA, 2015, pp. 1485–1494.
22. W.-H. Chong, E.-P. Lim, Tweet geolocation, Leveraging location, user and peer signals, in: Proceedings of the 2017 ACM on Conference on Information and Knowledge Management, 2017, pp. 1279–1288.
23. M. Lichman and P. Smyth, "Modeling human location data with mixtures of kernel densities," in Proc. 20th ACM SIGKDD Int. Conf. Knowl. Discovery Data Mining, New York, NY, USA, 2014, pp. 35–44.
24. E. Raad, A. Dipanda, and R. Chbeir, "User profile matching in social networks," in 2010 13th International Conference on Network-Based Information Systems(NBIS), Washington, DC, USA, 09 2010, pp. 297–304.
25. Y. Li, Y. Peng, W. Ji, Z. Zhang, and Q. Xu, "User identification based on display names across online social networks," IEEE Access, vol. 5, pp. 17 342–17 353, 2017.
26. X. Zhou, X. Liang, X. Du, and J. Zhao, "Structure based user identification across social networks," IEEE Transactions on Knowledge and Data Engineering, vol. 30, no. 6, pp. 1178–1191, 2018.
27. Y. Li and Z. Su, "A comment on "cross-platform identification of anonymous identical users in multiple social media networks"," IEEE Transactions on Knowledge and Data Engineering, vol. 30, no. 7, pp. 1409–1410, 2018.
28. R. Zafarani and H. Liu, "Connecting users across social media sites: A behavioral-modeling approach," in Proceedings of the 19th ACM SIGKDD International Conference on Knowledge Discovery and Data Mining, New York, NY, USA, 2013, pp. 41–49.
29. Li, Y., Peng, Y., Zhang, Z. et al. Matching user accounts across social networks based on username and display name. World Wide Web 22, 1075–1097 (2019).
30. Yongjun Li, Zhaoting Su, Jiaqi Yang, Congjie Gao: Exploiting similarities of user friendship networks across social networks for user identification. Inf. Sci. 506: 78–98 (2020).

CI in Robotics and Automation

ATIAS: A Model for Understanding Intentions to Use AI Technology

Farhana Faruqe, Larry Medsker, and Ryan Watkins

Abstract The interdisciplinary quantitative research method presented in this chapter is used to investigate people's trust in, and intention to use, AI systems. ATIAS (AI Trust and Intention to use AI Systems) is a hybrid model that combines AI ethics variables with technology acceptance model (TAM) variables. The approach is appropriate for surveys of large populations of consumers and other decision-makers to collect data on their levels of trust in AI and their intentions to choose and use AI systems. In this chapter, ATIAS is applied to the healthcare domain, where AI is increasingly being used. ATIAS is used to examine the impact of known technology acceptance factors and AI ethical factors on users' trust in and positive attitudes toward AI. The method uses Partial Least Squares Structural Equation Modeling (PLS-SEM) as the data analysis method. ATIAS addresses the gap in the current research on human trust in AI systems, which tends to focus on either ethical factors or technology acceptance factors. By combining both types of factors in a hybrid model, the approach aims to provide a more comprehensive understanding of why people use AI systems. ATIAS may prove valuable for policymakers, AI system designers, and healthcare providers who need to understand the factors that influence users' trust in AI systems. By identifying the factors that are most important in shaping users' attitudes toward AI, the method may inform the development of more effective AI systems that are trusted and accepted by users.

Keywords AI technology · AI ethics · Technology acceptance · Trust AI · TAM · PLS-SEM

F. Faruqe (✉) · L. Medsker · R. Watkins
The George Washington University, Washington, DC, USA
e-mail: faruqe@gwu.edu

L. Medsker
e-mail: lrm@gwu.edu

R. Watkins
e-mail: rwatkins@gwu.edu

F. Faruqe
University of Virginia, Charlottesville, VA, USA

K. Daimi et al. (eds.), *Cutting Edge Applications of Computational Intelligence Tools and Techniques*, Studies in Computational Intelligence 1118,
https://doi.org/10.1007/978-3-031-44127-1_5

1 Introduction

The rapid advancement in AI has led to the development of increasingly sophisticated and autonomous systems that have the potential to transform many aspects of human life. However, it is essential to ensure that these systems are trustworthy and can interact effectively with humans to achieve their intended goals. One critical area of focus is developing user-centered design principles that ensure that the technology is intuitive and easy to use, which can improve the overall user experience. Another important consideration is ensuring that these systems are transparent, so users can understand how they make decisions and how they use data to inform those decisions. Building trust between humans and machines is crucial. This involves establishing clear expectations about how the system will be used, what information it will collect, and how that information will be used. Ensuring that humans can intervene or override decisions made by the machine when necessary is essential. As AI systems become more pervasive in our lives, human-centered design and trust building are required to ensure that the technology is beneficial to society and is not viewed as a threat to human well-being.

Trust is not an internal quality of an AI system like accuracy. Instead, trust has multidimensional facets that affect the human-machine relationship. Understanding the depth of trust in AI systems requires detailed research. Organizations are coming together to provide guidelines and principles to build AI systems using ethical frameworks that enable the AI systems to be trustworthy and reliable. Leading examples are: (1) the European Commission has provided very detailed ethical guidelines for building AI systems, (2) AI ethics journals such as *AI and Ethics* (Springer Nature) have started to promote discussions regarding ethics, policy, and regulation for AI development, and (3) Private organizations (Google, IBM, Microsoft, and others) provide guidelines to operationalize ethical AI to make AI trustworthy.

Research in trust in AI systems aims to fill this gap through research on users' intent to use AI systems based on their level of trust in and positive attitude in terms of technology acceptance and AI ethical factors. The focus here is on the domain of AI used for healthcare and AI users in the first layer of the AI literacy pyramid shown in Fig. 1 [1]. "AI literacy as a set of competencies that enables individuals to critically evaluate AI technologies; communicate and collaborate effectively with AI;" [2], p 2. The levels of competency within the larger area of AI Literacy (and related AI ethics) among people in different groups within our society vary across a spectrum—from AI researchers and developers to workers and consumers (See Fig. 1). As a result, AI Literacy is also essential for policymakers as they make important decisions across this spectrum so that laws and regulations have the desired effects so that individuals and societal groups are protected.

Though various groups in the continuum can be considered, the focus for this work comprises the (1) consumers as someone who uses the outputs of AI to improve their work or life, (2) co-worker as someone knows the basics of how the AI systems work and uses AI outputs in the work, (3) collaborator as someone who works alongside one or more AI systems to improve a worker's performance, and (4) creator as someone

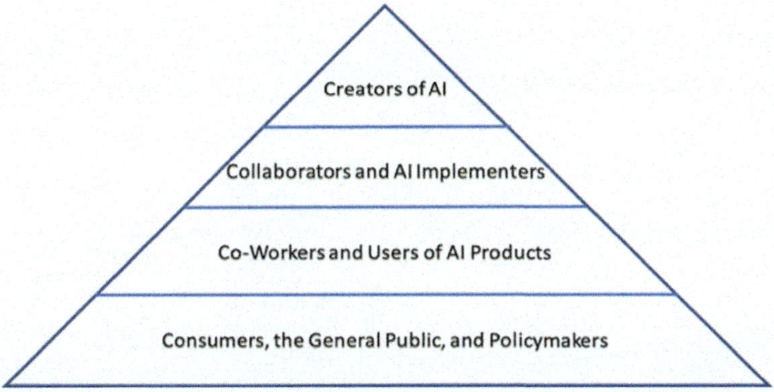

Fig. 1 Levels of AI literacy

who develops and tests new AI systems and underlying models (see Fig. 1). Each of the groups thereby requires a different level of AI Literacy knowledge and skills, and they each have different needs in establishing trust in AI systems and products.

The study addresses the existing gaps in the literature regarding the application of the Technology Acceptance Model (TAM) to AI systems, which often overlooks the consideration of AI ethics factors. Many studies of TAM applied to AI systems treat them as a technology to be accepted and do not also consider AI ethics factors. Studies investigating trust in AI systems are variable, and few (e.g., [3]) considered the relative impact of technology acceptance—that it might outweigh the ethics issues. Furthermore, studies on trust in AI systems rarely explore the relative impact of technology acceptance and ethics issues. With only a limited number of published articles found in the field, this research aims to fill these research gaps and propose a model called AI Trust and the Intention to use AI Systems (ATIAS). The ATIAS model combines AI ethics impacts with TAM, focusing on the relationships between AI ethics attributes, intention to use AI systems, and trust in AI systems. The five components of this model are (1) AI ethics factors, (2) technology acceptance factors (user's perception in this case), (3) trust in AI systems, (4) attitude towards AI usage, and (5) intention to use an AI system as illustrated in Fig. 2. By investigating the relationships among AI ethics, technology acceptance, trust, and intention to use AI systems, this research contributes to a better understanding of AI technology adoption and usage.

An important question is the following: How is user trust in AI affected by the combination of technology acceptance factors and AI ethical factors such that users will intend to use an AI system? The answer involves AI ethics and technology acceptance. Trust can be viewed from an ethical perspective and the literature has been rapidly expanding. On the other hand, technology acceptance is an additional well-established consideration that has not been addressed in combination with the ethical issue. In terms of current AI Ethics research, the most important factors that influence users' trust in AI systems are (1) perceived explainability, (2) perceived

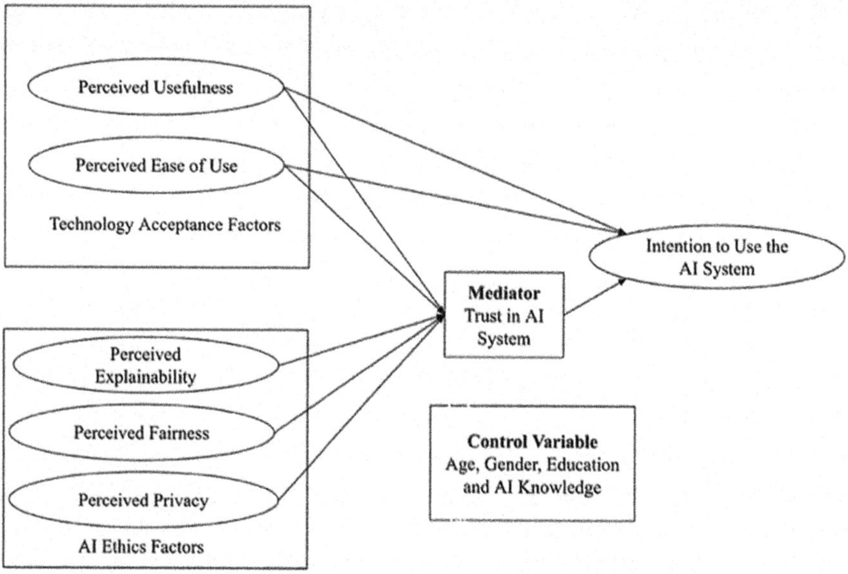

Fig. 2 Elaborated model of AI trust and the intention to Use AI systems (ATIAS)

fairness, and (3) perceived privacy; and for Technology Acceptance, the most important factors that influence the intention to use AI systems are (1) perceived usefulness, and (2) perceived ease of use. Research in this area requires data collection to test the theory of trust and intention to use AI systems.

In the case example in this chapter, data was collected from 233 students from a large private university. Among the participants, 90% are 18 to 23 years old, and the rest in the age range is 24 to 39 years old. The majority (70%) of participants are male. There are no restrictions on participants' educational and technical backgrounds if they can understand English. The survey is the primary data collection method for this study. To ensure data quality, a pilot version of the survey wase used to collect data in an Amazon MTurk survey in several batches to test and improve the survey and ensure a high-quality data collection process. After each set of data collection, the outcome was inspected and update the survey questions accordingly. Since SEM (structural equation model) was the primary method for data analysis, a crucial step is to ensure better measurement for all the model constructs. In this research study, the five constructs or exogenous (independent) variables are (1) perceived explainability, (2) perceived fairness, (3) perceived privacy, (4) perceived usefulness, and (5) perceived ease of use. The one endogenous (dependent) variable is the intention to use an AI system and one mediator, trust in AI systems.

2 Background and Theoretical Foundation

The term "ethics" is derived from the Greek word "ethos," which can mean custom, habit, character, or disposition. At its simplest, ethics is a system of moral principles. Ethics is concerned with what is good for individuals and society and is also described as moral philosophy. AI systems or machine learning (ML) algorithms are used to make decisions in everyday life, such as selecting movies, navigating (best route to take to avoid traffic), whom to connect to in social media, which product to buy, and in many more ways. AI systems are used to make life-changing decisions for others, such as granting loans, diagnosing conditions, employing individuals, granting college admissions, and making key decisions within the justice system [4–8]. At this point, society has become conscious of whether these decisions are fair or non-discriminatory to an individual or a group. Some research shows that considering ethics in AI helps to bring trust in the AI system [9–15].

2.1 Trust and Its Components

Hoff and Bashir [16] identified that trust has three common components: trustor, trustee, and involvement of some sort of task, which needs to be performed by the trustee. The performed task should have a risk factor associated with it. Though there are some similarities between interpersonal (human to human) trust and human-automation (human-machine) trust, interpersonal trust is not the same as trust in automation according to the authors. Human automation trust is based on the purpose or performance of the machine [17]. On the other hand, interpersonal trust is based on the trustee's honesty [18]. Hall and McQuany [19] review the literature from various disciplines to understand the concept of trust and trustworthiness since these concepts are heavily applied in human and automation interactions. Authors define trust and trustworthiness in an iterative manner based on similarities from various disciplines such as automation, psychology, economics, and more. With respect to trust in automation, two types of trust applicable to human-machine interaction have been identified: dispositional trust (trust reflected in the first interaction for another person or machine) and history-based trust (trust established through interactions between human-machine or human-human). Trust (T) and trustworthiness (TW) are not the same, even though these terms are frequently mixed up. "TW is a property of Y (but in relation to a potential evaluator/partner X (the trustor)); while T is a property of X (but in relation to Y (the trustee))" [20], p. 47. The authors have also mentioned that TW is a multidimensional profile of Y (trustee), and it will require multiple measures for evaluation. Trust can be defined as an attitude, rather than as a belief, intention, or behavior. Trust can be defined as "the attitude that an agent will help achieve an individual's goals in a situation characterized by uncertainty and vulnerability." [21].

Trust is the foundation of any interpersonal (human-human) relationship and is also critical in maintaining a happy relationship [22]. From the lens of psychology, an interpersonal trust situation can be defined between two individual partners when they need to rely on each other based on their mutually beneficial decisions and on other emotional attributes. For my research project, it would be useful to transform the "mutual beneficial decision" to a form of matrix where components will play a role to measure trust. In this article, Simpson [22] presented this concept in a model, "The Dyadic Model of Trust in Relationships." This model consists of two components: "normative component" and "individual-difference." The latter component represents the dispositions of individual partners in a relationship and their link to one of the five normative constructs. The model represents an iterative process; it starts with when individuals are able to trust each other and feel secure in their relationship, then they will enter the trust situation again and this loop restarts. Though there are some similarities between interpersonal trust and trust in automation (human-machine), interpersonal trust is not the same as trust in automation [16, 21]. At the early stage of the human-machine relationship, there may be little known information about the machine's performance. As such, trust can depend on the machine's purpose [21] and human-machine trust can be based on either purpose or performance of the machine based on the stage of relationship [16, 17]. On the other hand, human to human trust develops gradually. It starts with a foundation of performance, then moves to dependability and then finally develops into faith [21]. These dimensions (purpose, process, performance) of trust are considered as "attributional abstractions," which can be achieved by understanding the purpose, process or performance of the system of trust development [21].

2.2 *Trust in Human-Machine Interaction (HMI)*

In HMI, there are generally three components identified in the domain of trust: trustor (human), trustee (machine) and situation and there is always a risk of trust violation (Meritt and Ilgen 2008) [19]. A similar concept is described by Hoff and Bashir [16] in which definitions of trust generally have three common components: trustor, trustee, and the involvement of some sort of task that needs to be performed by the trustee. With respect to trust in HMI, two types of trust have been identified: dispositional trust (trust reflected in the first interaction) and history-based trust (trust established over interactions) (Meritt and Ilgen 2008) [19]. Since history-based trust can change as a function of human-machine summative interactions, it can be used for dynamic measures of trust between human-machines. A similar concept is carried over by Hoff and Bashir [16] who constructed a three-layered framework that captures the trust variability: situational trust, learned trust, and dispositional trust. Descriptions of these components have been taken from Hoff and Bashir [16] (p. 413–421): (1) Dispositional trust: "represents an individual's overall tendency to trust automation, independent of context or a specific system. We use the term dispositional trust to refer to long-term tendencies arising from both biological and environmental influences."

Hoff and Bashir [16], p. 413. Factors influencing dispositional trust: culture, age, gender, and personality traits. (2) Situational Trust: "the development of trust as well as its significance to behavior varies greatly depending on the situation." Hoff and Bashir [16], p. 415. Factors influencing situational trust: (1) internal variability: self-confidence, subject matter experience, mood, and attentional capacity, (2) external variability: type of system, system complexity, task difficulty, workload, received risks, received benefits, organizational setting, and framing of task, (3) learned Trust: "represents an operator's evaluation of a system drawn from past experiences or the current interaction. This layer of trust is directly influenced by the operator's preexisting knowledge and the automated system's performance." Hoff and Bashir [16], p. 420. Hoff and Bashir [16] also considered the different factors that are capable of influencing human-machine trust and reliance. The authors' concluding remark is that human-machine interactions are increasing, as humans are getting more dependent on automation systems. This is critical to the formation of trust in automation to ensure proper use and minimize machine related accidents [16]. This definition of trust can be used in real-life scenarios to build trust-sensor models and to measure trust level in a human when the human is interacting with a machine (machine, in this prospectus, refers to any man-made autonomous system).

The European Union (EU) Trustworthy AI framework and Trust Theory by Catelfranchi and Falcone is a theoretical foundation used by researchers, and its primary components of trust in AI systems is used in this research. To promote Trustworthy AI the EU has provided Ethics guidelines for Trustworthy AI, which consist of three components: lawful, ethical, and robust. The framework for Trustworthy AI includes seven components, covering human agency and oversight, technical robustness and safety, privacy and data governance, transparency, diversity, non-discrimination, and fairness, societal and environmental well-being, and accountability [23]. Each component is equally important in making an AI system trustworthy. However, the framework does not explicitly address lawful AI, but offers guidance on ethical and robust AI.

Trust and trustworthiness, as well as their applications in interpersonal and human-automation interactions, are distinct concepts. Trust is a property of the trustor, while trustworthiness is a property of the trustee. Trustworthiness is a multidimensional construct that requires multiple measures to evaluate, as it involves factors such as reliability, competence, benevolence, and integrity [20].

Human-automation trust differs from interpersonal trust in various ways, including the attribution process and the basis for trust. Trust in automation typically starts with faith in the technology's capabilities, followed by its dependability and predictability. In contrast, interpersonal trust usually begins with a foundation of reliability, which then develops into dependability and ultimately involves faith. Trust in automation is based on the machine's purpose or performance [17], whereas interpersonal trust is based on the trustee's honesty Meyer et al. [18]. The involvement of a task with a risk factor is also a common component of both interpersonal and human-automation trust. Overall, these distinctions between trust and trustworthiness, as well as the differences between interpersonal and human-automation trust, can have significant implications for designing and evaluating systems that involve trust-based interactions.

2.3 Technology Acceptance Model

User acceptance is critical for the successful adoption of AI systems, and TAM is one of the common models that has been used to determine acceptance among users. This model was introduced by Davis in 1989, and the model has several versions. In the technology domain, "attitudes" are regarded as psychological tendencies that signify the degree to which individuals like or dislike a particular entity [24]. Attitude, considered as a moderator of beliefs and intentions, exists in both the theories of TAM and the theory of planned behavior (TPB) [25]. Users' attitude towards AI-based systems is an important contributor for user intent to use the AI-based system (TAM). Previous research shows that user adoption has a relationship between human behavior and attitude [26]. Perceived usefulness (PU) and perceived ease of use (PEU) factors are identified as influential factors for users' technology acceptance. Years of efforts and research have been on the technology acceptance model, its expansion TAM2, TAM3 [27, 28], and the unified theory of acceptance and use of technology [29].

TAM has been used as a theoretical foundation to understand the intention to use and this popular theory has several versions [27, 28, 30]. The two factors of this the case here, perceived usefulness and perceived ease of use, are two main components of the original TAM in terms of influencing the intention to use the technology. Perceived usefulness (PU) can be defined as "the degree to which a person believes that using a particular system would enhance his or her job performance." and perceived ease of use (PEOU) if used as "the degree to which a person believes that using a particular system would be free of effort." [27], p 5. Many studies have explained PE and PEOU using TAM to understand the public acceptance and intention to use AI based systems in various domains such as recruitment, healthcare, education and autonomous vehicles, education [25, 31–37].

Leading AI ethics factors that influence trust in AI systems are (1) explainability (2) fairness and (3) privacy. The focus of this study is to understand the user's perception of the explainability, fairness, and privacy of an AI system combined with the main factors derived from TAM that could be influencing factors for users' intentions to use AI systems: (1) perceived usefulness and (2) perceived ease of use.

2.4 ATIAS Components

(1) **Perceived Explainability** (PE) can be defined in various ways, and none is accepted by everyone. According to DARPA [38], p. 1 Explainable AI (XAI) aims to "produce more explainable models while maintaining a high level of

learning performance (prediction accuracy),and enable human users to understand, appropriately, trust, and effectively manage the emerging generation of artificially intelligent partners". FAT/ML (Fairness, Accountability, and Transparency in Machine Learning) stated the goal of explainability is to "ensure that algorithmic decisions, as well as any data driving those decisions, can be explained to end-users and other stakeholders in non-technical terms." [39, 40] has done a linguistic search to identify the key concepts of explainable AI and created a visual including Important terms. The appearance of the terms represents the frequency of the terms in their surveyed articles, and it shows that "Explainable AI" and "Interpretable AI" are the most frequently used terms according to the study by [40] and the term "Explainable" is used more than the "interpretable" in public settings. In this case example, the public and users are the focus for perceived explainability.

(2) **Perceived Fairness** (PF) can be defined as "ensure that algorithmic decisions do not create discriminatory or unjust impacts when comparing across different demographics (e.g. race, sex, etc.)." [39]. In various domains, fairness, social bias, or discrimination is identified as an important factor in trusting the AI system [41, 42]. Perceived fairness can be defined as the degree to which a person believes that data and algorithms are fair and not discriminatory to a certain group. Example: minority groups. Fairness, bias, and discrimination are the terms used frequently these days in the context of AI systems that make decisions in the everyday lives of individuals and groups. The following section will provide the details about it.

(3) **Perceived Usefulness** (PU) is "The degree to which a person believes that using a particular system would enhance his or her job performance" [34], p. 320. Venkatesh and Davis defines measurements items for this construct (2000) as, using the AI product would useful in my daily work, using AI products would enhance effectiveness in my daily work. "Perceived usefulness is most frequently affected by job relevance, complexity, and social influence" [43], p. 5497)

(4) **Perceived Ease of Use** (PEU) According to Davis et al. [44], perceived ease of use is mostly affected by attitude, anxiety, perceived enjoyment, trialability, and anxiety. There are several criteria relevant to AI-based technology acceptance [45]; these criteria can be defined as, attitude: the degree to which a person likes or dislikes the object; anxiety: an individual's apprehension or even fear, when she/he is faced with the possibility of using computers; trialability: the degree to which an innovation may be experimented with before adoption, and perceived enjoyment: the extent to which the activity of using a specific system is perceived to be enjoyable in its own right, aside from any performance consequences resulting from system usage.

3 Research Method

The research method is interdisciplinary, combining both AI ethics and TAM perspectives. By using a hybrid model, the study aims to capture the complex interplay between technology acceptance factors and AI ethical considerations. This approach may provide a more comprehensive understanding of how users perceive and evaluate AI systems. The research is also quantitative in nature, relying on survey data and data analysis methods to draw conclusions. Quantitative research involves collecting numerical data and analyzing it using mathematical methods, particularly statistics, to explain phenomena [46].

The data analysis method aims to identify the relative importance of technology acceptance and AI ethical factors in influencing users' trust and intention to use AI systems. Overall, the proposed research aims to contribute to a better understanding of the factors that influence users' trust and intention to use AI systems and the relationship. By taking an interdisciplinary approach, the study may shed light on the complex interplay between technology acceptance and AI ethics considerations and provide insights that can inform the development of more trustworthy and ethical AI systems.

Trust in AI systems is crucial for successful human-AI system interaction [47]. Additionally, suboptimal AI ethics, manifested through the system's biases or discriminatory behavior can lead to negative use. AI ethics is closely connected to user trust in the AI system experiences, which may limit use. Research has been conducted that explores the ethical factors that influence trust in AI systems [3, 48–50]. The body of research shows that the various factors influencing trust in AI systems can be explainability, fairness, and privacy.

3.1 Research Design

To establish a sound model, a pilot study was conducted to test the approach. For the pilot study, Qualtrics and Amazon Mturk were used to design, collect, and store the data. The main study was then conducted in a similar setting but in person; Qualtrics for survey design, creation, distribution, and data collection; and SmartPLS 4 software to analyze the data and SEM (structural equation model).

The current study's research design is depicted in Fig. 3, which consisted of five main stages. The first stage, Design, involved developing survey questions by reviewing previous research studies and testing them on small groups of participants. The second stage, Data Collection, included creating the survey using Qualtrics and distributing it to participants, with data collected from small groups of Amazon MTurk participants for the pilot study and final data collected from university students. In the Data Preparation stage, the data was processed to prepare it for analysis, including keeping necessary data points, cross-checking entry quality, and removing records with missing values. The Data Analysis stage used SmartPLS and

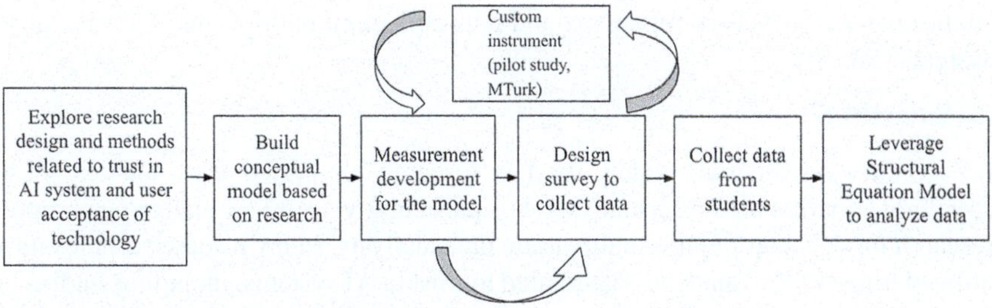

Fig. 3 An illustrative research method map

structural equation model (SEM) as a statistical technique to analyze quantitative analysis, which involved evaluating the measurement model, reflective measurement models, and structural model. Finally, in the Data Interpretation stage, the results were explained statistically and generally, with key findings presented along with study limitations and recommendations for future research. Details of the software and tools used in the study, such as Qualtrics, Amazon MTurk, and SmartPLS4, are included in the following sections.

(1) **Qualtrics**

This study used Qualtrics to create surveys and collect data. It is a powerful cloud-based software. It is a popular tool for conducting research and surveys in both academic and business settings. Qualtrics offers a range of features that make it easy to create and customize surveys, distribute them to respondents, and analyze the results. One of the key features of Qualtrics is the drag-and-drop interface that allows users to create surveys quickly and effectively. In addition, Qualtrics offers a range of distribution options, such as email invitations, website embedding, social media sharing, and QR codes. The software also supports mobile devices, making it easy for respondents to take surveys on their mobile devices. Qualtrics is known for its robust security features, which include encryption, multi-factor authentication, and regular security audits. The software complies with various data protection regulations, such as GDPR and HIPAA. Many research articles [51, 52] have used Qualtrics as a tool for survey creation and data collection.

(2) **Amazon Mechanical Turk (MTurk)**

MTurk is an online platform that allows researchers to hire workers, known as "Turkers," to complete tasks such as surveys, data entry, and image labeling. This platform is often used in academic research as a convenient and cost-effective way to collect data from a large and diverse sample. Researchers can set qualifications for the workers, such as their location, age, or previous experience on the platform, to ensure that they meet the desired criteria for the study. Despite some limitations, such as potential issues with data quality and participant selection bias, MTurk has been shown to produce reliable and valid results in a variety of research areas. Several studies have used MTurk for survey and data collection in the domain of trust in AI systems [53, 54].

(3) **SmartPLS**

It is a software program widely used in research for partial least squares (PLS) structural equation modeling analysis. It is particularly useful for analyzing complex research models that require multivariate data analysis. Many research articles have utilized SmartPLS to analyze data related to trust in AI systems, including studies on explainable AI, privacy concerns, and ethical implications. An example is the study by Siau and Wang [49], who used SmartPLS to examine the factors influencing users' trust in autonomous systems. More studies used SmartPLS to analyze data related to various aspects of user behavior such as technology acceptance, mobile internet usage, and purchase intentions [55].

(4) **Structural Equation Modeling (SEM)**

It is a statistical technique used to analyze the relationships between multiple variables. It is a versatile tool that can be applied to a wide range of research questions and is commonly used in fields such as psychology, social sciences, and business. SEM involves two main components: a measurement model and a structural model. The measurement model specifies how the observed variables (indicators) are related to the underlying constructs of interest, while the structural model specifies how the constructs are related to each other. Partial least squares structural equation modeling (PLS-SEM) is a variant of SEM that is particularly useful for analyzing complex models with many latent variables and small to medium-sized samples. PLS-SEM is a non-parametric method that does not require strict assumptions about the distribution of the data and can handle both reflective and formative measurement models. PLS-SEM is also robust to non-normality, outliers, and missing data. It is a popular tool in fields such as marketing, management, and information systems, where it is used to model complex relationships between constructs such as customer satisfaction, and technology adoption. PLS-SEM has gained popularity [56] due to its ability to handle complex models with many latent variables and small sample sizes, making it a valuable tool for researchers in various fields.

3.2 *Research Questions and Hypotheses*

The research strategy for this study is to conduct a quantitative study to clarify the roles of trust in AI systems and consumers' intention to use AI systems. The study focuses on one research question and multiple hypotheses. The research question is "How are users' trust in AI affected by technology acceptance factors and AI ethical factors such that users will intend to use an AI system" There are several hypotheses that will be tested in this study. The first three hypotheses focus on direct effects, where H1 suggests that user trust in AI has a positive impact on the intention to use the AI system, H2 suggests that the user's perception of the usefulness of the AI has a positive impact on the intention to use the AI system, and H3 suggests that the user's perception of ease of use of the AI has a positive impact on the intention to use the AI system. The remaining hypotheses focus on mediated effects, where H4–H8 suggests that trust mediates the positive relationship between various factors and the intention to use the AI system. Specifically, H4 suggests that trust mediates the positive relationship between usefulness and intention to use, H5 suggests that trust mediates the positive relationship between ease of use and intention to use, H6 suggests that trust mediates the positive relationship between explainability and intention to use, H7 suggests that trust mediates the positive relationship between fairness and intention to use, and H8 suggests that trust mediates the relationship between privacy and intention to use.

RQ: How are users' trust in AI affected by technology acceptance factors and AI ethical factors such that users will intend to use an AI system?

Hypotheses: Direct Effects

H1: User trust in AI has a positive impact on intention to use the AI.
H2: User perception of usefulness of the AI has a positive impact on intention to use the AI.
H3: User perception of ease of use (or user friendliness) of the AI has a positive impact on intention to use the AI.

Hypotheses: Mediated Effects

H4. Trust mediates the positive relationship between Usefulness and Intention to Use.
H5. Trust mediates the positive relationship between Ease of Use and Intention to Use.
H6. Trust mediates the positive relationship between Explainability and Intention to Use.
H7. Trust mediates the positive relationship between Fairness and Intention to Use.
H8. Trust mediates the relationship between Privacy and Intention to Use.

3.3 *Measurement Development*

Measurement development is a crucial step in SEM, as it involves the creation of a reliable and valid measurement instrument for the constructs of interest. This process typically involves multiple steps, including defining the construct, selecting appropriate items, assessing the reliability and validity of the measurement instrument, and refining the instrument as necessary. Once the measurement instrument is developed, it can be used to test the relationships between constructs in the structural model. For this study, measurements are guided by best practices in previous research, such as in [25, 28], and Esmaeilzadeh [41]. The descriptions of the measurements for each factor are listed in Table 1 along with the variables used in the analytical system.

Table 1 Measurement of the factors of the model

Factors	Measurement items for this study
Perceived usefulness	PU1: helps me understand the diagnosis PU2: improves my health management PU3: recommends reliable treatment options PU4: suggests effective care planning
Perceived ease of use or user-friendly	FRN1: the system is easy for me to understand and operate FRN2: my interaction with the system is easy FRN3: I will be able to explain the system to someone else FRN4: I become skillful using this app quickly
Perceived explainability	EX1: helps me understand how this AI-based app works EX2: shares the decision-making factors with me EX3: provides justification for suggested decisions EX4: explains everything in simple language
Perceived fairness	FR1: use data for a wide range of demographic cases FR2: have been checked for built-in bias that could influence decisions FR3: make decisions specific to my situation FR4: reach the best outcome for anyone who uses it
Perceived data privacy	DP1: collect my personal information only with my consent DP2: protect my information from unauthorized uses DP3: prevent access to my health information by outsiders DP4: use my health record information only with my consent
Intention to use AI based systems	IU1: diagnose my medical condition IU2: create my treatment plan IU3: predict my future medical situations IU4: manage my healthcare
Perceived trust in AI based systems	TR1: prescribe medicine for my critical illness TR2: predict my health condition TR3: manage my parents' healthcare TR4: prescribe medicine for common illness

(1) **Measurement Scale**

Likert scale was used to measure responses from the participants. The Likert scale is a tool commonly used in social science research to measure attitudes or opinions. The scale involves individuals indicating their level of agreement or disagreement with a statement on a scale ranging from "strongly agree" to "strongly disagree." Developed by psychologist Rensis Likert in the 1930s, the Likert scale has become a widely used tool in survey research to measure various attitudes, including opinions on products, political views, or job satisfaction. It is important to note that the reliability and validity of the Likert scale as a measurement tool depends on the careful design of questions and the appropriate analysis of responses. The Likert scale is commonly used for measuring attitudes in social science research and includes ways for researchers to enhance the reliability and validity of their Likert scale data. The case study used Likert scale to measure responses where 1 = strongly disagree, 2 = somewhat disagree, 3 = neutral (neither disagree nor agree), 4 = somewhat agree, and 5 = strongly agree.

(2) **Evaluation of Reflective Measurement Models**

Hair et al. [57] explained the rules of thumb to assess the reflective measurement models: a valid reflective measurement model must have (1) internal consistency reliability, (2) indicator reliability, (3) construct reliability, (4) convergent validity, and (5) discriminant validity.

Internal Consistency Reliability can be assessed using Cronbach's alpha, which estimates the reliability based on how the observed variables are intercorrelated [57]. However, in PLS-SEM, indicators are prioritized based on their individual reliability, which is not accounted for by Cronbach's alpha. Additionally, Cronbach's alpha tends to underestimate internal consistency reliability and is sensitive to the number of items on a scale. A more appropriate measure in PLS-SEM is composite reliability, which considers the different outer loadings of the indicator variables. The interpretation of composite reliability is similar to Cronbach's alpha, where higher values indicate higher reliability. Acceptable values range from 0.60 to 0.70 for exploratory research and 0.70 to 0.90 for more advanced stages of research [58].

Indicator Reliability refers to how well each indicator variable measures the underlying construct. This can be measured by looking at the outer loadings, or the correlation between the indicator and the construct, which should ideally be 0.708 or higher. The variance extracted from an item refers to how much of the indicator's variation is explained by the construct. A rule of thumb is that at least 50% of the indicator's variance should be explained by the construct. In some cases, the outer loading may be weaker than 0.70, especially in social science studies. Effects need to be checked for these weaker indicators on both composite reliability and content validity before eliminating them. Removing an indicator that improves the composite reliability could be considered, but weaker indicators may be kept if they contribute to the content validity of the construct. Indicators with very low outer loadings (below 0.4) should always be removed from the construct, as recommended by Hair et al. [57].

Construct Reliability and construct validity are two important aspects of scale development. Construct reliability refers to the consistency of a measurement tool, while construct validity refers to how well the tool measures what it claims to measure. Internal consistency reliability is a commonly used method to assess construct reliability, which evaluates the consistency of responses across multiple items within a scale (Malhotra 2010). On the other hand, construct validity ensures that the measurement tool accurately measures the intended construct, and it is determined by comparing the scores of the tool with a gold standard or other existing measures. The measurements used in this case study are should have coherence between the conceptual and operational definitions of the constructs.

Convergent Validity of reflective constructs considers the outer loading of the indicators and the average variance extracted (AVE) [57]. Validity is the degree to which a measure correlates positively with other measures of the same construct. Higher outer loadings on a construct indicate stronger associations with the indicators, meaning that they capture more of the construct. Indicator reliability should be at least 0.50, as recommended by Hair et al. [57].

Discriminant Validity is crucial in ensuring that the measures being used in the study are distinct from other constructs in the model [57]. This helps researchers to determine whether the structural paths in the model are real or just a result of chance or discrepancies. By demonstrating a lack of correlation among differing constructs, researchers can be more confident in the validity of their findings and conclusions. Three measures of discriminant validity are commonly used by researchers. The first approach, cross-loadings, assesses the outer loading values of each indicator on the associated constructs. For discriminant validity, the outer loadings should be higher than any cross-loadings on other constructs. The second, the Fornell-Larcker criterion, is widely used to assess discriminant validity [57]. It compares the square root of the AVE values with the latent variable correlations. More specifically, discriminant validity is obtained when the square root of the AVE is higher than the absolute value of the correlation shared between any of the other constructs [57]. Third, the Heterotrait-Monotrait Ratio (HTMT) is much more conservative and is more reliable than the Fornell-Larcker Criterion [59]. Assessing discriminant validity helps ensure that the measures used in the study are distinct from other constructs in the model and helps increase confidence in the results obtained from the study.

(3) **Evaluation of the Structural Model**

To test the model's predictive capabilities, researchers can use various measures, such as the coefficient of determination (R2), the predictive relevance (Q2), and the effect size (f2) [57]. The R2 measures the amount of variance in the dependent variable that is explained by the independent variable(s) in the model. To assess the relationships between constructs, researchers can examine the path coefficients, which indicate the strength and direction of the relationships between the latent variables in the model. Additionally, researchers can also look at the significance levels of the path coefficients and the bootstrapping confidence intervals to determine the statistical significance and reliability of the relationships. Overall, the evaluation of the structural model should provide insights into the fit and usefulness of the proposed theoretical framework and the relationships between the constructs being studied. Having established the reliability and validity of measurements of the latent variables in the previous section, all conditions are met for evaluating the structural model. The next step, following [57], is to assess the PLS-SEM structural model results. This is done by testing the model's predictive capabilities and the relationships between constructs.

The study focused on the users, regardless of their technical expertise and education. Data has been collected from 233 students from a large, western (USA), private university. Among the participants, 90% are from 18 to 23 years, and the rest 10% age range is 24 to 39 years old. About 80% have some college, but they are still in undergraduate programs. Most participants are male, about 70%, and the rest of them are female and others. Out of 233 participants, about 80% (209) passed the attention check questions. Regarding the level of understanding of AI, 56% of the participants can understand articles on AI in the popular media and 39% can explain the basic AI concepts to others, and 5% have no knowledge of AI. Seventeen records have been excluded to ensure data quality because of the incompleteness of their survey responses. We discarded the data that did not pass the attention checker.

4 Findings

The research model for this study is shown in Fig. 2, five constructs or exogenous (independent) variables are (1) perceived explainability, (2) perceived fairness, (3) perceived privacy, (4) perceived usefulness, and (5) perceived ease of use. The one endogenous (dependent) variable is the intention to use an AI system. Trust in the AI system is the one mediator. The data collected was in Likert scale ordinal form. As mentioned earlier, for data analysis, SmartPLS4 is used and applied PLS-SEM and bootstrapping to the data set. The steps in the data analysis follow established standards [57]. A valid reflective measurement model must have.

- Internal consistency (composite reliability)
- Convergent validity (average variance extracted)
- Discriminant validity

Step 1: Internal consistency (composite reliability) "is a form of reliability used to judge the consistency of results across items on the same test. It determines whether the items measuring a construct are similar in their scores (i.e. if the correlations between the items are large)" [57], p. 320. "Composite reliability should be higher than 0.70 (in exploratory research, 0.60 to 0.70 is considered acceptable). Consider Cronbach's alpha as the lower bound and composite reliability as the upper bound of internal consistency reliability" (p. 112). Cronbach's alpha and composite reliability are calculated to test the construct reliability; and the value of these should be >0.70 (see Table 1 for our results).

Step 2: Convergent Validity "is the extent to which a measure correlates positively with alternative measures of the same construct" [57], p. 112. To evaluate the convergent validity for reflective constructs, the outer loading of the indicators and the average variance extracted (AVE) should be considered [57]. Convergent validity is achieved when AVE >0.5.

Step 3: Discriminant Validity is "the extent to which a construct is truly distinct from other constructs by empirical standards" [57], p. 115. This study uses the HTMT (heterotrait-monotrait ratio) criterion to assess discriminant validity in PLS-SEM. Based on simulation and previous research, [59] recommend that HTMT values should not exceed 0.90 if the path model includes constructs that are conceptually similar. The outcomes for all these steps are included next.

Table 2 shows the reliability and validity of measurements of the latent variables. Comparing the listed critical values in the previous section with the following table shows that the outcome meets the evaluation criteria. The two values less than the published thresholds are still above acceptable targets for exploratory research (Hair et al. 2010).

Table 2 Evaluation of the reflective measurements

Latent variable	Indicators	Convergent validity		Internal consistency		Discriminant validity
		Indicator reliability (outer loadings)	Average variance extracted	Cronbach's alpha	Composite reliability (rho_c)	HTMT
		≥0.70	≥0.50	≥0.70	≥0.70	<0.90
Fairness	FR1	0.600	0.552	0.731	0.829	Yes
	FR2	0.726				
	FR3	0.833				
	FR4	0.791				
Data privacy	DP1	0.869	0.819	0.786	0.9	Yes
	DP4	0.939				
Explainability	EX1	0.686	0.583	0.681	0.805	Yes
	EX2	0.716				
	EX3	0.875				
Intention to use	INT1	0.720	0.593	0.772	0.853	Yes
	INT2	0.828				
	INT3	0.721				
	INT4	0.805				
Trust in AI system	TR1	0.809	0.604	0.78	0.859	Yes
	TR2	0.725				
	TR3	0.844				
	TR4	0.724				
Usefulness	USE1	0.612	0.634	0.811	0.872	Yes
	USE2	0.799				
	USE3	0.894				
	USE4	0.85				
User-friendly	FRN1	0.716	0.540	0.725	0.825	Yes
	FRN2	0.745				
	FRN3	0.738				
	FRN4	0.741				

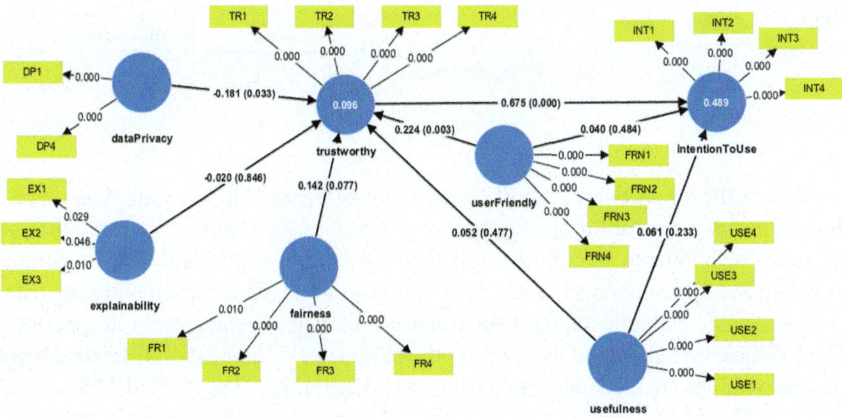

Fig. 4 PLS-SEM bootstrap model (SmartPLS4)

The study aims to assess the significance and relevance of the structural model relationships. This is achieved by running the PLS algorithm to estimate the structural model relationships as shown in Fig. 4. This outcome helps to validate the hypotheses of this study and determine the significance of the effects, discussed next. The outcome shows that user trust in AI has a high significant positive impact on intention to use the AI with a p value <0.001 (H1). User perception of usefulness (H2) and user friendliness (H3) did not show any evidence to have a positive impact on intention to use the AI system while accounting for trust. For the mediated effects, the following hypotheses were identified as significant, with p values <0.05.

H5: Trust mediates the positive relationship between user friendliness and Intention to Use.
H8: Trust mediates the relationship between Privacy and Intention to Use negatively. This can be interpreted as people who are less concerned about data privacy will have more trust in AI systems.
The following hypotheses were not supported:
H4: Trust mediates the positive relationship between Usefulness and Intention to Use.
H5: Trust mediates the positive relationship between user friendliness and Intention to Use.
H6: Trust mediates the positive relationship between Explainability and Intention to Use.
H7: Trust mediates the positive relationship between Fairness and Intention to Use.

Figure 4 shows the outcome generated from SmartPLS 4. Bootstrap is a resampling technique used in partial least squares structural equation modeling (PLS-SEM) to evaluate the reliability and validity of the model estimates. In the PLS-SEM context, the bootstrap resampling procedure is used to determine the accuracy of the parameter estimates and to calculate the standard errors and t-values for each of the parameters in the model.

Table 3 R-square values

	R-square	R-square adjusted
Intention to use	0.489	0.482
Trust in AI system	0.096	0.074

In SmartPLS4 software, the PLS-SEM Bootstrap Model is a statistical method that performs a resampling procedure to generate many samples from the original data set. The bootstrap procedure is repeated many times in this case, 5000 to obtain a distribution of parameter estimates. The mean and standard deviation of this distribution are used to estimate the parameter values and their standard errors, respectively. The t-values are calculated by dividing the parameter estimate by its standard error, and the p-values are then calculated from the t-distribution. The PLS-SEM Bootstrap Model in SmartPLS4 provides a robust and efficient way to estimate the model parameters and to evaluate the significance of the relationships between the latent variables and their indicators. The samples are generated by randomly selecting observations from the data set with replacements. Each sample is analyzed using the PLS-SEM algorithm to estimate the model parameters, such as path coefficients, loadings, and R-squared values (as shown in Table 3).

It is important to understand the model's predictive power, for which in this study R-squared values were calculated. Hair et al. [57] explain that R-squared is a measure of the model's predictive power and that "the coefficient represents the exogenous latent variables' combined effects on the endogenous latent variable. R-squared represents the amount of variance in the endogenous constructs explained by all of the exogenous constructs linked to it" (p. 198). The R^2 value for intention to use is 0.489 as shown in Fig. 4 and Table 6 which can be interpreted as about 50% of the variability observed in the target variable (intention to use AI system) is explained by the model. About 10% of the variability observed in the trust is explained by the model.

The case study employed Partial Least Squares Structural Equation Modeling (PLS-SEM) using SMARTPLS4 to conduct a Multicollinearity Assessment. The assessment aimed to examine the presence of multicollinearity among the variables in the model. Multicollinearity assessment is crucial in research because it helps to evaluate the relationships between predictor variables and identify potential issues that can affect the reliability and validity of statistical analyses. Multicollinearity refers to a high degree of correlation between two or more predictor variables in a regression or structural equation model. Assessing multicollinearity helps to ensure the accuracy and stability of coefficient estimates. In the presence of multicollinearity, the coefficient estimates may become unstable, leading to inflated standard errors and less precise estimates. This makes it challenging to determine the true impact of each predictor variable on the dependent variable. The Variance Inflation Factor (VIF) values were calculated for this research to understand the extent to which the variables are correlated with each other. For this research the results reveal that VIF values for all the variables IntentionToUse, Trust, DataPrivacy, Explainability,

Fairness, Usefulness, and UserFriendly are below 5, suggesting the absence of multicollinearity in the inner model. This finding signifies that the variables included in the study are not highly interrelated, ensuring the reliability of the obtained results and the independence of the predictor variables in explaining the dependent variables.

Post-hoc analysis with a control variable is an important step in statistical analysis, including Partial Least Squares Structural Equation Modeling (PLS-SEM). Control variables are additional factors included in the analysis to account for potential confounding effects or to test the influence of specific variables on the relationships being examined [57]. In this study, four control variables–age, gender, education, and basic understanding about AI–were included to ensure the accuracy and validity of the relationships between the independent and dependent variables by controlling for potential spurious effects. The latest version of SmartPLS introduces an enhanced Process function that allows for the inclusion of control variables in mediation and moderation analyses. Using this feature, a Bootstrap model was constructed to assess the influence of these additional variables on the relationships of interest. In PLS-SEM with control variables, the R-square value represents the proportion of variance in the endogenous latent variables that is explained by the model, while considering the impact of control variables. This value helps evaluate the model's predictive ability, taking into account both the variables of interest and the control variables.

The R-square value for Intention to Use is 0.528, indicating that the model, considering ethical factors, TAM factors, and the control variables, explains 52% of the variance in users' intention to use AI systems. This suggests that the included variables, along with the controlled factors, account for a substantial portion of the observed variability in users' intention to use AI systems. The R-square value for Trust in AI systems is 0.214, indicating that the model, considering ethical factors, TAM factors, and the control variables, explains 21% of the variance in users' trust in AI systems. A higher R-square value suggests that the model, when control variables are taken into account, provides a better fit to the data and has a stronger predictive capability for the endogenous latent variable. By comparing these outcomes to the original analysis, it can be determined whether the relationships between the main variables of interest change or remain significant after controlling for the influence of the control variable. In this study, the outcomes remain similar, indicating that the relationships between the main variables remain significant even after accounting for the influence of the control variables.

5 Discussion

The result of this study is an in-depth analysis of quantitative data including examining both reflective measurements, as well as evaluating the structural model, and analyzing the research models and hypotheses analysis. This interdisciplinary research project combines AI ethics and technology acceptance factors to develop a hybrid model called ATIAS, which could be used to investigate the impact of these factors on users' trust toward AI systems in the healthcare domain. Surveys were conducted to collect data, and this study used Partial Least Squares Structural Equation Modeling (PLS-SEM) as the data analysis method. To address the gap in current research, the study aims to provide policymakers, AI system designers, and healthcare providers with a more comprehensive understanding of the factors that influence users' trust in AI systems, which can inform the development of more effective and accepted AI systems.

The research question that guided this study is how users' trust in AI is affected by technology acceptance factors and AI ethical factors, such that users will intend to use an AI system. In addition, the study proposes several hypotheses, including direct effects of users' trust, usefulness, and ease of use on their intention to use an AI system, as well as mediated effects where trust mediates the relationship between various ethical factors and users' intention to use the AI system. The study aims to contribute to the development of effective and trustworthy AI systems by identifying the factors that influence users' intention to use them.

Table 4 shows an overview of the hypotheses of this study and their corresponding p-values and results. The study aimed to investigate the impact of different factors on the intention to use AI based systems. The hypotheses were related to the positive impact of user trust, usefulness, ease of use, explainability, fairness, and data privacy on the intention to use AI. The results indicate that user trust and user perception of ease of use have a significant positive impact on the intention to use AI, while the user perception of usefulness did not show a significant impact. The hypotheses related to the mediating effect of trust on the relationship between usefulness, explainability, fairness, and data privacy with intention to use AI had mixed results, where trust did not mediate the relationship in some cases, while it did mediate the relationship in others.

5.1 Interpretation of the Findings and Research Question

As mentioned earlier, this study has one research question that guided this research: how are users' trust in AI affected by technology acceptance factors and AI ethical factors such that users will intend to use an AI system? The findings support that user trust in AI and ease of use influence the intention to use AI. It also supports the hypothesis that trust mediates the connection between data privacy (AI ethical factor) and the intention to use AI systems. However, the findings do not support the hypothesis that trust mediates the connection between usefulness (a technology acceptance factor), explainability (an AI ethical factor), and intention to use AI. However, if the study considers a 90% confidence level (which is appropriate for exploratory research), it finds a relationship between the AI ethical factor fairness and the mediator trust in AI systems. Overall, trust impacts the intention to use AI systems in the study's model.

Table 4 Result for Each Hypothesis in This Study

	Hypotheses	P-Values	Results
Trust--> IntentionToUse	H1: User trust in AI has a positive impact on intention to use the AI	0.000	Supported
Usefulness--> IntentionToUse	H2: User perception of usefulness of the AI has a positive impact on intention to use the AI	NS	Not supported
EaseOfUse--> IntentionToUse	H3: User perception of Ease of Use (or user friendliness) of the AI has a positive impact on intention to use the AI	0.004	Supported
Usefulness--> Trust--> IntentionToUse	H4. Trust mediates the positive relationship between usefulness and intention to use	NS	Not supported
EaseOfUse--> Trust --> IntentionToUse	H5. Trust mediates the positive relationship between ease of use and intention to use	0.005	Supported
Explainability--> Trust--> IntentionToUse	H6. Trust mediates the positive relationship between explainability and intention to use	NS	Not supported
Fairness--> Trust--> IntentionToUse	H7. Trust mediates the positive relationship between fairness and intention to use	0.067	~Supported when ($p < 0.01$)
DataPrivacy--> Trust --> IntentionToUse	H8. Trust mediates the relationship between privacy and intention to use	0.040	Supported

5.2 *Limitations and Next Steps*

Artificial intelligence (AI) is rapidly transforming many industries, including healthcare. AI-based systems can help clinicians make more accurate diagnoses, predict treatment outcomes, and improve patient outcomes. However, for AI-based systems to be effective, they must be trusted and accepted by users, including clinicians and patients. This study addresses the research gaps in understanding users' acceptance of AI systems while considering ethical factors, trust, and technology acceptance factors. The study used a survey to gather data on users' interactions with AI-based healthcare systems and found a strong relationship between trust and intention to use the system. The study also found that perceived fairness and data privacy were significant factors in users' trust in AI systems.

This study does not cover "actual usage" because of the difficulty and time required for a longitudinal study. Instead, we measure the intention to use, which has been found in other studies to be strongly related. Another limitation is that for a complete understanding of the intention to use an AI system, additional extraneous decision factors may be important. Also, to simplify and focus the surveys, the context for the participants is that they have only the choice to use or not use the AI system presented to them. In terms of AI ethical factors, three factors (explainability, fairness, and privacy) have been considered among all the available AI ethical factors; several research studies consistently suggest that those three factors are the most impactful. User's acceptance of the AI system is a complex study, and one of the important factors is the domain or situation in which the user is using an AI system. Based on the scenario the response of the user can vary, so the concept of users' acceptance of AI systems cannot be generalized—the conclusion depends on the specific domain. The context for data collection and analysis is limited to the healthcare domain, and future research should explore additional domains. The data collected for this study was from college students, so it would be useful for future research to investigate how the ATIAS model performs with data collected from more diverse backgrounds.

The practical implications of ATIAS are equally significant by providing valuable insights for policymakers, AI system designers, and healthcare providers interested in creating trustworthy and accepted AI systems. By understanding the importance of trust in promoting the use of AI-driven healthcare technology, policymakers can make informed decisions about the regulation and implementation of these systems. Healthcare providers can use the study's insights to develop strategies to promote adopting AI-driven healthcare technology and improve patient outcomes while reducing healthcare costs. Users' trust in AI systems is based on the ethical factors fairness and data privacy and the ease of using the system. Further cases could extend our understanding of users' acceptance of AI-based systems and give insights into how to design AI systems that are trustworthy and effective.

Appendix

Definition of Key Terms

Human intelligence encompasses the cognitive capacity to acquire knowledge through experience, adjust to novel circumstances, comprehend abstract ideas, and employ acquired knowledge to influence one's surroundings [60, 61].

Artificial Intelligence (AI) could be defined in several ways, such as it is a field combining large datasets and the computational and learning power from the computer science domain to solve problems. AI could also be defined as a type of intelligence generated from machine learning, and the intelligence is not natural like human or animal intelligence; instead, it is artificial intelligence.

The relationship between human intelligence and AI is illustrated by AI being great in learning and problem-solving but far behind in emotional knowledge, intuition, and creativity. The new field of Human-Centered AI (HCAI) advocates unique roles for humans concerning AI systems and advocates the human user as central in designing AI systems [62].

Trust is a concept used in this case example as defined by Lee and See's [46]. The authors characterize trust as an attitude, which therefore has the components belief, feeling, and behavior. They describe trust as the attitude that an agent will assist in the attainment of an individual's objectives in a situation characterized by uncertainty and vulnerability [21].

Trust in an AI System means that users of an AI system can rely on and have confidence in the outcome of the AI system as demonstrated by users choosing or intending to use particular AI systems and products.

Technology Acceptance Model (TAM) is one of the common models that has been used to determine acceptance among users. This model was introduced by Davis in 1989, and the model has several versions.

Interpretability "it is defined as the ability to explain or to provide the meaning in understandable terms to a human." [63].

Explainability has two aspects: (1) the logic behind how the AI system makes certain decisions and (2) information about the decision that is understandable by the users regardless of their technical background or experience.

AI Transparency is a property of an application regarding how possible it is to understand a system's inner workings "in theory". It can also mean the way of providing explanations of algorithmic models and decisions that are comprehensible for the user. This deals with the public perception and understanding of how AI works. Transparency can also be taken as a broader socio-technical and normative ideal of "openness". Open questions focus on transparency versus explainability and what level of transparency is sufficient for different stakeholders. Depending on the specific situation, the precise meaning of "transparency" may vary. It is an open scientific question whether there are several different kinds or types of transparency. Moreover, transparency can refer to different things, whether the purpose is to analyze the legal significance of unjust biases or to discuss them in terms of features of machine learning systems (Rusanen & Nurminen, 2022).

Privacy No unwarranted disclosure of consumers' personal information—from data collection, processing, storing, and sharing.

Fairness The outcome of the AI system contribution does not favor any individual or group based on irrelevant demographic data.

MTurk, or Mechanical Turk, is a popular Amazon product for crowdsourcing to collect data from a diverse population (Amazon Mechanical Turk 2022). Study shows that "MTurk workers are notably effortful regardless of the inclusion of IMCs (Instructional Manipulation Checks) in a given study, and regardless of whether or not they have recently completed hundreds of surveys on the platform." (Anson 2018, p 6).

References

1. Faruqe, F., Watkins, R., & Medsker, L., " Competency model approach to AI literacy: Research-based path from initial framework to model," arXiv preprint arXiv:2108.05809, 2021.
2. Long, D., & Magerko, B., "What is AI literacy? Competencies and design considerations," In Proceedings of the CHI conference on human factors in computing systems (pp. 1–16), 2020.
3. Emaminejad, N., North, A. M., & Akhavian, R., "Trust in AI and Implications for the AEC Research: A Literature Analysis," arXiv preprint arXiv:2203.03847, 2022.

4. Jiang, F., Jiang, Y., Zhi, H., Dong, Y., Li, H., Ma, S., ... & Wang, Y. "Artificial intelligence in healthcare: past, present and future," Stroke and vascular neurology, 2(4), 2017.
5. Yu, K. H., Beam, A. L., & Kohane, I. S., " Artificial intelligence in healthcare. Nature biomedical engineering," 2(10), 719–731, 2018.
6. Albert, E. T., "AI in talent acquisition: a review of AI-applications used in recruitment and selection. Strategic HR Review," 18(5), 215–221, 2019.
7. Fujita, H., "AI-based computer-aided diagnosis (AI-CAD): the latest review to read first. Radiological physics and technology," 13(1), 6–19, 2020.
8. Köchling, A., Wehner, M. C., & Warkocz, J., "Can I show my skills? Affective responses to artificial intelligence in the recruitment process," Review of Managerial Science, 1–30, 2022.
9. Floridi, L., "Establishing the rules for building trustworthy AI. Nature Machine Intelligence,"1(6), 261–262, 2019.
10. Smuha, N., "Ethics guidelines for trustworthy AI. In AI & Ethics," Brussels (Digityser), Belgium, 2019.
11. Mezgár, "From Ethics to Standards; an Overview of AI Ethics in CPPS. IFAC-PapersOnLine, 54(1), 723–728.
12. Siau, K., & Wang, W., "Artificial intelligence (AI) ethics: ethics of AI and ethical AI," Journal of Database Management (JDM), 31(2), 74-87, 2020.
13. Hickman, E., & Petrin, M., "Trustworthy AI and Corporate Governance: the EU's ethics guidelines for trustworthy artificial intelligence from a company law perspective," European Business Organization Law Review, 22(4), 593-625, 2021.
14. Kaur, D., Uslu, S., & Durresi, A., "Requirements for trustworthy artificial intelligence–a review," In International Conference on Network-Based Information Systems (pp. 105–115). Springer, Cham, 2020, August.
15. Larsson, S., "On the governance of artificial intelligence through ethics guidelines," Asian Journal of Law and Society, 7(3), 437–451, 2020.
16. Hoff, K. A., & Bashir, M., 'Trust in Automation: integrating empirical evidence on factors that influence trust. Human Factors," The Journal of the Human Factors and Ergonomics Society, 57(3), 407-434. doi:https://doi.org/10.1177/0018720814547570, 2015.
17. Lee, J., & Moray, N., "Trust, control strategies and allocation of function in human-machine systems," Ergonomics, 35(10), 1243-1270, 1992.
18. Meyer, R. C., Davis, J. H., & Schoorman, F. D., "An integrative model of organizational trust," The Academy of Management Review, 20(3), 709–734, 1995. https://doi.org/10.2307/258792
19. Hall, S., & McQuay, W., " Review of trust research from an interdisciplinary perspective - psychology, sociology, economics, and cyberspace," Proceedings of the IEEE 2010 National Aerospace & Electronics Conference. doi:https://doi.org/10.1109/naecon.2010.5712918, 2010.
20. Castelfranchi, C., & Falcone, R., "Trust theory: A socio-cognitive and computational model." John Wiley & Sons, 2010.
21. Lee, J. D., & See, K. A, "Trust in automation: Designing for appropriate reliance," Human factors, 46(1), 50-80, 2004.
22. Simpson, J. A., "Psychological foundations of trust," Current Directions in Psychological Science, 16(5), 264–268, 2007. https://doi.org/10.1111/j.1467-8721.2007.00517.x
23. Eurpoean Commission, "Ethics guidelines for trustworthy AI,", European Commission, Brussels, Dec, 2018.
24. Eagly, A. H., & Chaiken, S., "The psychology of attitudes," Harcourt brace Jovanovich college publishers, 1993.
25. Sohn, K., & Kwon, O., "Technology acceptance theories and factors influencing artificial Intelligence-based intelligent products," Telematics and Informatics, 47, 101324, 2020.
26. Zhang, B., & Dafoe, A., "Artificial intelligence: American attitudes and trends," Available at SSRN 3312874, 2019
27. Davis, F, "Perceived usefulness, perceived ease of use, and user acceptance of information technology," MIS Quarterly, 13(3), 319–340. DOI: https://doi.org/10.2307/249008, 1989.

28. Venkatesh, V., & Davis, F. D., "A theoretical extension of the technology acceptance model: Four longitudinal field studies" Management Science, 46(2), 186-204, 2000.
29. Venkatesh, V., Morris, M. G., Davis, G. B., & Davis, F. D., Unified theory of acceptance and use of technology (UTAUT) [Database record]. APA PsycTests, 2003. https://doi.org/10.1037/t57185-000
30. Venkatesh, V., & Bala, H., Technology acceptance model 3 and a research agenda on interventions. Decision sciences, 39(2), 273-315, 2008.
31. Han, J. H., & Sa, H. J., "Acceptance of and satisfaction with online educational classes through the technology acceptance model (TAM): The COVID-19 situation in Korea," Asia Pacific Education Review, 1–13, 2021.
32. Libert, K., Mosconi, E., & Cadieux, N., "Human-machine interaction and human resource management perspective for collaborative robotics implementation and adoption," In Proceedings of the 53rd Hawaii international conference on system sciences, 2020.
33. Dünnebeil, S., Sunyaev, A., Blohm, I., Leimeister, J. M., & Krcmar, H., "Determinants of physicians' technology acceptance for e-health in ambulatory care," International journal of medical informatics, 81(11), 746-760, 2012.
34. Buckley, L.; Kaye, S.A., "Pradhan, A.K. Psychosocial factors associated with intended use of automated vehicles: A simulated driving study. Accid. Anal. Prev," 115, 202–208, 2018.
35. Walter, Z., & Lopez, M. S., Physician acceptance of information technologies: Role of perceived threat to professional autonomy. Decision Support Systems, 46(1), 206–215, 2008.
36. BenMessaoud, C., Kharrazi, H., & MacDorman, K. F., Facilitators and barriers to adopting robotic-assisted surgery: contextualizing the unified theory of acceptance and use of technology. PloS one, 6(1), e16395, 2011.
37. Zmud, J., Sener, I. N., & Wagner, J. (2016). "Self-driving vehicles: determinants of adoption and conditions of usage," Transportation Research Record, 2565(1), 57–64, 2016.
38. Gunning, D., "Explainable artificial intelligence (xai). Defense advanced research projects agency (DARPA)," nd Web, 2(2), 1., 2017.
39. Diakopoulos, N., Friedler, S., Arenas, M., Barocas, S., Hay, M., Howe, B., ... & Zevenbergen, B., "Principles for accountable algorithms and a social impact statement for algorithms," FAT/ML, 2017.
40. Adadi, A., & Berrada, M., "Peeking inside the black-box: a survey on explainable artificial intelligence (XAI)," IEEE access, 6, 52138-52160, 2018.
41. Esmaeilzadeh, P., "Use of AI-based tools for healthcare purposes: a survey study from consumers' perspectives," BMC medical informatics and decision making, 20(1), 1–19, 2020.
42. Angerschmid, A., Zhou, J., Theuermann, K., Chen, F., & Holzinger, A., "Fairness and explanation in AI-informed decision making," Machine Learning and Knowledge Extraction, 4(2), 556–579, 2022.
43. Laurim, V., Arpaci, S., Prommegger, B., & Krcmar, H., "Computer, whom should I hire?-acceptance criteria for artificial intelligence in the recruitment process," In Proceedings of the 54th Hawaii international conference on system sciences (p. 5495), 2021.
44. Davis, F. D., Bagozzi, R. P., & Warshaw, P. R., "User acceptance of computer technology: A comparison of two theoretical models," Management science, 35(8), 982–1003, 1989.
45. Laurim, V., Arpaci, S., Prommegger, B., & Krcmar, H., "Computer, whom should i hire?-acceptance criteria for artificial intelligence in the recruitment process, 2021.
46. Aliaga, M., & Gunderson, B., "Interactive statistics," Prentice Hall, 1999.
47. Asan, O., Bayrak, A. E., & Choudhury, A., "Artificial intelligence and human trust in healthcare: focus on clinicians," Journal of medical Internet research, 22(6), e15154, 2020
48. Faruqe, F., Watkins, R., & Medsker, L., "Monitoring Trust in Human-Machine Interactions for Public Sector Applications," arXiv preprint arXiv:2010.08140, 2020
49. Siau, K., & Wang, W., "Building trust in artificial intelligence, machine learning, and robotics," Cutter business technology journal, 31(2), 47-53, 2018
50. Ashoori, M., & Weisz, J. D., "In AI we trust? Factors that influence trustworthiness of AI-infused decision-making processes," arXiv preprint arXiv:1912.02675, 2019.

51. Liu, F., & Tan, T., "What factors influence the public's trust in artificial intelligence: A survey-based study," Computers in Human Behavior, 120, 106723. doi: https://doi.org/10.1016/j.chb.2021.106723, (2021.
52. Rallapalli, V. K., & Singh, N., "A conceptual framework and empirical investigation," Journal of Business Research, 131, 614-623. doi: https://doi.org/10.1016/j.jbusres.2021.01.038, 2021.
53. Bonnefon, J. F., Shariff, A., & Rahwan, I, "The social dilemma of autonomous vehicles," Science, 352(6293), 1573–1576, 2016
54. Shafti, A., Derks, V., Kay, H., & Faisal, A. A., "The response shift paradigm to quantify human trust in AI recommendations," arXiv preprint arXiv:2202.08979, 2022.
55. Moon, J. W., & Kim, Y. G., "Extending the TAM for a world-wide-web context," Information & management, 38(4), 217–230, 2001.
56. Cheng, E. W., Chu, S. K., & Ma, C. S., "Students' intentions to use PBWorks: A factor-based PLS-SEM approach," Information and Learning Sciences, 120(7/8), 489-504, 2019.
57. Hair Jr, J. F., Matthews, L. M., Matthews, R. L., & Sarstedt, M., "PLS-SEM or CB-SEM: updated guidelines on which method to use," International Journal of Multivariate Data Analysis, 1(2), 107–123, 2017.
58. Nunnally, J. C., & Bernstein, I. H., "Psychometric theory (3rd Ed.). New York: McGraw-Hill," (1994).
59. Henseler, J., Ringle, C. M., & Sarstedt, M., "A new criterion for assessing discriminant validity in variance-based structural equation modeling," Journal of the academy of marketing science, 43, 115-135, 2015.
60. Sternberg, R. J., & Detterman, D. K., "Human intelligence," 1979.
61. Lohman, D. F., "Human intelligence: an introduction to advances in theory and research," Review of Educational Research, 59(4), 333–373, 1989.
62. Shneiderman, B., "Human-centered AI," Oxford University Press. 2022. https://hcil.umd.edu/human-centered-ai/
63. Arrieta, A. B., Díaz-Rodríguez, N., Del Ser, J., Bennetot, A., Tabik, S., Barbado, A., ... & Herrera, F., "Explainable Artificial Intelligence (XAI): concepts, taxonomies, opportunities and challenges toward responsible AI,"Information fusion, 58, 82–115, 2020.
64. Kazim, E., Koshiyama, A. S., Hilliard, A., & Polle, R., "Systematizing audit in algorithmic recruitment," Journal of Intelligence, 9(3), 46, 2021.
65. Merritt, S. M., & Ilgen, D. R, " Not all trust is created equal: Dispositional and history-based trust in human-automation interactions," Human factors, 50(2), 194–210, 2008.
66. Shin, D, "The effects of explainability and causability on perception, trust, and acceptance: Implications for explainable AI," International Journal of Human-Computer Studies, 146, 102551, 2021.
67. Shin, D., "User perceptions of algorithmic decisions in the personalized AI system: perceptual evaluation of fairness, accountability, transparency, and explainability," Journal of Broadcasting & Electronic Media, 64(4), 541–565, 2020.
68. Streiner, D. L., "The reliability and validity of the Likert scale for the measurement of attitudes," In B. J. Bowers & T. D. Christensen (Eds.), Health Services Research Methods: A Guide to Best Practice (pp. 147–162). John Wiley & Sons, 2003.

Electronics Engineering Perspectives on Computer Vision Applications: An Overview of Techniques, Sub-areas, Advancements and Future Challenges

Yu Xun Zheng, K.-W. (G. H.) A. Chee, Anand Paul, Jeonghong Kim, and H. Lv

Abstract This chapter provides a strategic overview of applications in the computer vision domain. We initially introduce the etymology of computer vision, main tasks, key techniques, and algorithms. Traditional feature extraction methods and deep learning techniques, including prominent algorithms like Region-Based Convolutional Neural Network (R-CNN) and You Only Look Once (YOLO), are explored. We discuss important sub-areas such as image classification, object detection, and image semantic segmentation. The versatility of computer vision is showcased, particularly in autonomous vehicles, healthcare, and surveillance. Furthermore, we delve into the challenges and potential of computer vision, highlighting the necessity for advanced algorithmic methodologies, efficient hardware, robust privacy protections, and conscientious ethical considerations. We also explore upcoming trends, including cross-modal learning, sophisticated 'vision GPT' models, and unified models that share architecture and parameters across different tasks. These future directions indicate a transformative impact across various sectors, encompassing autonomous driving, healthcare imaging, and e-commerce. Additionally, we outline the future challenges and trends in the field, underscoring the significance of continuous research and development to address issues such as data scarcity, model interpretability, and privacy concerns. By effectively addressing these challenges and capitalizing on emerging trends, computer vision stands poised to make profound advancements with far-reaching implications. This comprehensive overview aims to provide a solid foundation for understanding the field of computer vision and its potential impact across multiple industries and applications.

Yu Xun Zheng and K.-W. (G. H.) A. Chee—These authors contributed equally and share first authorship.

Y. X. Zheng · K.-W. (G. H.) A. Chee (✉) · A. Paul · J. Kim
Kyungpook National University, Daegu 41566, Republic of Korea
e-mail: aghjuee@bh.knu.ac.kr

H. Lv
School of Information Science and Electronic Engineering, Zhejiang University, Hangzhou 315100, China

K. Daimi et al. (eds.), *Cutting Edge Applications of Computational Intelligence Tools and Techniques*, Studies in Computational Intelligence 1118,
https://doi.org/10.1007/978-3-031-44127-1_6

Keywords Object detection · Deep learning · Image classification · Scientific datasets · Neural networks

1 Introduction

Various applications of computational intelligence, such as in autonomous vehicles, healthcare, and surveillance, have emerged and developed over the past decades thanks to computer vision. Computer vision is a multidisciplinary field that aims to enable machines to interpret and understand visual information from their surrounding environment. Drawing upon techniques from neuroscience, computer science, and artificial intelligence, it seeks to replicate humans' ability to process and analyze visual data, ultimately allowing for the development of intelligent systems capable of performing tasks such as object recognition, scene understanding, and image matching. The evolution of computer vision has been marked by numerous milestones, ranging from foundational neural studies to breakthroughs in machine learning and deep learning, which together have propelled the field forward and expanded its applications across various industries.

In this chapter, we first introduce the origin and historical development of computer vision, the main tasks, key techniques, algorithms and sub-areas. Importantly, we demonstrate the versatility of computer vision in application scenarios such as autonomous driving, medical diagnosis, unmanned aerial vehicle (UAV) monitoring, and face recognition. Furthermore, we discuss the trends and future challenges, including research and development, in the various applications of computer vision.

1.1 History (Key Events)

The progress of computer vision has been marked by numerous significant milestones that have shaped the field's development. In 1959, Hubel and Wiesel's seminal study on the visual cortex neurons of cats laid the foundation for understanding the processing of visual information at the neural level, which has been crucial for the advancement of the field [1]. Subsequently, in 1963, Roberts' groundbreaking paper, "Machine Perception of Three-Dimensional Solids [2]," addressed the extraction of three-dimensional information from two-dimensional images, providing a cornerstone for the development of computer vision algorithms and techniques. During the 1970s, Marr's computational theory revolutionized the field by offering a comprehensive perspective on how humans interpret three-dimensional information from two-dimensional images [3]. The 1980s witnessed the incorporation of machine learning techniques to tackle computer vision problems, marking a substantial milestone in the discipline's evolution.

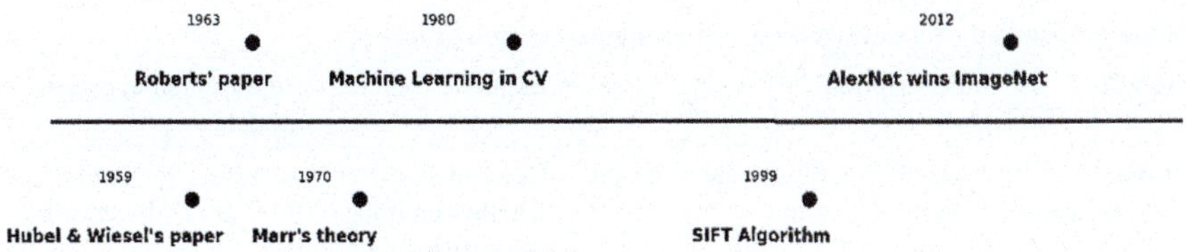

Fig. 1 Timeline of key events in computer vision

A pivotal development in object recognition and image matching occurred in 1999 with the introduction of the scale invariant feature transform (SIFT) algorithm by David Lowe. This robust feature detector and descriptor significantly impacted the field [4]. The deep learning revolution, which began in the 2010s and was exemplified by AlexNet's success in the ImageNet Large Scale Visual Recognition Challenge in 2012, heralded a new era of breakthroughs in computer vision and deep learning. This revolution drove the field forward and expanded the range of its applications [5]. Figure 1 provides a visual representation of these milestones.

1.2 Computer Vision Main Tasks

Computer vision covers a wide array of tasks aimed at interpreting and understanding visual information from the environment. Essential tasks include image classification, which categorizes images based on their content; object detection, identifying and locating specific targets within images; and semantic segmentation, assigning class labels to each pixel in an image. Additional tasks consist of instance segmentation, differentiating between separate instances of the same class; pose estimation, determining the position and orientation of objects; 3D reconstruction, recovering 3D structures from 2D images; and optical flow estimation, calculating object motion in image sequences. Table 1 provides a detailed overview of each of these tasks.

These tasks enable various applications across diverse domains, such as autonomous driving, intelligent surveillance, medical image analysis, virtual reality, and augmented reality. With the advancements in deep learning technologies, computer vision has made remarkable progress, with its performance on certain tasks approaching or even surpassing human levels. However, computer vision still faces numerous challenges, such as dealing with occlusions, lighting changes, and viewpoint variations in images. Additionally, real-world scenes are often highly complex, requiring computer vision algorithms to identify objects of interest and extract relevant information within these intricate environments.

To address these challenges, researchers in the field of computer vision have employed various techniques and methods. These include feature extraction, which processes images to extract features useful for object recognition, such as color, texture, and shape. Traditional feature extraction methods like SIFT [25], speeded-up robust features (SURF) [26], and histogram of oriented gradients (HOG) [27] have

Table 1 Detailed overview of each of the computer vision tasks

Task	Description	Example	Representative papers/ models
Image classification	Categorize input images based on their content	Determine whether an image contains a cat or a dog	AlexNet [5], VGG [6], ResNet [7], Inception [8]
Object detection	Detect specific objects within images and localize them using bounding boxes	Locate and draw bounding boxes around all pedestrians in an image	R-CNN [9], YOLO [10], SSD [11], Faster R-CNN [12]
Semantic segmentation	Assign a class label to each pixel in an image, segmenting the image into distinct regions corresponding to different object classes	Distinguish between different areas in an image, such as roads, buildings, and vegetation	FCN [13], DeepLab [14], U-Net [15]
Instance segmentation	Segment and identify individual instances of objects within the same class, differentiating between separate occurrences	Separate and identify two cars in an image, even if they are the same make and model	Mask R-CNN [16]
Pose estimation	Estimate the position and orientation of objects or keypoints within an image, often in the context of articulated objects like humans or animals	Estimate the coordinates of human body keypoints, such as shoulders, wrists, and knees	OpenPose [17], PIFPAF [18]
3D reconstruction	Recover the 3D structure and geometry of objects from 2D images, often using multiple images or depth information	Create a 3D model of a building using multiple images taken from different angles	Multi-View Stereo [19], SFM [20], KinectFusion [21], COLMAP [20]
Optical flow estimation	Compute the apparent motion of objects in an image sequence by estimating the displacement of pixels between consecutive frames	Estimate the speed and direction of moving objects from consecutive video frames	Lucas-Kanade [22], Horn-Schunck [23], Farneback [24], Dense optical flow Gunner-Farneback [24]

been utilized. Machine learning techniques, leveraging algorithms such as support vector machines (SVM) [28], decision trees [29], and random forests [30], have been applied to learn from the extracted features to perform tasks like image classification and object detection. Deep learning, using deep neural networks like convolutional neural networks (CNNs) and recurrent neural networks, directly learns feature representations from raw images. This approach has achieved remarkable success in computer vision and has become the dominant method. Multi-task learning trains a single model to handle multiple computer vision tasks simultaneously, such as image classification, object detection, and semantic segmentation. This can improve model generalization, reduce training time, and minimize computational resource consumption. Reinforcement learning enables computer vision systems to interact with their environment and make correct decisions in different scenarios, such as facilitating autonomous navigation for drones in complex environments.

Although computer vision has witnessed considerable advancements, the field still harbors significant potential yet to be discovered. Researchers tirelessly strive to propel its development by tackling these challenges and investigating novel techniques. As a result, computer vision technology will continue to evolve and expand, ultimately giving rise to more sophisticated and versatile applications across diverse domains.

2 Key Techniques and Algorithms in Computer Vision

2.1 Key Techniques

2.1.1 Traditional Feature Extraction Method

In the exploration of the captivating realm of computer vision, it is imperative to comprehend the fundamental techniques that have profoundly influenced this field over the course of its development. Initially, emphasis was placed on the formulation of traditional feature extraction methods, designed to capture distinctive visual attributes from images. These pioneering methods, including the SIFT, SURF, and HOG, established a solid foundation for early computer vision applications and catalyzed subsequent advancements. Consequently, it is essential to scrutinize these seminal methods, which emerged during the early 2000s, with SIFT making its debut, followed by SURF and HOG.

- SIFT, developed by David Lowe in 1999, is a landmark algorithm in computer vision. It extracts scale-invariant key points from images, providing robustness against rotation, scaling, and illumination changes. SIFT features have been widely used in image matching and object recognition tasks. However, the computational complexity of SIFT can be a drawback, especially in real-time applications.
- SURF, proposed by Herbert Bay and others in 2006, addresses the speed limitations of SIFT. It employs integral images and approximations of the Hessian matrix to speed up the feature extraction process. While similar to SIFT in many ways, SURF is computationally more efficient, making it an attractive alternative for applications where speed is crucial.

- HOG, introduced by Navneet Dalal and Bill Triggs in 2005, is a feature extraction method that describes the gradient direction information of local regions in an image. HOG is particularly effective at capturing shape and texture information, making it popular in pedestrian detection and vehicle recognition tasks. HOG features are generally faster to compute than SIFT and SURF but may have lower robustness and descriptiveness.

The comparison of these feature extraction methods can be summarized in Table 2.

SIFT, SURF, and HOG are significant feature extraction methods in computer vision that have evolved over time to address various challenges. SIFT and SURF are known for their robustness and descriptiveness, while HOG is valued for its speed and effectiveness in capturing shape information. The choice of which method to use depends on the specific application and dataset requirements.

Traditional feature extraction methods exhibit limitations due to their reliance on manually designed features, necessitating human expertise, and understanding for their extraction. Consequently, these methods prove inadequate in handling minor variations within images, such as alterations in illumination, angle, and scale, thereby lacking adaptability across diverse tasks. In contrast, deep learning techniques, notably CNNs, offer a viable solution to these challenges. Through the process of training, deep learning models autonomously acquire and extract effective features from extensive datasets. These learned features possess the ability to capture intricate and nuanced patterns present within images, rendering them more resilient and versatile. Thus, while traditional feature extraction methods retain utility in specific scenarios, the rapid advancement of deep learning methodologies, coupled with the availability of abundant annotated data, has facilitated the surpassing of traditional approaches by deep learning in performance across numerous computer vision tasks. Consequently, deep learning stands as the preeminent method for image comprehension and analysis in contemporary practice.

Table 2 Comparison of three feature extraction methods

Feature extraction method	Speed	Robustness	Descriptiveness	Main applications
SIFT (scale invariant feature transform)	Slower	High	High	Image matching, object recognition
SURF (speeded up robust features)	Faster	High	High	Image matching, object recognition
HOG (histogram of oriented gradients)	Fast	Moderate	Moderate	Pedestrian detection, vehicle recognition

2.1.2 Deep Learning Methods

In the 1980s, the development of the multilayer perceptron model [31] demonstrated the remarkable capabilities of computers in digit recognition. However, due to limitations in computing power, particularly CPU and storage resources, the data scale that could be processed was relatively small, and the models' expressive power was limited, making them incapable of handling complex image problems. To better address these issues, different types of deep learning models, including CNNs [32], recurrent neural networks (RNN) [33], and generative adversarial networks (GAN) [34], later emerged. To better understand the differences and respective strengths of these models, we provide a detailed comparative analysis in Table 3. By referring to Table 3, readers can gain a clearer understanding of the characteristics and application scenarios of these models and choose the appropriate network model for research and application according to specific tasks.

Table 3 Comparison of three types of neural network methods

Model	CNN	RNN	GAN
Purpose	Image processing, object detection, and classification	Sequence processing, time-series data, and natural language processing (NLP) tasks [35]	Generating new data samples, data augmentation, and image synthesis
Structure	Comprises convolutional, pooling, and fully connected layers	Consists of a sequence of hidden layers with recurrent connections	Comprises two sub-models: a generator and a discriminator
Input data	Typically fixed-size, grid-like structures (e.g., images)	Sequential data or data with temporal dependencies (e.g., text, speech)	Random noise and data samples from the true distribution
Training	Supervised learning using backpropagation	Supervised learning using backpropagation through time	Adversarial training: generator and discriminator models are trained simultaneously in a zero-sum game framework
Key features	Effective in learning spatial hierarchies and local features	Can handle variable-length input/output sequences and capture long-range dependencies	Can generate high-quality samples by learning the true data distribution
Applications	Image classification, object detection, segmentation	Language modeling, machine translation, speech recognition	Image-to-image translation, super-resolution, generating art, and data augmentation

2.1.3 Other Technologies

In conjunction with traditional feature extraction methods and deep learning techniques, several other noteworthy technologies have emerged as significant contributors to the field of computer vision. These technologies, namely Transfer Learning, Zero-shot Learning (ZSL), and Multimodal Learning, offer distinct yet complementary approaches to tackle the challenges inherent in various computer vision tasks. In the following sections, we will delve into the details of each of these technologies.

- Transfer learning is a method that leverages pre-trained models' knowledge to solve new tasks. In the computer vision field, it helps train high-performance models on limited datasets quickly. Initially proposed by Microsoft researcher Donaldson in 1992, transfer learning has been widely used in deep learning, particularly in image classification and object detection tasks. Classic applications include using pre-trained CNNs for feature extraction and training new classifiers with those features.
- ZSL aims to recognize categories do not present in the training data. By learning the relationship between categories and attributes in training data, ZSL can identify new categories without additional annotated data. First proposed by Christopher Kanan and Ilya Narsky in 2013, ZSL has been successfully applied to images, audio, and video domains.
- Multimodal learning involves processing data from multiple modalities, such as images, text, and audio. By fusing information from different modalities, multimodal learning can improve the performance of computer vision tasks like image captioning and visual question answering. First introduced by Li Fei-Fei's team in 2010, multimodal learning has been widely adopted in deep learning and has become an essential research direction in computer vision and natural language processing fields.

The above key technologies and algorithms provide robust support for the development of the computer vision field. With continuous technological advancements, the performance of computer vision systems will keep improving, offering powerful support for various application scenarios.

2.2 *Key Algorithms*

2.2.1 Region-Based Convolutional Neural Networks and Their Variants

Transitioning from key techniques, the exploration of pivotal algorithms that have propelled the field of computer vision forward becomes imperative. These algorithms have revolutionized image recognition, object detection, and semantic understanding, offering vital contributions to a wide range of applications, from autonomous driving to medical image analysis. Among these algorithms, the R-CNN framework, initially proposed by Ross Girshick and colleagues in 2014 [9], has emerged as a significant

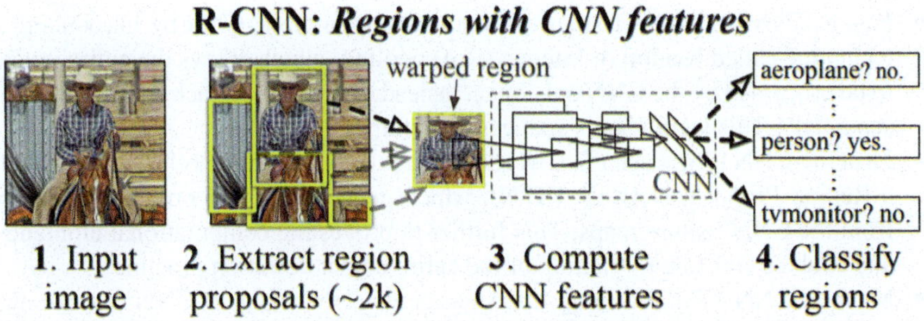

Fig. 2 Visual representation of the R-CNN algorithm's process. Reprinted with permission from the arxiv version of [9] with Attribution—Non-Commercial 4.0 International license

breakthrough in object detection. By seamlessly integrating region proposals and CNNs, R-CNN represents a substantial leap in enhancing the accuracy of object detection, surpassing traditional sliding window methods. Figure 2 provides a visual depiction of the R-CNN algorithm's intricate process. The R-CNN object detection algorithm's workflow are well illustrated, and how the different components interact with each other are depicted. To gain a deeper understanding, it is essential to explore the underlying mechanisms and steps of the R-CNN framework, which can be broadly categorized as follows:

1. Region Proposals: Given an input image, the R-CNN framework first generates a set of region proposals that may contain objects. This is typically done using selective search or other region proposal algorithms. These algorithms analyze the image for possible object locations based on color, texture, and size. The generated region proposals are usually in the form of bounding boxes.
2. Feature Extraction: Each of the region proposals is then processed by a pre-trained CNN to extract features. The region proposals are typically resized to a fixed size (e.g., 224 × 224 pixels) to be compatible with the CNN input requirements. The output of this step is a fixed-length feature vector for each region proposal, which captures the visual information of the potential object within the bounding box.
3. Classification: The extracted features are then fed into a classifier, such as a Support Vector Machine (SVM), to determine the object class. The classifier is trained on features extracted from labeled object bounding boxes in the training dataset.
4. Bounding Box Regression: To improve the localization accuracy of the object bounding boxes, R-CNN also employs a bounding box regression step. This step refines the coordinates of the predicted bounding boxes to better match the ground truth bounding boxes in the training dataset.

Following R-CNN, several improvements have been proposed to address its limitations, including computational efficiency and end-to-end training. These improvements include:

- Fast R-CNN [12]: This framework improves upon the R-CNN by introducing a technique called Region of Interest (RoI) pooling, which allows the entire image to be processed by the CNN only once, instead of processing each region proposal separately. This significantly reduces the computational cost.
- Faster R-CNN [36]: Faster R-CNN substitutes the selective search algorithm with a Region Proposal Network (RPN), which generates region proposals directly from the CNN feature maps. This further improves the computational efficiency and enables end-to-end training of the entire object detection pipeline.
- Mask R-CNN [16]: Building upon Faster R-CNN, Mask R-CNN extends the framework to perform instance segmentation, where each object in the image is not only detected but also precisely segmented from the background. It achieves this by adding a parallel branch for predicting object masks along with the existing branches for classification and bounding box regression.

These advancements have significantly improved the efficiency and performance of object detection and segmentation tasks, making R-CNN and its variants widely adopted in various computer vision applications.

Nevertheless, despite the remarkable success and far-reaching impact of the R-CNN framework and its derivatives, notable limitations remain inherent in these methodologies. One crucial drawback pertains to their computational speed. While advancements such as Fast R-CNN and Faster R-CNN have substantially improved computational efficiency, they still fall short in meeting the real-time requirements of numerous applications, particularly those demanding real-time or near real-time processing, such as autonomous driving or video surveillance. This limitation predominantly arises from the computationally intensive and time-consuming two-stage process involving region proposal and subsequent classification.

Moreover, the R-CNN framework and its variants necessitate a significant memory allocation to store region proposals and their corresponding features. Additionally, the reliance on region proposal mechanisms and separate classifiers contributes to the complexity and computational expense of the training process. Notably, this architecture lacks true end-to-end integration, as distinct components are essential for training.

These challenges have spurred the development of more efficient and end-to-end architectures for object detection. Among these, the You Only Look Once (YOLO) algorithm has emerged as a prominent solution, which will be elaborated upon in the subsequent section.

2.2.2 You Only Look Once (YOLO)

In 2016, Joseph Redmon and his colleagues introduced You Only Look Once (YOLO) as the first official single-stage object detector in the deep learning era. This method completely abandoned the two-stage detection mode of "region proposal + regression" and resized the input image to a uniform size, dividing it into multiple grids. Based on the grid containing the object center, it predicted the object category and

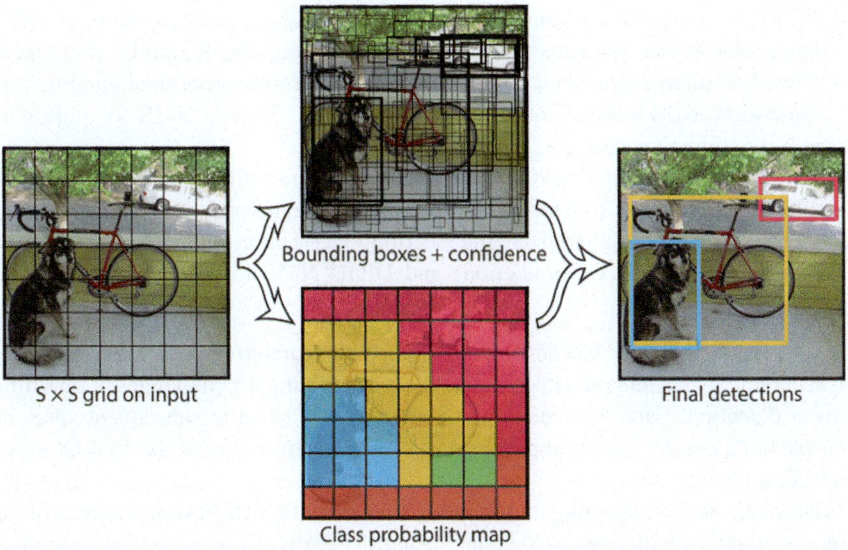

Fig. 3 Visual representation of the YOLO algorithm's process. Reprinted with permission from the arxiv version of [10] with Attribution—Non-Commercial 4.0 International license

output the detection results in the last convolutional layer. Figure 3 shows the visual representation of the YOLO algorithm's process This process can be understood as completing feature extraction, bounding box regression, and object classification tasks using only one CNN, saving a significant amount of computational cost.

The YOLO family of object detectors has evolved significantly since its inception, with each version introducing new features and improvements in accuracy and speed. The YOLO series has demonstrated improvements in terms of accuracy, speed, and robustness, effectively addressing the challenges of real-time object detection. As a result, YOLO and its variants have become popular choices for various computer vision applications that require efficient and accurate object detection.

The updates for each version are as follows:

- YOLOv2 [37] employed the Darknet-19 backbone, introduced anchor boxes to better handle different object sizes, and added passthrough layers for finer-grained feature detection.
- YOLOv3 [38] switched to the Darknet-53 backbone for improved feature extraction, changed the input size for better object detection at different scales, and incorporated a Feature Pyramid Network (FPN) for multi-scale detection, which combined low-level features with high-level features to enhance the detection performance.
- YOLOv4 [39] further enhanced the architecture with the CSPDarknet53 backbone, which increased the learning capability and decreased the computation cost. It also introduced Mosaic data augmentation for better generalization, self-adversarial training (SAT) for improved robustness, Spatial Pyramid Pooling

(SPP) to capture multi-scale features, and Path Aggregation Network (PANet) layers for better information flow between layers. Additionally, it adopted Complete Intersection Over Union (CIOU) loss for more accurate bounding box regression and Distance Intersection Over Union (DIOU) NMS for improved non-maximum suppression.

- YOLOv5 [40] modified the CSPDarknet53 backbone with a Focus module, which allowed the network to learn more representative features by combining adjacent pixels. It utilized Generalized Intersection Over Union (GIOU) loss for more accurate bounding box predictions and DIOU NMS for better non-maximum suppression.

As the YOLO series has developed, it has demonstrated improvements in terms of accuracy, speed, and robustness, effectively addressing the challenges of real-time object detection. Each new module and feature introduced in subsequent versions has played a crucial role in improving the overall performance of the YOLO object detection framework.

However, notwithstanding their notable strengths, the YOLO series of algorithms may encounter challenges in specific scenarios. Particularly, in scenes characterized by numerous diminutive objects or necessitating precise object localization, the YOLO series may exhibit suboptimal performance owing to the inherent trade-off between speed and accuracy. Furthermore, detecting objects with irregular or highly variable shapes can pose difficulties for the YOLO series, as its reliance on bounding boxes as object detection representations may be less effective in such cases. Notwithstanding these limitations, the ongoing advancements and refinements within the YOLO series are progressively addressing these challenges, rendering it a robust and versatile choice for an extensive range of applications.

3 Main Sub-areas of Computer Vision

In the domain of computer vision, object detection, image classification, and semantic segmentation are among the core tasks, offering foundational and widely applicable value. These tasks encompass basic concepts of image understanding, such as recognizing objects, classifying scenes, and comprehending the relationships between objects and their environment. At the same time, they pose technical challenges that involve addressing issues like illumination variations, occlusions, viewpoint changes, and scene complexity. Therefore, this chapter primarily focuses on these three tasks.

3.1 Image Classification

Image classification is a vital application in the computer vision domain. It primarily entails assigning a suitable semantic category label to an input image based on its content, enabling computers to classify the image accordingly. Notable advancements in image classification using deep CNNs are apparent in the ImageNet ILSVRC challenge. Essential network models for this task include AlexNet, ZF-Net [41], GoogleNet [8], VGG, and ResNet.

The MNIST dataset, which consists of handwritten digits, has been a widely used benchmark in the field of image classification, serving as an important starting point for many researchers in the computer vision community. Figure 4 shows some actual results from the MNIST dataset. In addition to the MNIST dataset, other commonly used datasets for image classification include the ImageNet dataset, Caltech-101, Caltech-256, TinyImage, and SUN. Table 4 provides an overview of some popular datasets in the field of image classification, along with their essential information.

Image classification stands as a fundamental and pivotal task within the realm of computer vision, serving as a cornerstone benchmark for assessing the performance of a myriad of models. The evolutionary trajectory commenced with relatively straightforward assignments, exemplified by the recognition of grayscale handwritten digits across ten classes in the MNIST dataset. This developmental trajectory progressively advanced towards more intricate challenges, including the ten-class CIFAR-10 and the hundred-class CIFAR-100 tasks, ultimately culminating in the prestigious ImageNet challenge. The proliferation of expansive datasets has been

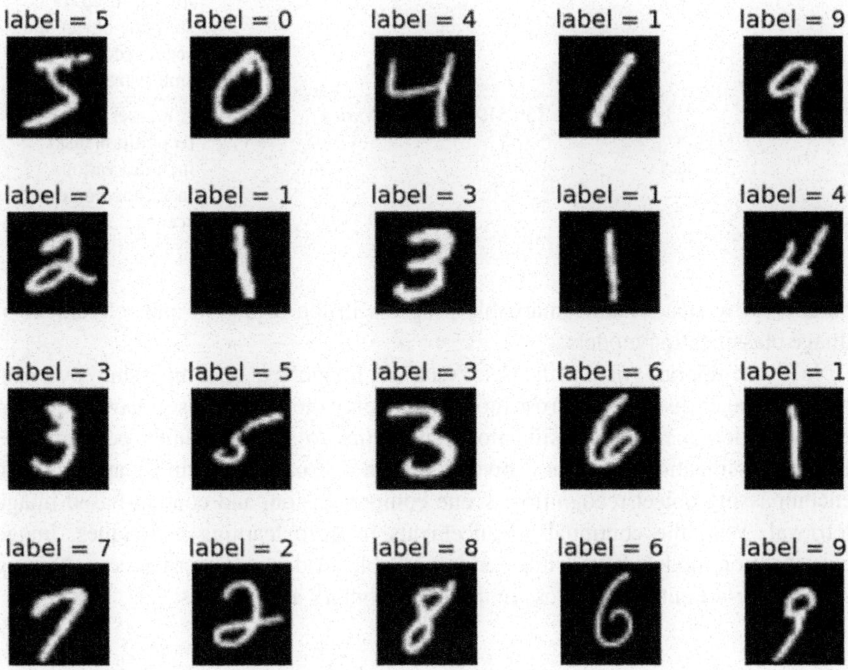

Fig. 4 Actual results from the MNIST dataset

Table 4 An overview of some popular datasets in the field of image classification

Dataset	Number of categories	Number of images	Resolution range	Description
ImageNet	1,000	14,197,122	Varies	Contains images of many categories, used for the large-scale visual recognition challenge (ILSVRC)
Caltech-101	101	9,144	~300 × 200	Each category contains 40 to 800 images, used for fine-grained classification and recognition
Caltech-256	256	30,607	~300 × 200	An extension of Caltech-101, includes more categories and images, used for more complex classification tasks
TinyImage	75,062	79,302,017	32 × 32	A large number of low-resolution images, used for studying object and scene recognition in small-sized images
SUN	397	130,519	Varies	A dataset for scene recognition tasks, includes various indoor and outdoor scenes

paralleled by steady and remarkable progress in the evolution and refinement of image classification models.

In contemporary times, datasets such as ImageNet, encompassing over ten million images distributed across more than twenty thousand classes, have propelled computer-driven image classification capabilities to surpass human performance. Image classification assumes a pivotal role in a broad spectrum of applications, encompassing object recognition, scene comprehension, and content-based image retrieval. With the continual advancements in deep learning techniques, image classification models have witnessed remarkable strides in performance, achieving state-of-the-art outcomes across multiple benchmark assessments.

3.2 *Object Detection*

Object detection is a fundamental and challenging task in the field of computer vision. It primarily focuses on locating specific objects within an image and determining their categories, making it a more complex image recognition problem compared to image classification. Traditional object detection algorithms usually rely on sliding window approaches and use handcrafted features.

The breakthrough of deep CNNs in image classification tasks in 2012 led many researchers to adopt Deep CNNs for object detection, resulting in significant improvements in detection accuracy. The R-CNN algorithm was one of the first to utilize this approach. It used Selective Search to extract candidate windows from the input image and then applied a deep CNN to extract features from these windows. Linear classifiers, such as SVM, were used to categorize the candidate windows into objects and backgrounds based on these features. Finally, NMS (non-maximum suppression) was employed to discard some candidate windows and obtain the final object localization results.

Despite achieving excellent results, selective search strategies and CNN-based object detection algorithms faced a speed bottleneck. In response, researchers developed various object detection methods, including Faster R-CNN, the YOLO series, EfficientDet [42], CenterNet [43], DETR [44], and Swin Transformer [45]. Each of these methods has its unique techniques and architectures to improve detection accuracy and optimize computational performance.

For example, Faster R-CNN replaced the traditional selective search strategy with a deep CNN-based Region Proposal Network (RPN), which shared convolutional layer features with the object detection network. The YOLO series provided an end-to-end, real-time object detection system with several improvements in speed and accuracy across its different versions (e.g., YOLOv2, YOLOv3, YOLOv4, and YOLOv5). Figure 5 shows the vehicle detection results of YOLO on the VisDrone dataset. EfficientDet combined the EfficientNet backbone network with a novel bidirectional feature pyramid network (BiFPN) and weighted box regression to achieve a good balance of accuracy and speed. CenterNet transformed the object detection task into a keypoint regression task, while DETR utilized the Transformer architecture to treat object detection as directly predicting pairs of bounding boxes and categories within images.

These emerging object detection algorithms have achieved impressive results in their respective application domains and datasets, making them suitable for various application scenarios, such as autonomous driving, security surveillance, and smart retail. As a result, they have significantly advanced the field of object detection, offering improved accuracy and computational performance.

Fig. 5 Vehicle detection results of YOLO on the VisDrone dataset

3.3 *Image Semantic Segmentation*

Semantic segmentation plays a pivotal role within the domain of computer vision due to its multifaceted nature. Unlike image classification, semantic segmentation not only identifies objects present in an image but also accurately delineates their precise boundaries. Traditional approaches to image semantic segmentation typically encompass three fundamental components. The initial stage primarily focuses on performing low-level image segmentation, effectively dividing the image into numerous subregions. Subsequently, the second stage extracts low-level features, such as color, texture, and shape, from these subregions. Lastly, the third stage involves learning the mapping from these low-level features to a high-level semantic space. Based on the acquired mapping model, the image is annotated, thereby recognizing the semantic categories of image regions and even individual pixels. A noteworthy example of such work is the TextonBoost method introduced by Shotton et al. [46], which leverages a boosted decision tree classifier. In this method, a conditional random field (CRF) encompassing all image pixels is utilized, where the single-point potential of each pixel is learned via texture layout filters. Additionally, smoothness constraints are imposed on the pixel-level semantic annotations within the random field, ultimately leading to the acquisition of pixel-level semantic annotations through energy minimization.

With the successful application of deep CNNs in image detection, classification, and other tasks, many researchers have applied deep CNN to the field of image semantic segmentation. For example, Long et al. [13] proposed a fully convolutional network (FCN) at CVPR 2015, which can obtain the target classification results of each pixel end-to-end. Unlike the classic CNN that require fixed-size image inputs and use fully connected layers after the convolutional layers to obtain fixed-length

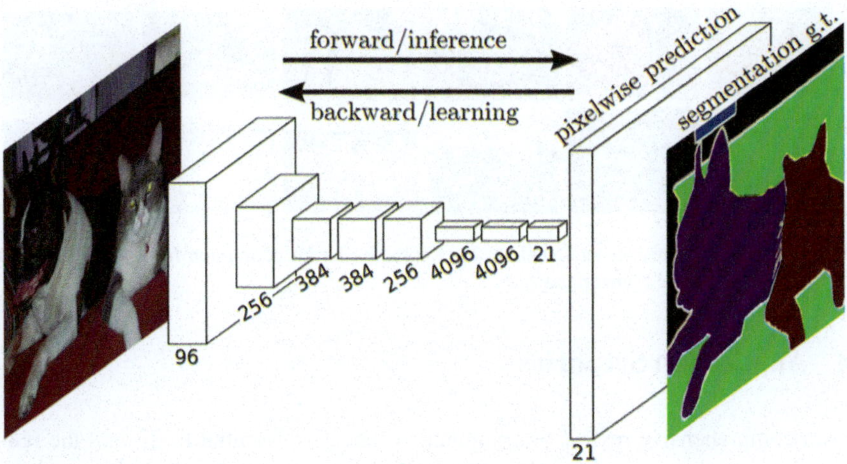

Fig. 6 Schematic diagram of the FCN structure. Reprinted with permission from the arvix version of [13] with Attribution—Non-Commercial 4.0 International license

feature vectors, FCN can accept input images of any size and use only convolutional layers. FCN uses deconvolution layers to upsample the feature maps of the last convolutional layer, making the feature maps the same size as the input image, thereby generating a semantic prediction for each pixel. This process retains the spatial information in the original input image, and finally calculates the softmax classification loss on the upsampled feature map pixel-by-pixel. Figure 6 shows the schematic diagram of the FCN structure used for semantic segmentation.

In recent years, several advanced algorithms have emerged as improvements to the original FCN approach, addressing some of its limitations. These algorithms include SegNet [47], U-Net, PSPNet [48], DeepLabv3 [49], and DeepLabv3+ [50]. Each of these methods offers unique improvements, such as encoder-decoder architectures, atrous convolutions, and pyramid pooling. By incorporating these techniques, these algorithms provide more accurate semantic segmentation results, especially when it comes to capturing fine details and handling various object scales. For example, you can refer to an image illustrating the results obtained using DeepLabv3+ (Fig. 7) to get a better understanding of the performance of this algorithm in semantic segmentation tasks. This image shows the original input image, the ground truth labels, and the predicted segmentation results from DeepLabv3+.

Fig. 7 Results obtained using DeepLabv3+. Reproduced with permission from [50]. Copyright 2018, Springer Nature Switzerland AG

4 Application Scenarios

As a comprehensive review of computer vision, it is essential to discuss the real-world application scenarios where computer vision technology has made significant contributions. The advancements in this field have resulted in a wide range of applications across various industries, demonstrating the far-reaching impact of computer vision. Here, we present a few notable examples of computer vision applications in different fields.

- Autonomous Driving: Computer vision technology has played a pivotal role in the development of autonomous vehicles by enabling them to identify traffic signs, pedestrians, other vehicles, and obstacles. This, in turn, contributes to improved road safety, reduced traffic congestion, and enhanced traffic efficiency.
- Medical Diagnosis: The medical field has witnessed remarkable achievements through the application of computer vision. By analyzing medical images, computer vision algorithms assist doctors in detecting and diagnosing diseases such as cancer and heart disease. These algorithms have been developed to analyze various types of medical imaging, including histopathological and fluorescence images. Furthermore, computer vision technology aids surgical procedures, like robotic surgery, to ensure higher accuracy and safety.
- UAV Monitoring: UAVs, or drones, have become increasingly popular in recent years, opening up new possibilities for computer vision applications. UAVs are being employed for monitoring, security, and emergency response purposes. Real-time image data analysis, terrain modeling, moving target tracking, and other high-level tasks become possible with computer vision technology, ultimately improving the efficiency and safety of UAVs. UAV monitoring leverages computer vision algorithms to analyze aerial images and videos captured by drones for various purposes, including environmental monitoring, disaster management, agriculture, surveillance, and infrastructure inspection.
- Face Recognition: Face recognition technology, driven by computer vision algorithms, has gained widespread adoption in security, payment, and social networking applications. By swiftly and accurately identifying faces in images and

matching them with database records, face recognition technology has become indispensable in various scenarios, including airport security, mobile phone unlocking, and online payment systems.

4.1 Autonomous Driving

Figure 8 illustrates the role of computer vision in autonomous driving. Computer vision in autonomous vehicles can lead to increased road safety, reduced traffic congestion, and improved traffic efficiency. By leveraging advanced techniques in object detection and semantic segmentation, autonomous vehicles can identify and track various objects in real-time, playing a crucial role in the perception and interpretation of their surroundings. This capability significantly enhances road safety, reduces traffic congestion, and improves overall traffic efficiency. Two key aspects of computer vision that contribute to the success of autonomous driving systems are object detection and semantic segmentation.

In the field of autonomous driving, traffic sign and road recognition are crucial for the success of computer vision systems. To evaluate and compare different approaches, researchers have developed several publicly available traffic sign and road recognition datasets. Table 5 provides a detailed comparison of some of these public datasets, including their names, the number of images, the number of classes, the year of release, the country, the annotation type, and the main features. By utilizing

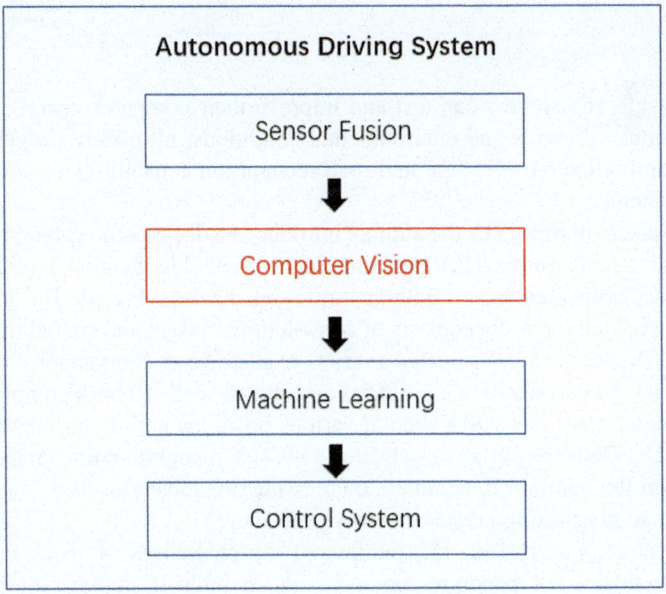

Fig. 8 An illustration of computer vision's role in autonomous driving

Table 5 Comparison of some public traffic sign and road recognition datasets

Dataset	Images	Classes	Year	Country	Annotations	Main features
GTSDB	900	43	2013	Germany	Bounding box	Traffic signs in various conditions; day and night
LISA	6,610	47	2011	USA	Bounding box	Diverse traffic signs; various lighting and occlusions
Cityscapes	25,000	30	2016	Germany	Pixel-level	Urban scenes; street-level imagery
Mapillary Vistas	25,000	66	2017	Global	Pixel-level	Street-level imagery; diverse scenes and countries
ApolloScape	140,000	26	2018	China	Pixel-level	Large-scale dataset; diverse scenes
BDD100K	100,000	19	2018	USA	Bounding box	Diverse driving scenarios; day and night
IDD	25,000	26	2018	India	Pixel-level	Diverse scenes; unique driving conditions in India

these datasets, researchers can test and improve their computer vision algorithms under various scenarios and environmental conditions, ultimately providing more accurate and reliable traffic sign and road recognition capabilities for autonomous driving systems.

For instance, in Ref. [51], the authors introduced a deep learning-based approach to traffic sign recognition. They performed a series of classification experiments on publicly available German and Belgian traffic sign datasets (Fig. 9). The deep neural network used in their study consists of convolutional layers and spatial transformer networks. The researchers evaluated a variety of adaptive and non-adaptive stochastic gradient descent optimization algorithms and investigated different combinations of Spatial Transformer Networks sited at various positions within the primary neural network. The CNN they propose achieved a 99.71% recognition rate on the German Traffic Sign Recognition Benchmark, outclassing previous state-of-the-art methods while also demonstrating greater memory efficiency.

Furthermore, Porzi et al. [52] addressed the challenges of using crop-based training strategies for panoptic segmentation on multi-megapixel images. They proposed a novel crop-aware bounding box (CABB) regression loss to improve the consistency of predictions with the visible parts of cropped objects, without

Fig. 9 German traffic sign detection benchmark (GTSDB) dataset

Fig. 10 Comparison of different methods on Mapillary Vistas validation set. Reprinted with permission from the arvix version of [52] with attribution—non-commercial 4.0 International license

over-penalizing them for extending outside of the crop. They also introduced a new data sampling and augmentation strategy to enhance generalization across scales by counteracting the imbalanced distribution of object sizes. By combining these contributions with a carefully designed, top-down panoptic segmentation architecture, the authors achieved state-of-the-art results on challenging datasets such as Mapillary Vistas Dataset (MVD), Indian Driving, and Cityscapes. Their approach surpassed the previous best method on the MVD dataset by 4.5% in Panoptic Quality (PQ) and 5.2% in mean Average Precision (mAP), demonstrating the effectiveness of their techniques in addressing scale variation and improving panoptic segmentation performance. Figure 10 showcases the comparison of different methods on Mapillary Vistas validation set.

4.2 Medical Diagnosis

In recent years, numerous publicly available datasets have been released to facilitate the development and evaluation of computer vision algorithms in medical diagnosis. Table 6 shows a comparison of common pathology datasets.

Table 6 Comparison of some public pathology datasets

Dataset	Images	Classes	Data type	Use cases
Camelyon16	400	2	Histopathological	Breast cancer detection
Camelyon17	1000	5	Histopathological	Breast cancer detection, metastasis localization
BreakHis	9097	8	Histopathological	Breast tumor classification
BreaKHis-VGG16	7909	8	Histopathological	Breast tumor classification
KIMIA Path24	24	24	Histopathological	Pathology image retrieval, classification
KIMIA Path960	960	20	Histopathological	Pathology image retrieval, classification
MoNuSeg	30	9	Fluorescence	Nuclei segmentation
TNBC	21	3	Histopathological	Triple-negative breast cancer classification

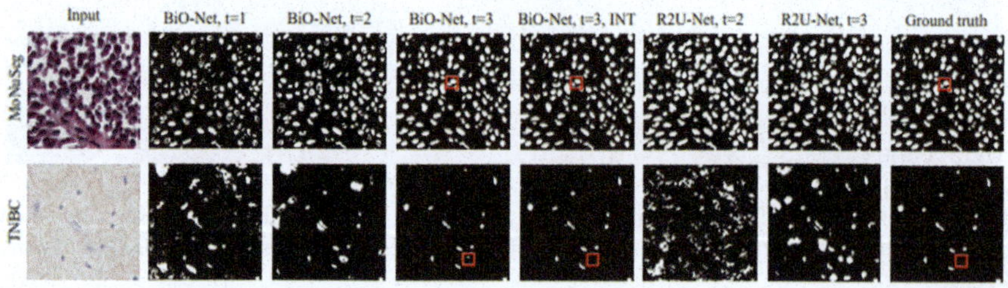

Fig. 11 Qualitative comparison between BiO-Net and the R2U-Net on the MoNuSeg testing set and TNBC. Reprinted with permission from [53]. Copyright 2020, Springer Nature Switzerland AG

In Ref. [53], the authors proposed a novel bi-directional O-shape network (BiO-Net) that enhanced the performance of U-Net-based models without increasing model complexity. Unlike previous U-Net variants [54], which mainly focused on modifying existing building blocks or developing new functional modules, BiO-Net utilizes the building blocks in a recurrent fashion, while avoiding the introduction of additional parameters. The proposed bi-directional skip connections can be readily incorporated into an encoder-decoder structure, enhancing its performance across different task domains. Figure 11 shows a comparison of the application of BiO-Net on multiple medical image analysis tasks.

Table 7 Comparison of some public UAV datasets

Dataset	Domain	Type	Images	Classes	Released
UAVDT	Traffic monitoring	RGB	80,000	3	2017
VISDrone	Object detection, tracking	RGB	10,209	12	2018
DOTA	Object detection in aerial images	RGB	2,806	15	2018
UCAS-AOD	Object detection in aerial images	RGB	1,510	3	2016
Stanford drone	Multi-task aerial video	RGB	N/A	8	2016

4.3 UAV Monitoring

As the field of UAV monitoring has grown, several datasets have been released to facilitate the development and evaluation of computer vision algorithms in this domain. Table 7 shows a comparison of common UAV datasets.

The core members of the Remote Sensing Image Analysis team (DH_RSIA) at Zhejiang Dahua Technology Co. Ltd., had achieved outstanding results in the field of aerial image analysis. Zhejiang Dahua Technology Co. Ltd. is a world-leading provider and operator of intelligent IoT solutions. The DH_RSIA team ranked first on the DOTA dataset (Fig. 12), utilizing an improved Cascade R-CNN detector inspired by the RoI Transformer algorithm. They employed ResNeXt101 with FPN as the backbone and applied data augmentation techniques, such as multi-scale, flip, and rotation, to the images sliced by the DOTA_devkit before training. Moreover, their method incorporated multi-scale training, multi-scale testing, and model merging to achieve optimal results [56–59].

4.4 Face Recognition

Face recognition algorithms have been developed to analyze facial features, detect landmarks, and perform robust matching, even under challenging conditions such as varying lighting or facial expressions. In recent years, numerous publicly available datasets have been released to facilitate the development and evaluation of computer vision algorithms in face recognition. Table 8 shows a comparison of common face recognition datasets.

In Ref. [60], the authors introduce a novel technique called face X-ray, which effectively detects forgery in face images. Figure 13 illustrates how face X-ray can expose blending boundaries present in manipulated face images, while producing a blank image when applied to genuine images. This approach can help to enhance the reliability and security of face recognition systems by identifying manipulated images.

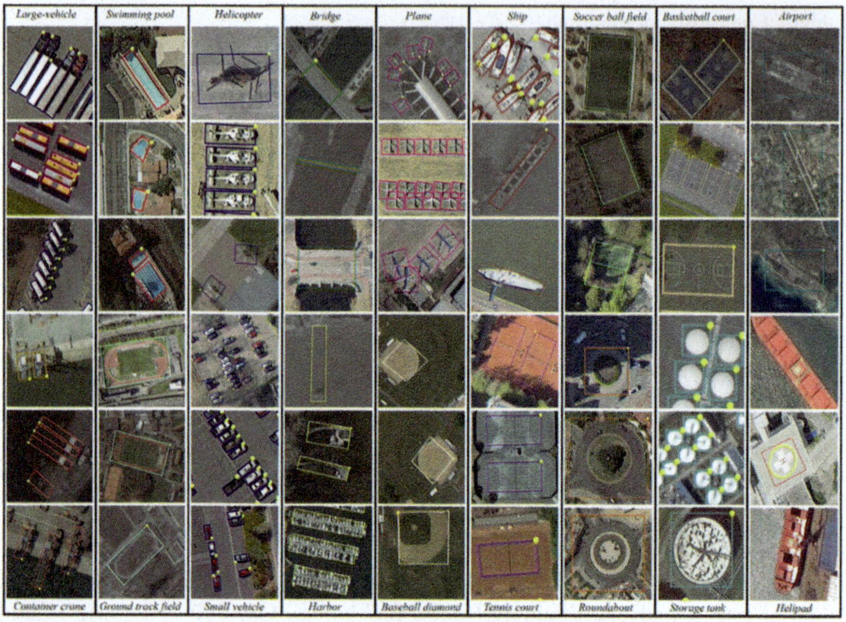

Fig. 12 Examples of annotated images on DOTA dataset. Reproduced from https://captain-whu.github.io/DOTA/index.html with permission from the authors on condition of citing their related paper [55]

Table 8 Comparison of some public datasets

Dataset	Number of images	Number of subjects
LFW	13,233	5,749
YTF	3,425	1,595
CASIA-WebFace	494,414	10,575
MS-Celeb-1 M	10,000,000	100,000
VGGFace	2,622,800	2,622
VGGFace2	3,314,679	9,131

In summary, the application of computer vision technology in autonomous driving, medical diagnosis, UAV monitoring, and face recognition has resulted in remarkable progress. As technology continues to develop and innovate, computer vision applications will expand further, enhancing convenience and value in people's daily lives and work.

Fig. 13 Face X-ray exposes the blending boundaries present in forged face images and produces a blank image when applied to genuine images. **a** A genuine image and its corresponding face X-ray; and **b** forged images and their related face X-rays. Reprinted with permission from the arvix version of [60] with Attribution—Non-Commercial 4.0 International license

5 Future Trends and Challenges

As a vibrant subfield of artificial intelligence, computer vision is advancing at an astonishing pace, holding the potential to redefine our perception and interaction with the surrounding world. Despite significant progress in recent years, the field is not without its challenges. Overcoming these hurdles is essential to fully unleash the capabilities of computer vision. Algorithm optimization stands as a prominent challenge. Although current algorithms have demonstrated impressive performance, there is still ample room for improvement. Deep learning models, while powerful, often demand extensive computational resources and memory. Thus, there is a continuous pursuit for more efficient optimization methods and model structures. Reinforcement learning and unsupervised learning methodologies hold promise in enhancing the adaptability and generalization of computer vision models.

Hardware requirements pose another significant challenge. As deep learning algorithms become increasingly complex and resource-intensive, traditional CPUs and GPUs may struggle to meet the demands. The future of computer vision hinges on the development of more efficient and energy-conserving hardware, such as dedicated AI chips and edge computing devices. These advancements could substantially reduce computational costs and enhance overall efficiency. Concurrently, data privacy emerges as a critical consideration. Numerous computer vision applications involve the processing of personal and sensitive data, necessitating robust privacy protection during data collection, storage, and analysis. Techniques such as differential privacy, homomorphic encryption, and distributed learning strategies like federated learning can play instrumental roles in safeguarding user privacy while enabling collaborative model training and optimization.

Ethical considerations are integral to the responsible deployment of computer vision technologies. Ensuring fairness, transparency, and accountability in algorithmic decision-making processes is paramount. Addressing algorithmic biases, advocating for fairness and interpretability, and enforcing relevant laws and policies are crucial steps in managing these ethical concerns. Despite these challenges, the future of computer vision presents promising trends. Cross-modal learning, which integrates visual information with other modalities like text and speech, is poised to gain prominence in research. Additionally, the advancements observed in NLP have set the stage for sophisticated computer vision models. Models like ChatGPT, Bard, and Bing developed by companies such as OpenAI, Google, and Microsoft, respectively, have made significant strides in this direction. ChatGPT, in particular, has garnered widespread recognition for its human-like text generation capabilities. This versatile language model, trained on diverse internet text, has demonstrated broad applicability across tasks such as email composition, complex question answering, language translation, and even video game dialogue generation. The success and adaptability of ChatGPT point towards an exciting trajectory for future advancements in computer vision. Thus, it is plausible to envision the emergence of "vision GPT" models that possess an unprecedented level of understanding and interpretation of visual data, potentially surpassing human-level performance across diverse computer vision tasks. The integration of unsupervised and self-supervised learning methodologies, as exemplified in ChatGPT, may increasingly feature in these sophisticated computer vision models.

Another emerging trend in computer vision is the development of unified models capable of performing multiple tasks, such as image classification, object detection, and segmentation. This innovative approach involves employing the same architectural framework or parameter set to accomplish various tasks, leading to more efficient and streamlined models. Furthermore, technologies like augmented reality and virtual reality are poised to propel computer vision advancements further. These immersive technologies offer richer experiences and are likely to stimulate the exploration of new methods and applications, enabling deeper understanding and manipulation of visual data. The profound impact of these advancements will extend to fields such as autonomous driving, healthcare imaging, surveillance, virtual reality, and e-commerce, revolutionizing how machines perceive, comprehend, and interact with the world beyond our current capabilities. However, to embark on this promising path, it is imperative that we navigate the complexities and challenges of advancing computer vision responsibly and ethically.

6 Conclusions

In conclusion, the field of computer vision, driven by recent advancements in artificial intelligence, stands on the precipice of a promising future. Despite persisting challenges in algorithm optimization, hardware development, data privacy, and ethical considerations, the potential for transformative progress is vast. The continuous advancement of AI models, exemplified by ChatGPT, and the potential emergence of their computer vision counterparts augur an expanding horizon of achievable milestones. Through the integration of diverse learning methodologies and cross-modal techniques, sophisticated models capable of interpreting and comprehending visual data with unparalleled accuracy and depth are poised to materialize.

Augmented reality (AR) and virtual reality (VR), as emerging technologies, will act as catalysts for further advancements in computer vision, enriching and enhancing user experiences. The repercussions of these developments will reverberate across a range of domains, encompassing healthcare, surveillance, autonomous driving, and e-commerce, fundamentally reshaping the way machines perceive, comprehend, and interact with the world. Nevertheless, it is imperative to tread the path of rapid technological progress with prudence and responsibility. Striking a delicate balance between the pursuit of cutting-edge innovation and safeguarding user privacy, ensuring data security, and upholding ethical standards will be a pivotal challenge. As researchers and practitioners, we bear the weighty responsibility of guiding the development and application of these technologies in a manner that is both beneficial and respectful, aligned with rigorous ethical principles.

In the grand tapestry of computer vision, the narrative is still unfolding. Each new discovery and innovation propels us closer to a future where machines possess the true ability to "see" and comprehend the world akin to human perception. It is an exhilarating journey filled with boundless opportunities, formidable challenges, and immense potential—an odyssey in which we are privileged to partake.

Acknowledgements Y. X. Zheng is greatly indebted to co-first author Professor K.-W. (G. H.) A. Chee for the conceptualization and the leading of the manuscript preparation, and the addition and editing of substantial technical content, data collation, and project supervision. This work was supported by the BK21 (Electronic Electric Convergence Talent Nurturing Education Research Center) funded by the Ministry of Education of Korea, the Dong-il Cultural Scholarship Foundation, and the Kyungpook National University Research Fund 2021.

References

1. Hubel, David H., and Torsten N. Wiesel. "Receptive fields of single neurones in the cat's striate cortex." *The Journal of physiology* 148.3 (1959): 574.
2. Roberts, Lawrence G. Machine perception of three-dimensional solids. Diss. Massachusetts Institute of Technology, 1963.
3. Marr, D. (1982). Vision: A Computational Investigation into the Human Representation and Processing of Visual Information. San Francisco: W.H. Freeman.

4. Lowe, David G. "Object recognition from local scale-invariant features." Proceedings of the seventh IEEE international conference on computer vision. Vol. 2. Ieee, 1999.
5. Krizhevsky, Alex, Ilya Sutskever, and Geoffrey E. Hinton. "Imagenet classification with deep convolutional neural networks." Communications of the ACM 60.6 (2017): 84–90.
6. Simonyan K, Zisserman A. Very deep convolutional networks for large-scale image recognition[J]. arXiv preprint arXiv:1409.1556, 2014.
7. He K, Zhang X, Ren S, et al. Deep residual learning for image recognition[C]//Proceedings of the IEEE conference on computer vision and pattern recognition. 2016: 770–778.
8. Szegedy C, Liu W, Jia Y, et al. Going deeper with convolutions[C]//Proceedings of the IEEE conference on computer vision and pattern recognition. 2015: 1–9.
9. Girshick R, Donahue J, Darrell T, et al. Rich feature hierarchies for accurate object detection and semantic segmentation[C]//Proceedings of the IEEE conference on computer vision and pattern recognition. 2014: 580–587.
10. Redmon J, Divvala S, Girshick R, et al. You only look once: Unified, real-time object detection[C]//Proceedings of the IEEE conference on computer vision and pattern recognition. 2016: 779–788.
11. Liu W, Anguelov D, Erhan D, et al. Ssd: Single shot multibox detector[C]//Computer Vision–ECCV 2016: 14th European Conference, Amsterdam, The Netherlands, October 11–14, 2016, Proceedings, Part I 14. Springer International Publishing, 2016: 21–37.
12. Girshick R. Fast r-cnn[C]//Proceedings of the IEEE international conference on computer vision. 2015: 1440–1448.
13. Long J, Shelhamer E, Darrell T. Fully convolutional networks for semantic segmentation[C]//Proceedings of the IEEE conference on computer vision and pattern recognition. 2015: 3431–3440.
14. Chen L C, Papandreou G, Kokkinos I, et al. Semantic image segmentation with deep convolutional nets and fully connected crfs[J]. arXiv preprint arXiv:1412.7062, 2014.
15. Ronneberger O, Fischer P, Brox T. U-net: Convolutional networks for biomedical image segmentation[C]//Medical Image Computing and Computer-Assisted Intervention–MICCAI 2015: 18th International Conference, Munich, Germany, October 5–9, 2015, Proceedings, Part III 18. Springer International Publishing, 2015: 234–241.
16. He K, Gkioxari G, Dollár P, et al. Mask r-cnn[C]//Proceedings of the IEEE international conference on computer vision. 2017: 2961–2969.
17. Cao Z, Simon T, Wei S E, et al. Realtime multi-person 2d pose estimation using part affinity fields[C]//Proceedings of the IEEE conference on computer vision and pattern recognition. 2017: 7291–7299.
18. Kreiss S, Bertoni L, Alahi A. Pifpaf: Composite fields for human pose estimation[C]//Proceedings of the IEEE/CVF conference on computer vision and pattern recognition. 2019: 11977–11986.
19. Furukawa Y, Hernández C. Multi-view stereo: A tutorial[J]. Foundations and Trends® in Computer Graphics and Vision, 2015, 9(1–2): 1–148.
20. Schonberger J L, Frahm J M. Structure-from-motion revisited[C]//Proceedings of the IEEE conference on computer vision and pattern recognition. 2016: 4104–4113.
21. Newcombe R A, Izadi S, Hilliges O, et al. Kinectfusion: Real-time dense surface mapping and tracking[C]//2011 10th IEEE international symposium on mixed and augmented reality. Ieee, 2011: 127–136.
22. Lucas B D, Kanade T. An iterative image registration technique with an application to stereo vision[C]//IJCAI'81: 7th international joint conference on Artificial intelligence. 1981, 2: 674–679.
23. Horn B K P, Schunck B G. Determining optical flow[J]. Artificial intelligence, 1981, 17(1–3): 185–203.
24. Farneback G. Two-frame motion estimation based on polynomial expansion[C]//Image Analysis: 13th Scandinavian Conference, SCIA 2003 Halmstad, Sweden, June 29–July 2, 2003 Proceedings 13. Springer Berlin Heidelberg, 2003: 363–370.

25. Lowe D G. Object recognition from local scale-invariant features[C]//Proceedings of the seventh IEEE international conference on computer vision. Ieee, 1999, 2: 1150–1157.
26. Bay H, Ess A, Tuytelaars T, et al. Speeded-up robust features (SURF)[J]. Computer vision and image understanding, 2008, 110(3): 346–359.
27. Dalal N, Triggs B. Histograms of oriented gradients for human detection[C]//2005 IEEE computer society conference on computer vision and pattern recognition (CVPR'05). Ieee, 2005, 1: 886–893.
28. Platt J. Sequential minimal optimization: A fast algorithm for training support vector machines[J]. 1998.
29. Quinlan J R. Induction of decision trees[J]. Machine learning, 1986, 1: 81–106.
30. Breiman L. Random forests[J]. Machine learning, 2001, 45: 5–32.
31. Rumelhart D E, Hinton G E, Williams R J. Learning representations by back-propagating errors[J]. nature, 1986, 323(6088): 533–536.
32. LeCun Y, Bottou L, Bengio Y, et al. Gradient-based learning applied to document recognition[J]. Proceedings of the IEEE, 1998, 86(11): 2278–2324.
33. Elman J L. Finding structure in time[J]. Cognitive science, 1990, 14(2): 179–211.
34. Goodfellow I, Pouget-Abadie J, Mirza M, et al. Generative adversarial networks[J]. Communications of the ACM, 2020, 63(11): 139–144.
35. Mikolov T, Chen K, Corrado G, et al. Efficient estimation of word representations in vector space[J]. arXiv preprint arXiv:1301.3781, 2013.
36. Ren S, He K, Girshick R, et al. Faster r-cnn: Towards real-time object detection with region proposal networks[J]. Advances in neural information processing systems, 2015, 28.
37. Redmon J, Farhadi A. YOLO9000: better, faster, stronger[C]//Proceedings of the IEEE conference on computer vision and pattern recognition. 2017: 7263–7271.
38. Redmon J, Farhadi A. Yolov3: An incremental improvement[J]. arXiv preprint arXiv:1804.02767, 2018.
39. Bochkovskiy A, Wang C Y, Liao H Y M. Yolov4: Optimal speed and accuracy of object detection[J]. arXiv preprint arXiv:2004.10934, 2020.
40. https://github.com/ultralytics/yolov5
41. Zeiler M D, Fergus R. Visualizing and understanding convolutional networks[C]//Computer Vision–ECCV 2014: 13th European Conference, Zurich, Switzerland, September 6–12, 2014, Proceedings, Part I 13. Springer International Publishing, 2014: 818–833.
42. Tan M, Pang R, Le Q V. Efficientdet: Scalable and efficient object detection[C]//Proceedings of the IEEE/CVF conference on computer vision and pattern recognition. 2020: 10781–10790.
43. Duan K, Bai S, Xie L, et al. Centernet: Keypoint triplets for object detection[C]//Proceedings of the IEEE/CVF international conference on computer vision. 2019: 6569–6578.
44. Carion N, Massa F, Synnaeve G, et al. End-to-end object detection with transformers[C]//Computer Vision–ECCV 2020: 16th European Conference, Glasgow, UK, August 23–28, 2020, Proceedings, Part I 16. Springer International Publishing, 2020: 213–229.
45. Liu Z, Lin Y, Cao Y, et al. Swin transformer: Hierarchical vision transformer using shifted windows[C]//Proceedings of the IEEE/CVF international conference on computer vision. 2021: 10012–10022.
46. Liao M, Shi B, Bai X, et al. Textboxes: A fast text detector with a single deep neural network[C]//Proceedings of the AAAI conference on artificial intelligence. 2017, 31(1).
47. Badrinarayanan V, Kendall A, Cipolla R. Segnet: A deep convolutional encoder-decoder architecture for image segmentation[J]. IEEE transactions on pattern analysis and machine intelligence, 2017, 39(12): 2481–2495.
48. Zhao H, Shi J, Qi X, et al. Pyramid scene parsing network[C]//Proceedings of the IEEE conference on computer vision and pattern recognition. 2017: 2881–2890.
49. Chen L C, Papandreou G, Schroff F, et al. Rethinking atrous convolution for semantic image segmentation[J]. arXiv preprint arXiv:1706.05587, 2017.
50. Chen L C, Zhu Y, Papandreou G, et al. Encoder-decoder with atrous separable convolution for semantic image segmentation[C]//Proceedings of the European conference on computer vision (ECCV). 2018: 801–818.

51. Arcos-García Á, Alvarez-Garcia J A, Soria-Morillo L M. Deep neural network for traffic sign recognition systems: An analysis of spatial transformers and stochastic optimisation methods[J]. Neural Networks, 2018, 99: 158–165.
52. Porzi L, Bulo S R, Kontschieder P. Improving panoptic segmentation at all scales[C]//Proceedings of the IEEE/CVF conference on computer vision and pattern recognition. 2021: 7302–7311.
53. Xiang T, Zhang C, Liu D, et al. BiO-Net: learning recurrent bi-directional connections for encoder-decoder architecture[C]//Medical Image Computing and Computer Assisted Intervention–MICCAI 2020: 23rd International Conference, Lima, Peru, October 4–8, 2020, Proceedings, Part I 23. Springer International Publishing, 2020: 74–84.
54. Alom, M.Z., Yakopcic, C., Taha, T.M., Asari, V.K.: Nuclei segmentation with recurrent residual convolutional neural networks based u-net (r2u-net). In: IEEE National Aerospace and Electronics Conference. pp. 228–233. IEEE (2018).
55. https://captain-whu.github.io/DOTA/index.html.
56. Cai Z, Vasconcelos N. Cascade r-cnn: Delving into high quality object detection[C]//Proceedings of the IEEE conference on computer vision and pattern recognition. 2018: 6154–6162.
57. Lin T Y, Dollár P, Girshick R, et al. Feature pyramid networks for object detection[C]//Proceedings of the IEEE conference on computer vision and pattern recognition. 2017: 2117–2125.
58. Ding J, Xue N, Long Y, et al. Learning roi transformer for oriented object detection in aerial images[C]//Proceedings of the IEEE/CVF Conference on Computer Vision and Pattern Recognition. 2019: 2849–2858.
59. Xie S, Girshick R, Dollár P, et al. Aggregated residual transformations for deep neural networks[C]//Proceedings of the IEEE conference on computer vision and pattern recognition. 2017: 1492–1500.
60. Li L, Bao J, Zhang T, et al. Face x-ray for more general face forgery detection[C]//Proceedings of the IEEE/CVF conference on computer vision and pattern recognition. 2020: 5001–5010.

CI in Manufacturing, Engineering, and Industry

Feature Importance Study for Biogas Production from POME Treatment Plants Using Out-of-Bag Permutation

Kishaan Ragu, Ong Qian Yee, Kiew Xin Yun, Hia Hung Yi, Kashwin Selvanathan, Sara Kazemi Yazdi, Chen Zhi Yuan, Chan Yi Jing, and Reza Godary

Abstract Methane capturing systems (MCS) for electricity generation in the palm oil mill effluent (POME) treatment process are emphasized to reduce methane emissions. However, the presence of carbon dioxide and hydrogen sulfide in POME biogas affect the heating quality of the biogas. Therefore, proper understanding on the parameters which could affect the emission of these impurities is necessary to curb their production. Due to limited available data, Synthetic Minority Oversampling Technique (SMOTE) was applied to expand the dataset for training purposes. In this study, a random forest based out-of-bag permutation feature importance study was conducted by assessing the influence of temperature, pH and organic loading rate (OLR), chemical oxygen demand (COD), total solids (TS), biological oxygen demand (BOD), suspended solids (SS), and hydraulic retention time (HRT) on methane, carbon dioxide and hydrogen sulfide emission. Temperature, pH and organic loading rate were found to be the most influential parameters for methane and carbon dioxide production, while pH was replaced by suspended solids the case of hydrogen sulfide. The final random forest machine learning model generated performance metrics for R^2 and RMSE with values of 0.98 and 0.131 and 0.99 and 0.061, respectively.

Keywords Feature importance · POME biogas · Machine learning · Random forest · SMOTE

K. Ragu · O. Q. Yee · K. X. Yun · H. H. Yi · K. Selvanathan · S. K. Yazdi (✉) · C. Y. Jing · R. Godary
Department of Chemical and Environmental Engineering, University of Nottingham, Semenyih, Malaysia
e-mail: sara.yazdi@nottingham.edu.my

C. Y. Jing
e-mail: Yi-Jing.Chan@nottingham.edu.my

C. Z. Yuan
Department of Computer Science, University of Nottingham, Semenyih, Malaysia
e-mail: zhiyuan.chen@nottingham.edu.my

K. Daimi et al. (eds.), *Cutting Edge Applications of Computational Intelligence Tools and Techniques*, Studies in Computational Intelligence 1118,
https://doi.org/10.1007/978-3-031-44127-1_7

1 Introduction

Palm oil is the most consumed edible oil in the world as the crop can produce more oil per acre of land used compared to other vegetable oil crop [1, 2]. In 2019, Malaysia contributed to 26% (19 million tons) of the global palm oil production [3]. In 2020, the agricultural sector in Malaysia contributed to 7.4% of the national GDP with 37.1% major gross value to the sector added by the palm oil industry [4]. Despite all its benefits, the palm oil industry has been under fire for being the main cause of biodiversity loss and deforestation within their plantation [5].

Figure 1 presents a simplified block diagram representing the overall POME process has been shown. During palm oil production, 90% of the total FFB forms biomass waste comprising of empty fruit bunches (EFB), palm oil mill effluent (POME), biomass sludge and others. POME is typically produced during sterilisation (36% total POME) and clarification (60% total POME) [6]. 5–7 tonnes of water are needed to produce one tonne of CPO, with roughly 50% of the water ending up as POME [7].

POME is a non-toxic effluent containing 90% water, 4–5% soil particles, 2–4% suspended solids and 0.6–0.7% residual oils, which is highly polluting and cannot be released to the environment without treatment POME has a 100 times higher polluting capability than municipal wastewater in terms of biological oxygen demand (BOD_5) and chemical oxygen demand (COD) concentrations [8]. This means that releasing untreated POME into water bodies will most definitely disrupt the ecosystem in ways like water oxygen depletion, eutrophication, and the death of aquatic organisms.

Anaerobic treatment (AD), membrane treatment (MD) and evaporation method (EM) are some of the available methods for POME treatment. According to [9], the anaerobic and aerobic ponding system using bacteria is the most extensively

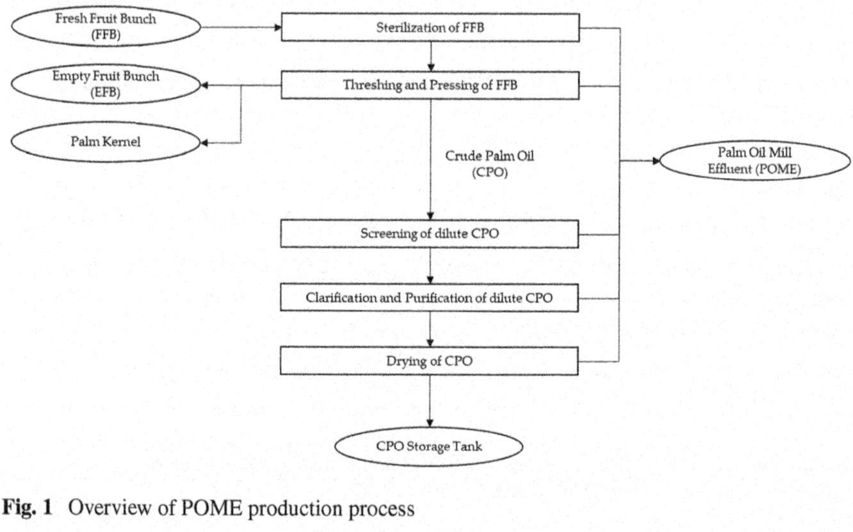

Fig. 1 Overview of POME production process

used POME treatment technique. This method of treatment is cost-effective, energy efficient, low maintenance, highly reliable and simple in design. In Malaysia, 85% of palm oil mills (POMs) utilize this method [10].

The four different stages of AD are shown in *Error! Reference source not found.*, namely hydrolysis, acidogenesis, acetogenesis and methanogenesis. Hydrolysis includes the conversion of complex organic matter into simple sugars to allow bacteria to further process the content. Next, acidogenesis converts soluble organic content into organic materials like CO_2, H_2, NH_3 and organic acids, which are then converted into acetic acids. During methanogenesis, methanogens convert the intermediate products into biogas.

During the anaerobic digestion of POME, organic matters in the effluent undergoes degradation under the absence of oxygen to form biogas consisting of 60–70% methane (CH_4), 30–35% carbon dioxide (CO_2), trace amounts of hydrogen sulfide (H_2S) and water vapor [11]. It is estimated that 1 ton of POME can produce roughly 28.13 m^3 of biogas [12].

According to the [13], a 20-year trend suggests that CH_4 has 80 times more global warming potential than CO_2 CH_4 is also the primary contributor to the formation of hazardous ground level ozone, which have caused a million premature deaths each year [13, 14]. In 2016, a total of 57,211 Gg CO_2 eq of CH_4 was emitted by Malaysia, with roughly 23.76% (13,593 Gg) coming from POME treatment [15]. Hence, the release of vast amount of CH_4 to the atmosphere is a relevant concern that must be addressed (Fig. 2).

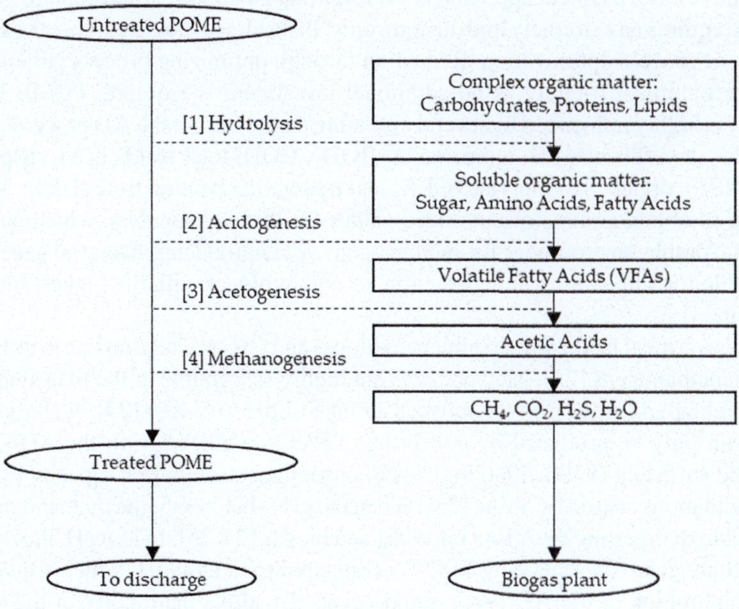

Fig. 2 Overview of different stages of AD process [12]

CH_4 has a higher heating value (50–55 MJ/kg) than coal or oil [16]. The combustion of CH_4 also produces half the amount of carbon that a coal fuel would [17]. In an MCS, CH_4 is captured and redistributed within the plant for heat and power generation purposes through combustion. The energy generated from biogas combustion can be used as a heat source in reboilers to produce low pressure steam for FFB sterilization, or even for a steam turbine to generate electricity. Typically, electricity produced via biogas is resupplied to the POM, making the industry self-sustainable. In certain plants, where the heat and power generation scheme is highly efficient, the electricity generated through biogas will exceed the electricity demand of the plant. This excess electricity can be connected and sold to the national grid to generate additional income [18]. With this scheme in place, the overall GHGs emission and carbon footprint of the palm oil industry can be reduced. Implementing energy recovery and transforming waste to wealth also kickstarts the industry's evolution to a circular economy.

This methane-rich biogas also contains impurities such as CO_2 (30–45%) and H_2S. The presence of CO_2 can reduce the volumetric heating value of the biogas, as the acid gas itself is inert to combustion [19]. H_2S is also highly acidic and in addition to that, toxic, which can cause pipeline erosion and catalyst poisoning [20]. To improve biogas quality, both CO_2 and H_2S contents must be reduced. As far as we are concerned, this can be achieved via two direct manners: purification of biogas, or optimization of the AD process.

The first option, biogas purification is extremely high in capital cost. In a biogas treatment process, H_2S must first be removed via adsorption in a hydro-desulfurizer. To remove CO_2, a typical approach is via installing an amine carbon capture system, which requires an extremely high heating duty. Instead, we should explore the second option of process optimization. Reduction through optimizing process parameter is more economical friendly as no additional investment is required. POME biogas quality is highly influenced by several operating parameters of the AD process, which includes the effluent's pH, temperature, BOD_5, COD, total solids (TS), suspended solids (SS), organic loading rate (OLR) and hydraulic retention time (HRT). We are aiming to obtain a range of optimized values for these parameters, which pose the most favorable environment for methanogens, CH_4 producing bacteria, and is less favorable for sulphate-reducing bacteria, to obtain biogas with the highest methane content.

pH is a critical factor which influences biogas and CH_4 production due to its impact on the methanogens [21]. Methanogens are highly susceptible to the pH value of its surrounding and only survives between a range of pH (6.5–7.8) [22]. As biogas and CH_4 can only be produced by converting VFAs, the alkalinity in the AD system must be sufficient (2500–5000 mgL^{-1}) to ensure adequate buffer capacity to avoid pH fluctuations caused by VFAs [23]. When the pH is below 6.5, the methanogens are incapable of digesting the VFAs into CH_4 and biogas [24, 25]. In low pH, the number of methanogens can reduce by 71–79% compared to neutral pH, which would lead to the inhibition of the AD process and negatively affect the quality of the biogas produced [26]. Cioabla et al. [27], Jayaraj et al. [28] and Vikrant et al. [29] found that the production of CH_4 by the methanogens is the enhanced between pH 6.5 and

7.5 that peaks at pH (6.8–7.2). This suggests that decrease in acidity or increase in alkalinity of the effluent will affect the CH_4 production negatively.

Like pH, methanogens are also sensitive to temperature. Temperature fluctuations during anaerobic digestion can affect the CH_4 production. AD is often conducted at mesophilic temperature (30–35 °C) or thermophilic temperature (50–65 °C) [30, 31]. Zinder et al. [32] reported that the *Methanosarcina* culture (a type of archaea that produces CH_4) experiences optimal growth at temperatures between 55 and 58 °C. When temperature reaches 65 °C, a halt in the microorganism growth and CH_4 production was also observed. Similarly, at temperatures between 40 and 50 °C, methanogenesis is observed to have significant improvement while sulfidogenesis, a process producing hydrogen sulfide was suppressed [33]. A decrease in temperature below 20 °C can cause a decline in biogas production and eventually halting the production when the temperature reaches 10 °C [27].

BOD_5 refers to the amount of oxygen required by microorganisms during the biochemical degradation of organic matter under anerobic conditions. These organic materials are crucial in supporting the growth of microorganisms (Kumar and [34]. Lower BOD_5 can result in lower count of microorganism which could affect the anaerobic digestion process negatively. CH_4 emissions can be calculated as factor of BOD_5. An estimation factor of 0.6 kg CH_4/kg BOD_5 is used as a general guideline [35]. Utami et al. [36] reported that the larger the AD HRT value, the greater the amount of BOD_5 removed, yielding more biogas. It is safe to say that increasing BOD_5 trend will result in higher CH_4 emissions [37].

COD is defined as the amount of oxygen equivalent consumed in the chemical oxidation of organic matter by a strong oxidant such as potassium dichromate. Putro [38] reported that one kg of COD from POME can contribute to 0.238 kg of CH_4, while the most recent Intergovernmental Panel on Climate Change (IPCC) Guideline for National Greenhouse Gas Inventories reported a default value of 0.25 kg CH_4/kg COD [35]. This suggests that COD has a huge influence on the CH_4 produced during treatment, as naturally, POME with higher COD will produce more CH_4 than that of a lower COD under the same conditions [37, 39]. In comparison to BOD_5, COD removal affects the CH_4 yield more significantly [36].

Suspended solids are defined as solid in water that can be trapped by a filter while dissolved solids are those that cannot be trapped by a filter [40]. Changes in the total solids can cause changes in microbial morphology affecting performance of anaerobic digestion especially, CH_4 production efficiency [41, 42]. Yan et al. [43] in their study on anaerobic co-digestion of dairy manure and maize stover, observed an increase in CH_4 production with an increase in TS value and associated the observation with the increase in OLR that led to increase in volatile fatty acids and CH_4 producing microbes. There were also reports on CH_4 production decreasing with increasing TS content [44] demonstrating a strong correlation, (R^2 = 0.96) between the TS concentrations and methanogenic activities and concluding that high TS content would reduce the overall yield of methane produced during digestion processes. This is because high TS content results in lower water content which directly affects the hydrolysis rate of bulk liquid bacteria present in AD systems that release enzymes necessary for the breakdown of the organic matter [45, 46].

HRT is average time interval substrate and cells are kept inside the digester [47]. Adequate HRT is required to produce maximum CH_4 as very short HRT will cause hydraulic overload resulting in microbial community especially methanogens being washed out causing lower CH_4 yield due to the decrease in the efficiency of nutrient removal [48]. The decrease in microbes in the system would lead to a decrease in conversion efficiency, lower CH_4 yield and instability of the AD system [21]. Gaby et al. [49] reported that CH_4 production from AD of food waste decreased with lower HRT. Long HRT is not necessarily good as well because it will result in huge capital cost due to the needs of huge reactor to store the substrate for a long time [50, 51].

OLR is defined as the amount of organic waste that is being fed per unit volume of the digester on a given day [52]. OLR plays an important role in the conversion of substrate and maintaining a balance between methanogenesis and acidogenesis (a process that provides volatile fatty acids to methanogens) [53]. Higher OLR relates to higher CH_4 released during the anaerobic digestion process as the concentration of substrate in the digester is high and the digester size can also be reduce [54, 55]. However, too high of an OLR value can reduce the CH_4 emissions due to disruption in the structure of the microbial community and inhibit the methanogenesis pathway. Accumulation of volatile fatty acid could also occur due to too high OLR which could lead to an irreversible failure in the anerobic system through a series of events. As such, maintaining an optimum OLR value is necessary in a wastewater treatment system to ensure continuous and stable CH_4 production [56].

These process parameters that were used as predictor variables in a parallel study conducted simultaneously by the authors of this paper on developing a CH_4, CO_2, and H_2S emission prediction tool for POME treatment have huge effect on the amount and quality of biogas produced at different stages of aerobic digestion. However, the level of impact of each of these variables on this extremely sensitive processes still needs further exploration and understanding [57]. As such, this study is aimed to determine the highest influencing parameters for each of these pollutant gases using a feature importance study.

Figure 3 presents an overview of the dependent and independent variables. To maximize methane content in biogas, the aim is to achieve a range of optimized values for these parameters to provide the optimal condition for methanogens and reduce the activity of sulfate-reducing bacteria.

The results of this study would aid in developing a better prediction tool by focusing on the important parameters that affect the emission of these pollutant gases. Besides that, the outcome of this study would give palm oil mill owners an idea on which parameters to focus during process optimization to maximize the production of biogas that is rich in CH_4 with low amounts of impurity.

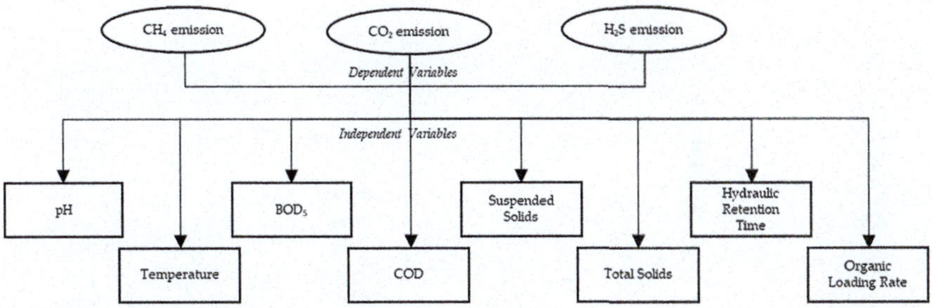

Fig. 3 Overview of dependent and independent variables

2 Materials and Methods

The overall approach to a feature importance study is shown in Fig. 4. Raw data is obtained and pre-processed prior to its insertion to a machine learning model. Once the model is successfully trained with satisfactory results, we can proceed with the feature importance study.

Step 1: Data Collection

Real scale industrial data of the industrial-scale covered lagoon POME anaerobic digestion unit used in this research were obtained from four different local plants in Malaysia over a period of 24 months (July 2019–June 2021). All the data provided were monthly average values from the plants. As shown in Fig. 5, this process

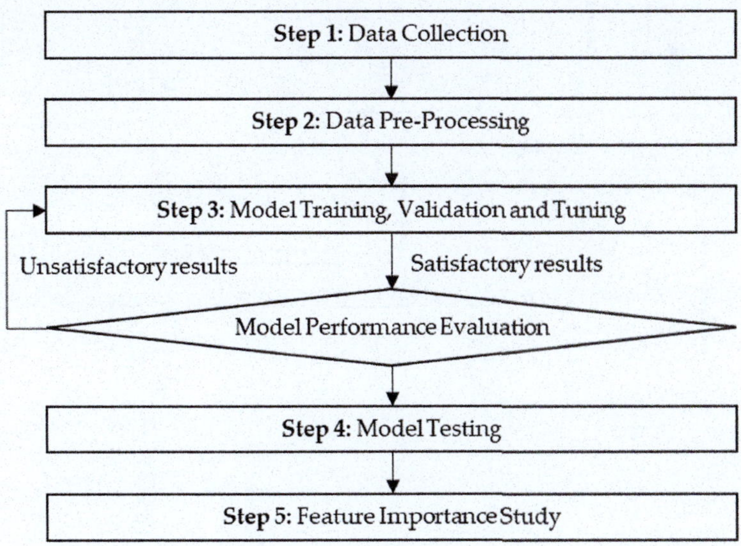

Fig. 4 Overview of the methodology

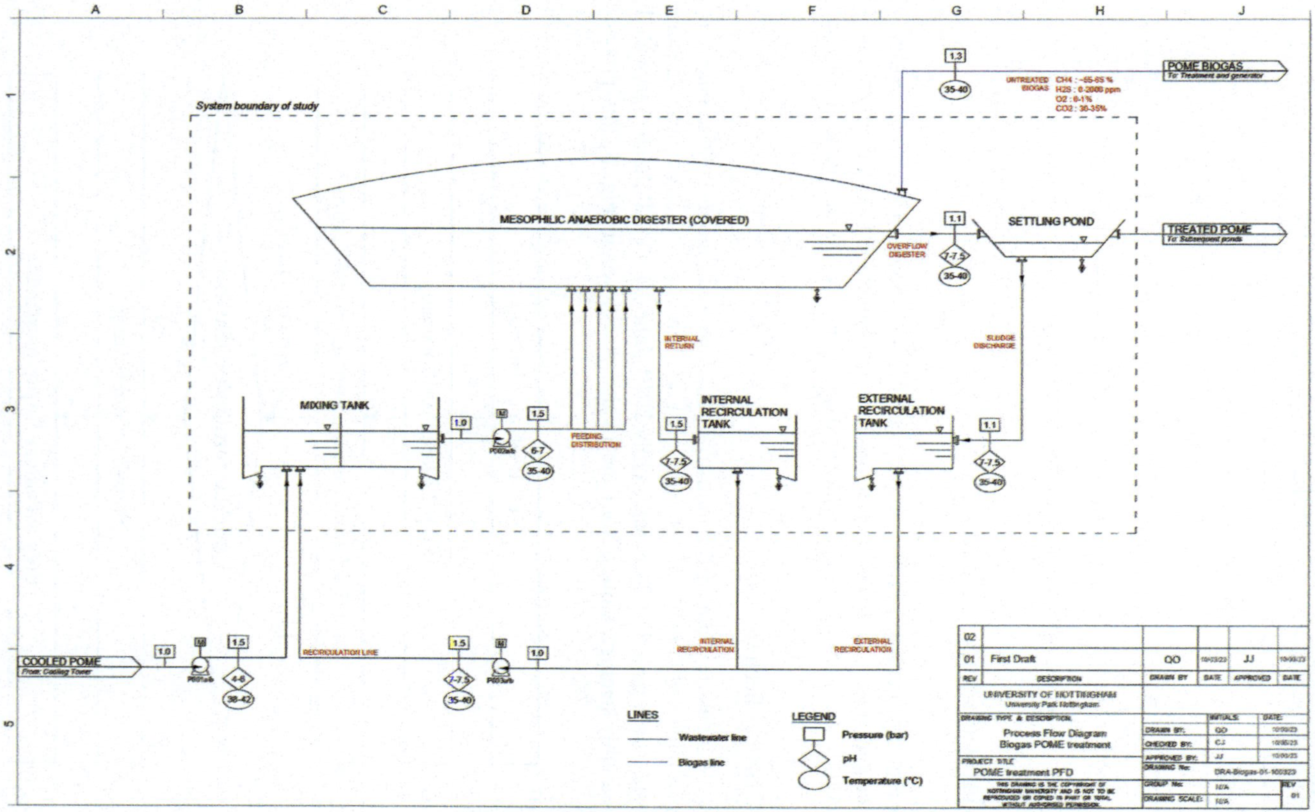

Fig. 5 Process flow diagram of an industrial covered lagoon anaerobic digester for POME

Table 1 Summary of the parameters

Parameters	Unit	Range
pH	–	4.20–5.23
Temperature	°C	46.80–62.39
BOD_5	mg/L	22,500.00–47,520.00
COD	mg/L	53,450.00–92,844.00
TS	mg/L	20,148.00–56,420.00
SS	mg/L	12,300.00–57,650.00
HRT	days	34.04–87.92
OLR	kg COD in/ m^3·day	0.85 – 1.79
CH_4 produced	Nm^3/month	11,340.08 – 295,479.86
CO_2 produced	Nm^3/month	6582.38–185,236.11
H_2S produced	Nm^3/month	4.51–709.90

considers mixing tank to be within the system boundary. The pH and temperature are obtained at the inlet of the mixing tank, as it is imperative to control the inlet conditions to ensure proper mixing. The BOD_5, COD, TS and SS levels are obtained at the outlet of the mixing stream entering the anaerobic digester, while the HRT and OLR are obtained via the monthly records of the plant. The collected dataset consists of 96 data points. A summary of the parameters with the range and units are shown in Table 1.

Step 2: Data Pre-processing

The data was pre-processed in a manner similar to the parallel study conducted by the authors to develop the prediction tool. In this chapter, the z-score data normalisation (also known as standardisation) technique was applied to the raw data. As shown in Eq. 1, the data is normalised according to its standard deviation and mean.

$$Z = \frac{x_i - \overline{X}}{\sigma} \tag{1}$$

where x_i is the data point in the dataset, $\overline{X}$ is the dataset mean, and σ is the dataset standard deviation. Upon applying Eq. (1), the dataset will be converted into a single, standardised data format where the new mean and standard deviation values are 0 and 1 [58]. Unlike Min–Max normalisation, the standardised values can be either positive or negative. Z-score normalisation also has the advantage of being able to minimise the effects of outliers in any dataset [59].

A major challenge faced in this particular dataset is that the raw data points obtained is insufficient to train an accurate machine learning model, as a minimum of 500 data points is required [60]. Therefore, the exploration of data expansion (or augmentation) to generate synthetic datasets based upon the original dataset is carried out. A common approach to this for regression-based problems is by implementing the Synthetic Minority Oversampling Technique (SMOTE). SMOTE can synthesize new observations based on the existing dataset using the k-nearest neighbour (KNN) approach. To execute this function, the desired number of samples, N is set to be 6 while the number of nearest neighbours, k is set to be 6. In other words, this means that after performing SMOTE, the total number of samples will be 7 times the original amount ($N_{ori} = 96$, $N_{syn} = 576$, $N_{total} = 672$). The 6 nearest datapoints to an origin will be identified, and a random point along the vector connecting the origin to the KNN point is selected as the synthesized point, as shown in Fig. 6.

Step 3: Model Training, Validation and Tuning

The Regression Learner Toolbox from MATLAB®2022a was utilized for the model development. Out-of-bag permutation feature importance method for random forest of regression trees was implemented in this study. Random forest is a powerful ensemble of trees method that uses independent framework which is capable of solving complicated nonlinear classification/regression problems efficiently. It uses a random tree as the base classifier with the bagging approach which improves the ensemble diversity [61]. The approach employs a collection of classification/regression trees, each of which is constructed from a bootstrap sample of the data that has been iteratively split into more homogeneous sections. At each split, a random subset of all independent variables (typically with a fixed number) is chosen to

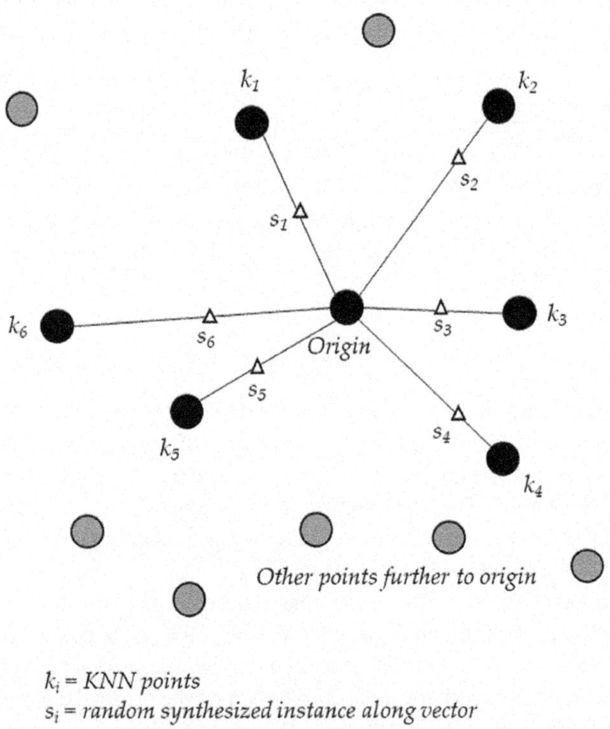

Fig. 6 Illustration of random point along the vector connecting the origin to the KNN points

identify the optimal strategy to divide the data at that node [62]. During training of a node, only a random subset of features is checked and sampled afresh for each node.

The bagged regression ensemble with tree learners was trained using MATLAB Code fitrensemble before computing the input variable importance score using MATLAB Code oobPermutedPredictorImportance. R^2 of the trained random forest model was obtained using the MATLAB code corr(X). A bar graph visualizing the importance of each variable was plotted.

Following that, a cross fold validation of 10-folds was used as it is a common practice used in most machine learning model development. Cross fold validation is a powerful resampling technique as the prediction performance of a model on concealed data is able to be tested. It also prevents the overfitting of data, which might affect the performance of any machine model. In this validation technique, the dataset is split randomly into 10 groups. 9 groups are used to train the model while the remaining 1 group is held out for validation of the trained model. To evaluate the model, the R^2 statistical indicator is used, as shown in Eq. 2.

$$R^2 = \frac{\sum_{i=1}^{n}(x_i - y_i)^2(y_i - \overline{y})}{\sum_{i=1}^{n}(x_i - \overline{x})^2\sum_{i=1}^{n}(y_i - \overline{y})^2} \tag{2}$$

As previously mentioned, due to the data generation method's incapability to generate a dataset with the same pattern and distribution as the existing dataset, SMOTE technique was applied to create synthetic observations for each dataset which involved oversampling the available datasets. Datasets that were synthesized and measured were divided and saved separately. In other words, only the synthesized datasets were employed for this iteration's model training, and the measured dataset was utilized for the model testing. This was accomplished to investigate the possibility of building a model that is solely based on synthetic data and closely resembles the pattern of measured data.

Step 4: Model Testing

Following the tuned model, the raw data representing the actual data from the POM was used to test and validate the legitimacy of the SMOTE based synthetic and trained models. For model testing scenarios where performance was assessed using the R^2 indicator, the evaluation metrics for each model were generated automatically by the regression learner in MATLAB. Generally speaking, high R^2 show how closely the predicted results match the actual data, while low RMSE shows how accurate a model is. The tuned models with comparable training performance were subjected to a test with the raw dataset to evaluate the models' capacity to handle real-world data.

Step 5: Feature Importance Study

According to [63], feature selection is defined as the process of extracting subsets from a feature set using a feature selection criterion that selects the dataset's relevant features. In addition to that, Miao and Niu [64] has also mentioned that feature selection helps in highlighting features more relevant to the response through importance scores thus providing insight to specific models. In this study, feature importance provides an insight on the impact of each input variable (process and effluent parameters) on the output variable (gas emissions). This would allow a better interpretation of the developed prediction model and aid in improving the overall performance of the model [65].

Permutation importance method or also known as mean decrease in accuracy (MDA) is a multivariate feature importance method whereby it takes into consideration the interactions between features used in the model to estimate the feature importance score. This model agnostic approach is extremely useful for real life applications as it is done at predict time. The predictor variable with the highest influence on the response variable will have a huge effect on the error when permuted and vice versa. The feature importance score is obtained by randomly permuting or reshuffling values from on feature in a selected dataset and calculating the difference between the benchmark score (the baseline model score) and the modified model score (trained using permuted dataset).

3 Results and Discussion

Table 2 displays the overall feature importance analysis results while Fig. 7a–c shows the figure depicting the feature importance estimate for each predictor variable for CH_4, CO_2, and H_2S emissions respectively. The importance score shows the influence of a predictor variable on the response variable. An influential variable will have a high score, thus will have bigger effect on the developed model and vice versa.

Temperature was found to be the most influential parameter for the production of all the gases. The significance of temperature influence on the production of CH_4 was found to be at least 40.73% greater than other parameters while in the case of CO_2, and H_2S the significance was slightly lower at 30.75% and 32.76% respectively. This is mainly because microorganisms involved in CH_4, CO_2, and H_2S production are largely sensitive to fluctuations in temperature which would affect their digestion capabilities. The multiple microorganisms that are responsible for producing these gases, digest waste efficiently at different temperature ranges. It becomes vital to maintain the temperature at which methanogen's activity is at peak and the sulfur

Table 2 Feature importance analysis results for CH_4, CO_2 and H_2S

Feature importance specification	Predictor variables							
	COD	OLR	HRT	pH	Temp	SS	TS	BOD
CH_4	1.3785	1.9518	1.1848	1.9293	3.2932	1.5697	0.9692	1.3202
CO_2	1.3541	2.3503	1.2068	2.1881	3.3939	1.7020	1.3079	0.9118
H_2S	1.0041	1.4628	1.3937	1.2857	2.5517	1.7157	1.0632	1.2875

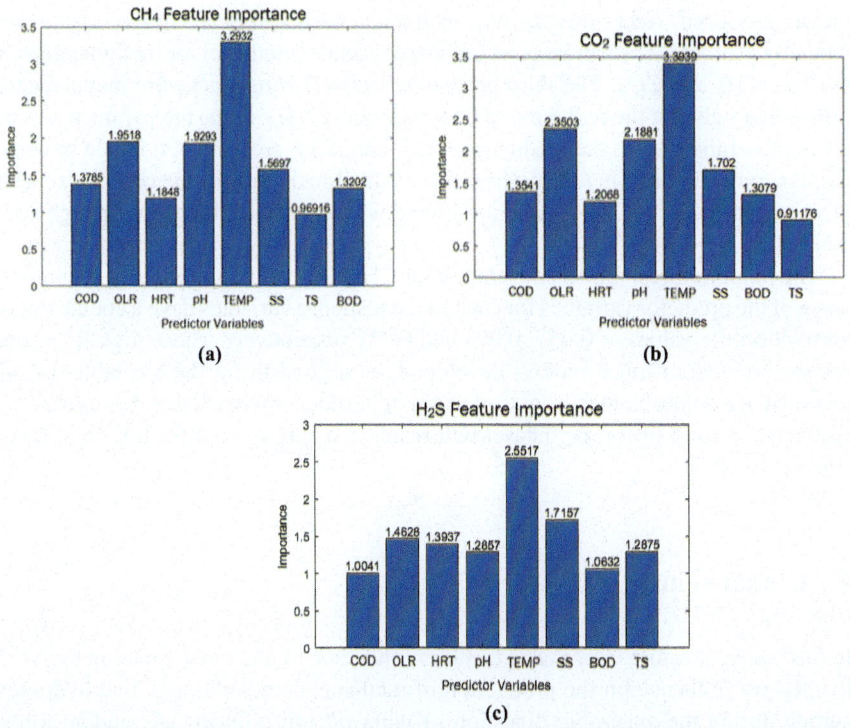

Fig. 7 **a** Predictor variables feature importance scores for methane emission; **b** Predictor variables feature importance scores for carbon dioxide emission; **c** Predictor variables feature importance scores for hydrogen sulfide emission

reducing bacteria's activity is inhibited to obtain biogas with the highest purity i.e. high methane content.

pH (only CH_4 and CO_2) and OLR can also be seen to have high influence on the production of these gases. This again comes down to creating a suitable environment for the methanogens to survive and having the optimum amount of nutrients to carry out anaerobic digestion to produce biogas. The top three parameters with the most influence on these emission gases have some major role to play in maintaining optimal microbial activity to maintain balance between methanogenesis and other processes (sulfidogenesis and acidogenesis) by creating a suitable environment for these microorganisms to thrive during anaerobic digestion.

On the other hand, TS scored the lowest for both CH_4 and CO_2, which indicates that this parameter has the least influence on the emission of these following gases. In other words, even if TS has not been monitored, it is still likely for the ML prediction model to be fairly accurate.

Reducing the H_2S by making the parameters more suited for methanogens is one way to obtain high purity biogas. Besides that, it also becomes crucial to reduce the amount of CO_2 produced during methanogenesis to increase the CH_4 content of the biogas produced. According to [66], methanogenesis can occur in three pathways whereby *acetoclastic* methanogens use either acetate, methanol, or methylamines to produce CH_4 and CO_2. The third process whereby *hydrogenotrophic* methanogens utilize hydrogen for the reduction of CO_2 to produce CH_4. Since the parameters with the highest influence on the production of CH_4 and CO_2 are similar, it would be beneficial to maintain them in a way where CO_2 reduction dominates the methanogenesis process compared to CO_2 production giving way to produce biogas with high CH_4 content.

The random forest models developed for CH_4, CO_2 and H_2S to assess the importance of the predictor variables for each of the response variables have a coefficient of correlation, R^2 values of 0.957, 0.973 and 0.957 respectively. These high R^2 values shows the random forest models developed are a good fit for the available dataset making the results obtained from the feature importance study reliable. Using a model that is not a good fit for the dataset will result in a sub-optimal feature importance study result [67].

4 Conclusions

In this study, a feature importance test was conducted to find out the parameters with the highest influence on the production of methane, carbon dioxide, and hydrogen sulfide during the anaerobic digestion of palm oil mil effluents. A random forest machine learning model is applied, obtaining R^2 values of 0.957, 0.973 and 0.957 for methane, carbon dioxide and hydrogen sulfide. The out of bag permutation feature importance technique has showed temperature, pH and OLR are the most influential parameters when it comes to methane and carbon dioxide production while for hydrogen sulfide it was temperature, OLR and SS. For future works, the optimization of identified AD conditions, such as temperature, OLR and pH, can be carried out to maximize methane production and biogas yield. Response surface methodology (RSM) is one of the potential process optimization method that can be employed in which it is a statistical forecasting model that is able to identify mathematical correlations and patterns based on the historical data. Furthermore, less sensitive input parameters such as HRT and TS can be voided from the prediction model to allow a simplified model with less training time to be developed. On the other hand, other feature analysis methods such as mean decrease in impurity (MDI) can be employed in future works to compare the results using OOB permutation. According to the number of samples it splits, MDI determines the relevance of each feature as the total of all the splits across all trees that include the feature. By comparing different feature analysis methods, the accuracy of the results can be further justified.

Author Contributions
Kishaan Ragu: Conceptualization, formal analysis and writing of the original draft of the manuscript; **Hia Hung Yi**: research methodology development and execusion; **Kashwin Selvanathan**: model development and validation and execusion; **Ong**

Qian Yee and **Kiew Xin Yun**: writing and editing of the review and the consecutive manuscript drafts and illustrations; **Sara Kazemi Yazdi** and **Chen ZhiYuan**: research methodology development, project supervision and administration; **Reza Godary**: reformatting of the manuscript's original structure and conducting an extended literature review upon receiving the first review feedback; **Chan Yi Jing**: original dataset curation and supervision of data collection. All authors have read and agreed to the published version of the manuscript.

Acknowledgements We are grateful to the University of Nottingham Malaysia for providing us with the software support required to conduct this research. We would also like to show appreciation for the engineers and managers of Lepar Hilir Palm Oil Mill, Adela Palm Oil Mill, Keratong Estate Oil Palm Mill and Felda Lok Heng Palm Oil Mill for providing the dataset required for this study.

Funding This research received no external funding.

Data Availability Statement Data will be made available upon request.

Conflicts of Interest The authors declare no conflict of interest.

References

1. Malaysian Palm Oil Board (2011) *Environmental Impact , Malaysian Palm Oil Board.*
2. WWF (2021) *Palm Oil Buyers Scorecard*. Gland, Switzerland.
3. McCarthy, N. (2020) *Which Countries Produce The Most Palm Oil? , Forbes.*
4. Department of Statistics Malaysia (2021) *Selected Agricultural Indicators, Malaysia, 2021, Department of Statistics Malaysia.*
5. Vijay, V., Pimm, S.L., Jenkins, C.N. and Smith, S.J. (2016) 'The Impacts of Oil Palm on Recent Deforestation and Biodiversity Loss', *PLOS ONE*, 11(7), p. e0159668. doi:https://doi.org/10.1371/JOURNAL.PONE.0159668.
6. Harsono, S.S., Grundmann, P. and Soebronto, S. (2014) 'Anaerobic treatment of palm oil mill effluents: potential contribution to net energy yield and reduction of greenhouse gas emissions from biodiesel production', *Journal of Cleaner Production*, 64, pp. 619–627. doi:https://doi.org/10.1016/j.jclepro.2013.07.056.
7. Ahmad, A.L., Ismail, S. and Bhatia, S. (2003) 'Water recycling from palm oil mill effluent (POME) using membrane technology', *Desalination*, 157(1–3), pp. 87–95. doi:https://doi.org/10.1016/S0011-9164(03)00387-4.
8. Kamyab, H., Din, M.F.M., Keyvanfar, A., Majid, M.Z.A., Talaiekhozani, A., Shafaghat, A., Lee, C.T., Shiun, L.J. and Ismail, H.H. (2015) 'Efficiency of Microalgae Chlamydomonas on the Removal of Pollutants from Palm Oil Mill Effluent (POME)', *Energy Procedia*, 75, pp. 2400–2408. doi:https://doi.org/10.1016/J.EGYPRO.2015.07.190.
9. Kamyab, H., Chelliapan, S., Din, M.F.M., Rezania, S., Khademi, T. and Kumar, A. (2018) 'Palm Oil Mill Effluent as an Environmental Pollutant', in *Palm Oil Mill Effluent as an Environmental Pollutant*. InTech, pp. 13–28. doi:https://doi.org/10.5772/INTECHOPEN.75811.
10. Kheang Loh, S., Mei Ee, L., Muzzammil Ngatiman, ;, Weng Soon, L., Yuen May, C., Zhang, Z. and Salimon, J. (2013) 'ZERO DISCHARGE TREATMENT TECHNOLOGY OF PALM OIL MILL EFFLUENT', *Journal of Oil Palm Research*, 25(3), pp. 273–281.
11. Shakib, N. and Rashid, M. (2019) 'Biogas Production Optimization from POME by Using Anaerobic Digestion Process', *Journal of Applied Science & Process Engineering*, 6(2), pp. 369–377. doi:https://doi.org/10.33736/JASPE.1711.2019.

12. A Aziz, M.M., Kassim, K.A., ElSergany, M., Anuar, S., Jorat, M.E., Yaacob, H., Ahsan, A., Imteaz, M.A. and Arifuzzaman (2020) 'Recent advances on palm oil mill effluent (POME) pretreatment and anaerobic reactor for sustainable biogas production', *Renewable and Sustainable Energy Reviews*, 119, p. 109603. https://doi.org/10.1016/j.rser.2019.109603.
13. United States Environmental Protection Agency (2021) *Importance of Methane, Global Methane Initiative.*
14. United Nations Environment Programme (2021) *Methane emissions are driving climate change. Here's how to reduce them., Climate Action.*
15. Ministry of Environment and Water Malaysia (2020) *MALAYSIA THIRD BIENNIAL UPDATE REPORT TO THE UNFCCC.* Putrajaya.
16. World Nuclear Association (2016) *Heat values of various fuels , World Nuclear Association.*
17. Energy Commission Malaysia (2020) *MALAYSIA ENERGY STATISTICS HANDBOOK 2020.* Putrajaya.
18. Ministry of Energy, G.T. and W.M. (2017) 'Waste', in *Green Technology Master Plan Malaysia 2017 - 2030.* Putrajaya: Ministry of Energy, Green Technology and Water Malaysia, pp. 109–130.
19. Tippayawong, N. and Thanompongchart, P. (2010) 'Biogas quality upgrade by simultaneous removal of CO2 and H2S in a packed column reactor', *Energy*, 35(12), pp. 4531–4535. doi:https://doi.org/10.1016/J.ENERGY.2010.04.014.
20. Islamiyah, M., Soehartanto, T., Hantoro, R. and Abdurrahman, A. (2015) 'Water Scrubbing for Removal of CO2 (Carbon Dioxide) and H2S (Hydrogen Sulfide) in Biogas from Manure', *KnE Energy*, 2(2), p. 126. doi:https://doi.org/10.18502/ken.v2i2.367.
21. Choong, Y.Y., Chou, K.W. and Norli, I. (2018) 'Strategies for improving biogas production of palm oil mill effluent (POME) anaerobic digestion: A critical review', *Renewable and Sustainable Energy Reviews*, 82, pp. 2993–3006. doi:https://doi.org/10.1016/J.RSER.2017.10.036.
22. Anderson, K., Sallis, P. and Uyanik, S. (2003) 'Anaerobic treatment processes', *Handbook of Water and Wastewater Microbiology*, pp. 391–426. https://doi.org/10.1016/B978-012470100-7/50025-X.
23. McCarty, P.L. (1964) 'Anaerobic Waste Treatment Fundamentals', *Public Works*, 95, pp. 91–94.
24. Akhbari, A., Kutty, P.K., Chuen, O.C. and Ibrahim, S. (2020) 'A study of palm oil mill processing and environmental assessment of palm oil mill effluent treatment', *Environmental Engineering Research*, 25(2), pp. 212–221. doi:https://doi.org/10.4491/EER.2018.452.
25. Poh, P.E. and Chong, M.F. (2009) 'Development of anaerobic digestion methods for palm oil mill effluent (POME) treatment', *Bioresource Technology*, 100(1), pp. 1–9. doi:https://doi.org/10.1016/J.BIORTECH.2008.06.022.
26. Singkhala, A., Mamimin, C., Reungsang, A. and O-Thong, S. (2021) 'Enhancement of Thermophilic Biogas Production from Palm Oil Mill Effluent by pH Adjustment and Effluent Recycling'. doi:https://doi.org/10.3390/pr9050878.
27. Cioabla, A.E., Ionel, I., Dumitrel, G.A. and Popescu, F. (2012) 'Comparative study on factors affecting anaerobic digestion of agricultural vegetal residues', *Biotechnology for Biofuels*, 5(1), pp. 1–9. doi:https://doi.org/10.1186/1754-6834-5-39/FIGURES/9.
28. Jayaraj, S., Deepanraj, B. and Velmurugan, S. (2014) 'STUDY ON THE EFFECT OF pH ON BIOGAS PRODUCTION FROM FOOD WASTE BY ANAEROBIC DIGESTION Solar heat pumps View project Domestic refrigerators View project', *International Green Energy Confrence*, 5(August), pp. 799–803.
29. Vikrant, U.D., Ajit, C.C. and Yogesh, V.A. (2015) 'Temperature, pH and loading rate effect on biogas generation from domestic waste', *2014 International Conference on Advances in Engineering and Technology, ICAET 2014* [Preprint]. doi:https://doi.org/10.1109/ICAET.2014.7105292.
30. Chin, K.K. and Wong, K.K. (1983) 'Thermophilic anaerobic digestion of palm oil mill effluent', *Water Research*, 17(9), pp. 993–995. doi:https://doi.org/10.1016/0043-1354(83)90039-8.
31. Kim, S.H., Choi, S.M., Ju, H.J. and Jung, J.Y. (2013) 'Mesophilic co-digestion of palm oil mill effluent and empty fruit bunches', *Environmental technology*, 34(13–16), pp. 2163–2170. doi:https://doi.org/10.1080/09593330.2013.826253.

32. Zinder, S.H., Anguish, T. and Cardwell, S.C. (1984) 'Effects of Temperature on Methanogenesis in a Thermophilic (58°C) Anaerobic Digestor', *Applied and Environmental Microbiology*, 47(4), p. 808. doi:https://doi.org/10.1128/AEM.47.4.808-813.1984.
33. Wu, J., Liu, Q., Feng, B., Kong, Z., Jiang, B. and Li, Y.Y. (2019) 'Temperature effects on the methanogenesis enhancement and sulfidogenesis suppression in the UASB treatment of sulfate-rich methanol wastewater', *International Biodeterioration & Biodegradation*, 142, pp. 182–190. doi:https://doi.org/10.1016/J.IBIOD.2019.05.013.
34. Kumar, R. and Kumar, A. (2005) 'WATER ANALYSIS | Biochemical Oxygen Demand', in Worsfold, P., Townshend, A., and Poole, C.B.T.-E. of A.S. (Second E. (eds) *Encyclopedia of Analytical Science*. Oxford: Elsevier, pp. 315–324. doi:https://doi.org/10.1016/B0-12-369397-7/00662-2.
35. IPCC (2019) 'Wastewater Treatment and Discharge', in *2019 Refinement to the 2006 IPCC Guidelines for National Greenhouse Gas Inventories*. Bangkok, Thailand: IPCC.
36. Utami, I., Redjeki, S., Astuti, D.H. and Sani (2016) 'Biogas Production and Removal COD – BOD and TSS from Wastewater Industrial Alcohol (Vinasse) by Modified UASB Bioreactor', *MATEC Web of Conferences*, 58, p. 01005. doi:https://doi.org/10.1051/MATECCONF/201658 01005
37. ICF (2019) *USER'S GUIDE FOR ESTIMATING METHANE AND NITROUS OXIDE EMISSIONS FROM WASTEWATER USING THE STATE INVENTORY TOOL*. U.S. Environmental Protection Agency.
38. Putro, L.H.S. (2022) 'Emissions of CH4 and CO2 from wastewater of palm oil mills: A real contribution to increase the greenhouse gas and its potential as renewable energy sources', *Environment and Natural Resources Journal*, 20(1), pp. 61–72. doi:https://doi.org/10.32526/ENNRJ/20/202100149.
39. Xu, Y., Jiang, Y., Chen, Y., Zhu, S. and Shen, S. (2014) 'Hydrogen Production and Wastewater Treatment in a Microbial Electrolysis Cell with a Biocathode', *Water Environment Research*, 86(7), pp. 649–653. doi:https://doi.org/10.2175/106143014x13975035525500.
40. Ismail, A.F., Khulbe, K.C. and Matsuura, T. (2019) 'RO Membrane Fouling', in *Reverse Osmosis*. Elsevier, pp. 189–220. doi:https://doi.org/10.1016/B978-0-12-811468-1.00008-6.
41. Wang, Z., Jiang, Y., Wang, S., Zhang, Y., Hu, Y., Hu, Z. hu, Wu, G. and Zhan, X. (2020) 'Impact of total solids content on anaerobic co-digestion of pig manure and food waste: Insights into shifting of the methanogenic pathway', *Waste Management*, 114, pp. 96–106. doi:https://doi.org/10.1016/J.WASMAN.2020.06.048.
42. Yi, J., Dong, B., Jin, J. and Dai, X. (2014) 'Effect of Increasing Total Solids Contents on Anaerobic Digestion of Food Waste under Mesophilic Conditions: Performance and Microbial Characteristics Analysis', *PLOS ONE*, 9(7), p. e102548. doi:https://doi.org/10.1371/JOURNAL.PONE.0102548.
43. Yan, J., Zhao, Yehua, He, H., Cai, Y., Zhao, Yubin, Wang, H., Zhu, W., Yuan, X. and Cui, Z. (2022) 'Anaerobic co-digestion of dairy manure and maize stover with different total solids content: From the characteristics of digestion to economic evaluation', *Journal of Environmental Chemical Engineering*, 10(3), p. 107602. doi:https://doi.org/10.1016/J.JECE.2022.107602.
44. Bujoczek, G., Oleszkiewicz, J., Sparling, R. and Cenkowski, S. (2000) 'High solid anaerobic digestion of chicken manure', *Journal of Agricultural and Engineering Research*, 76(1), pp. 51–60. doi:https://doi.org/10.1006/JAER.2000.0529.
45. Budiyono, B., Syaichurrozi, I., Suhirman, S., Hidayat, T. and Jayanudin, J. (2021) 'Experiment and Modeling to Evaluate the Effectof Total Solid on Biogas Production fromthe Anaerobic Co-Digestion of Tofu LiquidWaste and Rice Straw', *Polish Journal of Environmental Studies*, 30(4), pp. 3489–3496. doi:https://doi.org/10.15244/PJOES/127277.
46. Panico, A., D'Antonio, G., Esposito, G., Frunzo, L., Iodice, P. and Pirozzi, F. (2014) 'The effect of substrate-bulk interaction on hydrolysis modeling in anaerobic digestion process', *Sustainability (Switzerland)*, 6(12), pp. 8348–8363. doi:https://doi.org/10.3390/SU6128348.
47. Dong, R., Qiao, W., Guo, J. and Sun, H. (2022) 'Manure treatment and recycling technologies', *Circular Economy and Sustainability: Volume 2: Environmental Engineering*, pp. 161–180. doi:https://doi.org/10.1016/B978-0-12-821664-4.00009-1.

48. David, B., Federico, B., Cristina, C., Marco, G., Federico, M. and Paolo, P. (2019) 'Biohythane Production From Food Wastes', in *Biohydrogen*. 2nd edn. Elsevier, pp. 347–368. doi:https://doi.org/10.1016/B978-0-444-64203-5.00013-7.
49. Gaby, J.C., Zamanzadeh, M. and Horn, S.J. (2017) 'The effect of temperature and retention time on methane production and microbial community composition in staged anaerobic digesters fed with food waste', *Biotechnology for Biofuels*, 10(1), pp. 1–13. doi:https://doi.org/10.1186/S13068-017-0989-4/FIGURES/5.
50. Chen, S., Xie, J. and Wen, Z. (2021) 'Chapter Four - Microalgae-based wastewater treatment and utilization of microalgae biomass', in Li, Y. and Zhou, W.B.T.-A. in B. (eds). Elsevier, pp. 165–198. doi:https://doi.org/10.1016/bs.aibe.2021.05.002.
51. Shi, X.S., Dong, J.J., Yu, J.H., Yin, H., Hu, S.M., Huang, S.X. and Yuan, X.Z. (2017) 'Effect of Hydraulic Retention Time on Anaerobic Digestion of Wheat Straw in the Semicontinuous Continuous Stirred-Tank Reactors', *BioMed Research International*, 2017https://doi.org/10.1155/2017/2457805
52. Gautam, R., Nayak, J.K., Daverey, A. and Ghosh, U.K. (2022) 'Emerging sustainable opportunities for waste to bioenergy: an overview', *Waste-to-Energy Approaches Towards Zero Waste*, pp. 1–55. doi:https://doi.org/10.1016/B978-0-323-85387-3.00001-X.
53. Jung, S.P. and Pandit, S. (2019) 'Chapter 3.1 - Important Factors Influencing Microbial Fuel Cell Performance', in Mohan, S.V., Varjani, S., and Pandey, A.B.T.-M.E.T. (eds) *Biomass, Biofuels and Biochemicals*. Elsevier, pp. 377–406. doi:https://doi.org/10.1016/B978-0-444-64052-9.00015-7.
54. Grangeiro, L.C., Almeida, S.G.C. de, Mello, B.S. de, Fuess, L.T., Sarti, A. and Dussán, K.J. (2019) 'New trends in biogas production and utilization', *Sustainable Bioenergy: Advances and Impacts*, pp. 199–223. doi:https://doi.org/10.1016/B978-0-12-817654-2.00007-1.
55. Liu, C., Wang, W., Anwar, N., Ma, Z., Liu, G. and Zhang, R. (2017) 'Effect of Organic Loading Rate on Anaerobic Digestion of Food Waste under Mesophilic and Thermophilic Conditions', *Energy and Fuels*, 31(3), pp. 2976–2984. doi:https://doi.org/10.1021/ACS.ENERGYFUELS.7B00018.
56. Orhorhoro, E.K., Ebunilo, P.O. and Sadjere, G.E. (2018) 'Effect of organic loading rate (OLR) on biogas yield using a single and three-stages continuous anaerobic digestion reactors', *International Journal of Engineering Research in Africa*, 39, pp. 147–155. doi:https://doi.org/10.4028/WWW.SCIENTIFIC.NET/JERA.39.147.
57. Meegoda, J.N., Li, B., Patel, K. and Wang, L.B. (2018) 'A review of the processes, parameters, and optimization of anaerobic digestion', *International Journal of Environmental Research and Public Health*, 15(10). doi:https://doi.org/10.3390/ijerph15102224.
58. Han, J., Kamber, M. and Pei, J. (2012) 'Data Preprocessing', in *Data Mining*. Third. Elsevier, pp. 83–124. doi:https://doi.org/10.1016/B978-0-12-381479-1.00003-4.
59. Jayalakshmi, T. and Santhakumaran, A. (2011) 'Statistical Normalization and Back Propagationfor Classification', *International Journal of Computer Theory and Engineering*, pp. 89–93. doi:https://doi.org/10.7763/IJCTE.2011.V3.288.
60. Nugaliyadde, A., Wong, K.W., Sohel, F. and Xie, H. (2017) 'Reinforced Memory Network for Question Answering', in *Lecture Notes in Computer Science (including subseries Lecture Notes in Artificial Intelligence and Lecture Notes in Bioinformatics)*. Springer Verlag, pp. 482–490. doi:https://doi.org/10.1007/978-3-319-70096-0_50.
61. Kuncheva, L.I. and Whitaker, C.J. (2015) 'Pattern Recognition and Classification', *Wiley StatsRef: Statistics Reference Online*, pp. 1–7. doi:https://doi.org/10.1002/9781118445112.STAT06503.PUB2.
62. Liu, Y., Traskin, M., Lorch, S.A., George, E.I. and Small, D. (2015) 'Ensemble of trees approaches to risk adjustment for evaluating a hospital's performance', *Health Care Management Science*, 18(1), pp. 58–66. doi:https://doi.org/10.1007/S10729-014-9272-4/TABLES/7.
63. Cai, J., Luo, J., Wang, S. and Yang, S. (2018) 'Feature selection in machine learning: A new perspective', *Neurocomputing*, 300, pp. 70–79. doi:https://doi.org/10.1016/J.NEUCOM.2017.11.077.

64. Miao, J. and Niu, L. (2016) 'A Survey on Feature Selection', *Procedia Computer Science*, 91, pp. 919–926. doi:https://doi.org/10.1016/J.PROCS.2016.07.111.
65. Altmann, A., Toloşi, L., Sander, O. and Lengauer, T. (2010) 'Permutation importance: a corrected feature importance measure', *Bioinformatics*, 26(10), pp. 1340–1347. doi:https://doi.org/10.1093/BIOINFORMATICS/BTQ134.
66. Arelli, V., Juntupally, S., Begum, S. and Anupoju, G.R. (2022) 'Solid state anaerobic digestion of organic waste for the generation of biogas and bio manure', in *Advanced Organic Waste Management: Sustainable Practices and Approaches*. Elsevier, pp. 247–277. https://doi.org/10.1016/B978-0-323-85792-5.00023-X.
67. Kuhn, M. and Johnson, K. (2019) 'Feature Selection Overview', in *Feature Engineering and Selection : A Practical Approach for Predictive Models*. Florida: CRC Press LLC, pp. 227–240.

Convolutional Neural Networks for Part Orientation in Additive Manufacturing

Valentina Furlan, Kevin Castelli, Luca Scaburri, and Hermes Giberti

Abstract The industrial applications of additive manufacturing (AM) have seen noticeable growth in recent years, pushing the studies on the parameters affecting the aesthetic, functional, and structural characteristics of the produced component. A central role is attributed to the orientation of the geometry on the building platform and the resulting building direction, the choice of which usually relies on the expertise of the operator. This work aims to elaborate an algorithm able to predict autonomously the optimal positioning of the component through the innovative techniques of deep learning, employed for their ability to draw information from a set of examples and build complex models. A convolutional neural network (CNN) is developed that, starting from the tridimensional representation of an object, predicts the rotation angle pair that leads to the optimal printing configuration. Two approaches have been compared: the first one predicts the angles, represented as points on a unit circle, through a regression that minimizes the angular difference loss. The second one performs a classification over a set of discreet rotations. The algorithm is trained and validated on two different datasets; finally, the generalization capacity of the model is investigated, highlighting the limits linked to the choice of examples used during training.

Keywords Additive manufacturing · Part orientation · Building direction · Convolutional neural networks · Predictive algorithm

---

V. Furlan (✉) · K. Castelli · H. Giberti
Dipartimento di Ingegneria Industriale e Dell'Informazione, Università di Pavia, via A. Ferrata 5, 27100 Pavia, Italy
e-mail: valentina.furlan@unipv.it

K. Castelli
e-mail: kevin.castelli@unipv.it

H. Giberti
e-mail: hermes.giberti@unipv.it

L. Scaburri
Dipartimento di Ingegneria Meccanica, Politecnico di Milano, via G. La Masa 1, 20156 Milano, Italy
e-mail: luca1.scaburri@polimi.it

K. Daimi et al. (eds.), *Cutting Edge Applications of Computational Intelligence Tools and Techniques*, Studies in Computational Intelligence 1118,
https://doi.org/10.1007/978-3-031-44127-1_8

1 Introduction

The increasing diffusion in industrial applications of Additive Manufacturing (AM) processes of the last decades is due to the several benefits provided by these technologies, such as the possibility of producing complex shapes, topologically optimized structures, design flexibility, and customizability, reduced material consumption and time and cost efficiency for small batches production [1, 2]. For this reason, the research has been pushed in the direction of understanding the effects of the many parameters involved in the process on the quality and mechanical properties of the AM components [3]. One of the most important parameters is *part orientation* referring to the building platform. The *part orientation* affects the mechanical properties of the 3D part resulting in anisotropy problems. On the other hand, *part orientation* affects also the building height and the use of support material resulting in different volumes and material consumption. Furthermore, *part orientation* influences the setting of layer height driving part quality, and the staircase effect [4–10]. *Part orientation* is crucial for different AM technologies moving from powder-bed technologies to direct energy deposition to fused deposition modeling (FDM) or to binder and material jetting. One example is the case of metallic components printed with thermal technologies. In these cases part orientation and consequently build direction affect heat conduction. The heat dissipation and the cooling rate are critical aspects in some AM technologies (i.e. selective laser melting, laser metal deposition gas metal arc welding). AM involves strong thermal cycles and cooling rates that are in the range of $10^3 - 10^8$ K/s [11]. Different cooling rates affect the mechanical performance of 3D parts. In fact, the cooling rate is directly related to the microstructures of the components resulting in different hardness and static performance. Moreover, the part orientation affects surface roughness, residual stress, delamination and defect formation [12]. Despite its relevance, the industrial machine only suggests the positioning for support reduction or volume reduction, which does not necessarily correspond to a better solution. The definition of the part position and its orientation is still based on the expertise of the operator resulting in a non-automated and manual approach. In this field, machine learning (ML) is a powerful technique to elaborate an orientation prediction algorithm due to its ability to gather information from large scale data and to build a prediction model. Neural Networks (NNs) in particular have strong evaluating skills for representing complex, highly non-linear relationships between input and output features. It is an interesting choice for this case in which one parameter (i.e. part orientation) affects many different aspects. In literature, many examples of applications of ML to AM can be found, spacing from topological optimization to material design, creation of process maps, relating process parameters to quality indicators at mesoscale and macroscale levels, in-process defect monitoring through audio or video systems [13–20], showing promising results in this direction. However, to the authors' knowledge no one complete work addresses the problem of part orientation through ML; for this reason, this work proposes, for the first time, a preliminary approach to build an algorithm taking as input the component 3D digital representation and giving as output the rotation angles to optimally place it on the building platform. To do so, two datasets of computer-aided design (CAD) files are employed. The first dataset is constituted of CAD models collected by an open platform. The information from the second dataset is collected through research work [21]. Further details on them are provided in Sect. 4. Several works on ML and AM are present in the literature. Moreover, it is interesting to observe that to the authors' knowledge no one exhaustive work addresses the problem of part orientation. This work proposes, for the first time, a preliminary approach to building an algorithm to predict parts orientation. The input is the 3D digital representation of the component while the outputs are the rotation angles to optimally place it on the building platform. To do so, two datasets of CAD files are employed:

1. featuring AM-produced components with a wide geometrical shape variety;
2. focusing on one specific component shape.

The first dataset acts as a learning basis for different classes of objects with different shapes, different sizes, and different complexity factors. The dataset is composed of FDM objects, however, the AM technology is not critical for the general purpose of the present research. This dataset can be continuously updated to enhance the algorithm and increase its capabilities. The second dataset focuses on a specific component for which many varieties in shapes and sizes are available. The component will become the case study for validating the algorithm. The components are manually rotated in the base position, corresponding to the best building orientation, and then mixed with rotated examples labelled with the related rotation angles. This approach will be used to train the convolutional neural network (CNN) algorithm, which has proved to be the top choice for visual orientation tasks among the other neural network (NN) algorithms. In fact, CNN has the advantage in the capacity of convolution of leveraging three levels to improve machine learning systems: sparse interactions, parameter sharing and equivariant representations [22, 23]. Previous work uses the CNN algorithm, nevertheless, this research differs in the inputs and outputs. After the validation, the ability of the model to fit different components will be tested underlining limits and capabilities. The aim is to mimic/codify the expertise of the AM operator performing the choice based on his experience through a program that learns from the results and replicates them on new unseen components. Furthermore, this technique does not rely on other parameters outside the geometry of the components which is provided through a CAD file. Process parameters, as much as differences in AM technology are not considered making this algorithm very general in its initial purpose. Moreover, this generalization is useful ad adaptable to the different AM technologies.

The chapter is subdivided into five sections without considering the introduction. Section 2 presents a summary of previous research conducted on part orientation and the development of algorithms and strategies for its optimization. Section 3 discusses in depth the method and the ML algorithm used to predict the orientation, the objective function, and the performance metrics employed. Section 4 describes the datasets employed in this analysis. Finally, Sect. 5 presents the results obtained while Sect. 6 gives the conclusions and future perspectives for the research work.

2 State of the Art of Related Works

2.1 Part Orientation

Although no examples of direct application of ML algorithms, and in particular NN, to the direct determination of part orientation can be found in literature, however numerous works oriented to the optimization of the position of the component on the building platform can be found, and the directions followed can be divided into two main approaches: error objective function minimization and part decomposition. An exception to this classification is constituted by the work of Leutenecker-Twelsiek et al. [24]. This work is not focused on the positioning of the final part. It addresses the design phase by producing a set of design guidelines according to the *early determination of the part orientation* principle. This principle is based on dividing the concept design into several design elements and analysing them separately.

2.1.1 Objective Function Minimization

The main discriminant within the works in this category is constituted by the choice of the objective to minimize through part rotation: considering works in which only one aspect is considered, [5] minimized cylindricity error computed directly from CAD models through the intersection with planar surfaces simulating the deposited layers. In their work, [25] determine the optimal build orientation through the minimization of support structures, computing the amount of overhang of each potential build face using the convex hull principle. Masood et al. [7] propose an algorithm based on the minimization of *volumetric error*, computed intersecting the standard tesselation language (STL) model triangles with parallel planes at one layer distance and projecting the contour on the adjacent plane to compute volume difference.

Other works instead create a weighted average of different undesired features and minimized them at once to provide a more accurate decision: [6] extend the cylindricity error formulation to perpendicularity, parallelism, angularity, conicity, total runout, and circular runout, combining them with the support structures volume, producing a minimization of tolerances and supports. Pham et al. [8] have developed an *orientation advisor* for stereolithography (SL) which measures the overhanging area and support volume, estimates the build time through layer thickness, laser velocity, and setup time, estimates the total cost including also pre and post-processing time and considers also several problematic features (pipes, shell, holes, axes, and critical surfaces). These parameters are used to rank some candidate orientations according to the total cost, build time, optimally oriented features, support volume or overhanging area optimization, or a weighted combination of the previous ones. Finally, [26] create an objective function given by the weighted average of five normalized evaluation criteria: build height, staircase error factor, material utilization factor, part surface area in contact with support structures, and volume of the support structure. Whilst most of the parameters considered are common to previous works, the introduction of a term directly addressing material utilization constitutes a novelty, and it is calculated based on the hollowing of the part obtained through the 2D curve offsetting approach; this corresponds to the volume between the outer part contour and the offset inner curve, divided by the volume of the full part.

2.1.2 Part Decomposition

This approach is based on the decomposition of the component into sub-geometries easier to build without the need for support structures or guaranteeing a higher surface quality, this is because they can be singularly oriented in a more optimal way having less critical features. The main drawback of this technique is the need to assemble the final part during post-processing, leaving more noticeable welding marks or seams and increasing post-production time, and better suits the application in sheet lamination [27]. Demir et al. [28] create a decomposition algorithm to simultaneously reduce printing time, decrease material consumption and increase fidelity. Their approximate convex decomposition algorithm partitions the initial polygonal model seeking a low number of near-convex components with no near-horizontal faces (i.e. faces are either horizontal or have an angle greater than a printer-defined threshold), which allows for reducing the staircase effect improving surface quality.

2.2 Convolutional Neural Network

CNN has proved to be very effective in the field of computer vision due to their capacity of dealing with data in tensor or matrix form for image classification and orientation detection applications involving pictures represented through pixel grids [29, 30]. Extending the concept to 3D object representation through occupancy grid or voxel grids, 3D object classification CNN have been introduced like the VoxNet project [31]. Furthermore, [32] employed orientation prediction to boost the performance of object recognition by introducing examples rotated around the $z - axis$, leading to an improvement in the generalization capacity of the model and increasing the ability to extract features.

The work of Eranpurwala et al. [33] establishes a relationship between standard machining features and AM build orientation, using deep learning to predict the orientation angles of new parts. The combinatory machine learning algorithm uses STL files as input and converts them into voxels. A segmentation process identifies the machining features which are individually passed to a feature classification 3D CNN model, determining both their typology and build orientation. Finally, a random forest regression analysis is used to predict the build orientation angle for each part based on minimum support structures volume. This approach based on component segmentation differs from the one proposed in this work where the object is considered as a whole, allowing to take into account not only the effect on the final build direction of each feature and their orientations but also their positioning inside the part. Furthermore, in this work the best orientation is not determined through the optimization of a parameter. The aim is to create an algorithm that returns a component rotated back to the base position. The position has been defined by the user. The result is a decoupling between the orientation prediction problem and the computation of the optimal building direction, whose definition can be changed to include more aspects or whose choice can be experience-driven.

3 The Method

In this chapter, a CNN has been used to create a model able to provide the angular values to be used to rotate a component around the x and y axes to bring it into the best position for building with a specific AM technology. In the present work, FDM is used as a representative case. Nevertheless, the obtained evidence can be applied to different AM technologies moving from powder-bed approaches to direct approaches, and to binder jetting. However, the CNN should be trained considering the characteristics and constraints of the desired method. In this research, the network is trained and validated using two datasets of components labelled through their rotation angles around the two axes; more details on how the datasets are built are available in Sect. 4. Since the aim of the CNN is to predict the orientation of an unseen object based on a dataset of already oriented objects, there is no need, at this stage of the work, of considering the AM technology to which its production is destined. For the same reason, the nature of the object itself and its function have low to no relevance.

The core network architecture is based on the work of [32], which in turn is based on VoxNet [31]. The difference is the absence of dropout layers, which have revealed not providing any significative improvement in the performances both in regression and classification tasks, and for this reason, have been excluded from this work. The CNN takes a 3D voxel grid (occupancy grid) as input and contains two convolutional layers with 3D filters followed by two fully connected layers, details on the parameters are available in Table 1. Since this choice was revealed not to be optimal in some cases due to the limited availability of examples for training, a slightly shallower network with only one convolutional layer has been employed as well.

The data from both datasets are converted into occupancy grids using the *trimesh library* [34], which generates a hollow representation of the converted object, and a converter of our implementation based on the ray detection technique described in [35], providing a full representation of the object. Since no significant difference between the results obtained with the two converters has emerged, only the ones obtained through the former will be reported.

Both regression and classification tasks have been conducted using Tensorflow library v. 2.4.1 with Keras Sequential API [36], and further details will be presented in the following paragraphs. The structure of the algorithms for the two tasks differ only in the final dense layer: in the former case 16 outputs, corresponding to the 16

Table 1 Model parameters

	3DConv1	3DConv2	3DMaxPool
# of filters	32	64	–
Kernel size	$3 \times 3 \times 3$	$3 \times 3 \times 3$	$2 \times 2 \times 2$
Stride	2	1	None
Padding	0	0	–
Batch normalization	✓	✓	–
	Dense1	Dense2	
# of outputs	128	variable	
Batch normalization	×	×	

classes of the labels, are employed, and the `softmax` activation function is applied. In the latter, just 4 outputs, corresponding to the sine and cosine of the two rotation angles, are used without introducing any activation function. For both models, *Adam* optimizer has been used.

3.1 Regression Task

An attempt at continuous angular value prediction has been done through this task, which required further considerations for what concerns objective functions: *mean squared error (MSE)* and *mean absolute error (MAE)* cannot be used directly as they are based on Euclidean distance, which does not apply to angular values, representing them a non-Euclidean space. For this reason, two different objective functions have been introduced: *approach 1* and *approach 2*. The *approach 1* extends the concept of distance to angles, computing the angular difference between the true angle θ and the predicted angle $\hat{\theta}$. This can be done converting the angles into unit circle coordinates trough sine and cosine by $v = (\cos\theta, \sin\theta)$; and then computing the angular difference using the arctangent as:

$$\begin{aligned} \Delta\theta &= \text{arctan2}\left(\frac{\sin(\theta - \hat{\theta})}{\cos(\theta - \hat{\theta})}\right) = \\ &= \text{arctan2}\left(\frac{\sin\theta\cos\hat{\theta} - \cos\theta\sin\hat{\theta}}{\cos\theta\cos\hat{\theta} + \sin\theta\sin\hat{\theta}}\right) \end{aligned} \tag{1}$$

and applying it to MSE and MAE, obtaining:

$$MSE = \frac{1}{N}\sum_{i=1}^{N}\left[\arctan2\left(\frac{\sin\theta\cos\hat{\theta} - \cos\theta\sin\hat{\theta}}{\cos\theta\cos\hat{\theta} + \sin\theta\sin\hat{\theta}}\right)\right]^2 \tag{2}$$

$$MAE = \frac{1}{N}\sum_{i=1}^{N}\left|\arctan2\left(\frac{\sin\theta\cos\hat{\theta} - \cos\theta\sin\hat{\theta}}{\cos\theta\cos\hat{\theta} + \sin\theta\sin\hat{\theta}}\right)\right|. \tag{3}$$

The second objective function, identified as *approach* 2, is derived from the work of [29]. In fact, representing the orientation angles as points on a unit circle, it is possible to use the scalar product to create a cost function to determine the distance between the true angle v and the predicted $\hat{v}$:

$$\begin{aligned} L = \frac{1}{N}\sum_{i=1}^{N}[1 - \cos(\Delta\theta_i)] = \frac{1}{N}\sum_{i=1}^{N}\left(1 - \frac{v_i \cdot \hat{v}_i}{|v_i||\hat{v}_i|}\right) = \\ = \frac{1}{N}\sum_{i=1}^{N}\left(1 - \frac{x_i\hat{x}_i + y_i\hat{y}_i}{\sqrt{\hat{x}_i^2 + \hat{y}_i^2}}\right). \end{aligned} \tag{4}$$

In both approaches the objective functions are evaluated separately on the rotation angles around the two axes, θ_x and θ_y, and then are averaged together to create the final objective function:

$$L_{tot} = (1 - \gamma) \cdot L_x + \gamma \cdot L_y \tag{5}$$

where $\gamma = 0.5$.

3.1.1 Classification Task

In this task instead of a continuous angular value prediction, a discretization is introduced: only four possible values $\{0°, 90°, 180°, 270°\}$ are considered for the two rotation angles around the x and y-axis. This way only 16 possible angle pairs combinations can be made, and they correspond to the 16 classes used for classification.

The objective function employed is *sparse categorical cross-entropy*, a common choice for classification tasks that requires that a label is associated with each rotated component through an integer between 0 and 15.

4 The Datasets

The important choice of the dataset is fundamental since ML algorithms strongly relate to data to correctly develop the model. Two different datasets have been employed:

- *Thingiverse*: AM *parts dataset:* this dataset is made of STL files of components coming from the *Makerbot Thingiverse* design community at *www.thingiverse.com*, which collects parts and projects destined primarily to FDM production. This dataset is meant to provide a wide variety of object shapes, featuring all those features considered critical in AM as holes, pipes, conformal channels, and overhanging surfaces. The AM technologies are promising due to their ability to build complex features and shapes not achievable with traditional approaches. However, these elements increase the complexity of the whole component and the issues with its positioning. A high number of examples increases the possible learning cases. A total number of 313 examples are present. The components have been rotated manually to the best building orientation according to the expertise of the authors.
- *ModelNet10-Chairs dataset:* this dataset is made starting from the ModelNet10 dataset [21], and it was chosen to provide a reduced shape variety among the components for the algorithm validation. For this reason, only one object class has been considered for this dataset: a number of 989 chair models are present inside ModelNet10, already oriented according to a reference direction. These have been chosen to create a stand-alone dataset completed with their rotated copies labelled with the rotation angles employed. Moreover, the chair is an object that can present some interesting aspects considering the different models. A chair can be axial symmetrical or can present features that are common to components for different applications (e.g. mechanical or structural components). This permits a possible expansion of the results to other objects.

The STL meshes of both datasets have been converted into voxels using $50 \times 50 \times 50$ voxel grids and stored in memory using boolean tensors, where 1 indicates the presence of a voxel and 0 the absence. To grant that the conversion preserves the characteristics of the object, it is crucial to check for the STL meshes being manifold and to repair them if not; this applies in particular to conversion algorithms using the ray detection method, which strongly relies on the consistency of face orientation and absence of internal faces. Figure 1 reports an example of a conversion of a non-manifold mesh before and after repairing it. Then the objects are manually oriented in the best building position, considered as the base position and to which the rotation angles label $(\theta_x, \theta_y) = (0°, 0°)$ is associated; then, for each component rotated copies are created and labelled with the respective rotation angles.

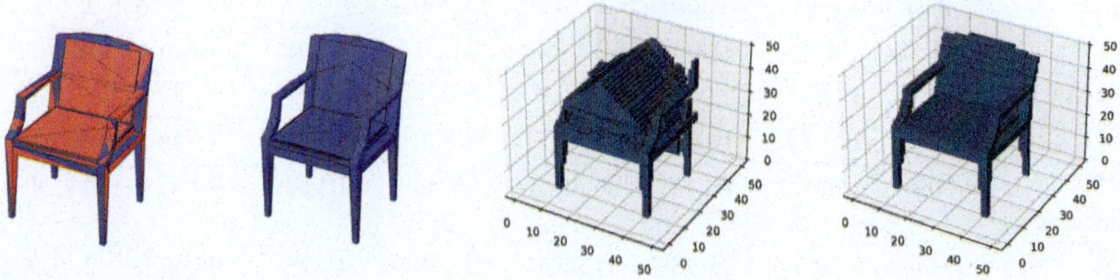

Fig. 1 Effect of non-manifold mesh repair on voxel conversion. In the first picture, the red faces present wrong orientation

For the classification task, the four rotation angles {0°, 90°, 180°, 270°} have been considered. From these four angles, 16 pairs of rotation angles around the x and y axes are obtained and used to rotate each one of the components of the two datasets.

Regarding the regression task, the need for a higher number of rotated components to provide a sufficient number of angles to be interpolated sensitively increases the size of the dataset. For this reason, angular values spaced by 20° have been chosen {0°, 20°, 40°, ..., 340°}, and from this set pairs have been generated using a random extraction function. To set a limit to the size of the final dataset, each element of the two starting datasets has been rotated using 35 different pairs randomly generated.

Finally, both the datasets have been divided into 80% and 20% subsets for training and validation, respectively.

5 Results

5.1 Regression Task

In order to evaluate the performance of the CNN in regression, MAE as defined in Eq. (3) has been used as a metric together with *accuracy-22.5* and *accuracy-45*, which are extensions of the concept of accuracy for regression tasks and defined as the ratio of samples whose predicted orientation is within 22.5° and 45° from the ground truth, respectively. In the regression task, neither one of the two datasets was allowed to reach convergence, presenting a situation of overfitting with the objective functions of both approaches, as it can be seen in Table 2, where the difference between the results obtained on the validation set outcomes significantly the one on validation set; furthermore, Fig. 2 displays the trend of MAE during training epochs with *approach 1* objective function, which is analogous to the results obtained with *approach 2*. In both cases the MAE evaluated over the training set shows the expected decreasing behaviour, whilst the one evaluated over the validation set is steady or lightly increases, never showing a reduction, denoting a clear situation

Table 2 Performance metrics for regression model—chairs dataset

	AM parts dataset				Chairs dataset			
	Approach 1		Approach 2		Approach 1		Approach 2	
	Train	Valid.	Train	Valid.	Train	Valid.	Train	Valid.
MAE	4.82°	89.66°	6.57°	90.53°	5.97°	85.31°	5.48°	84.51°
SD	7.47°	103.75°	10.07°	104.60°	8.88°	100.51°	8.25°	99.95°
45°-accuracy	99.60%	6.93%	99.44%	6.88%	99.53%	10.57%	99.57%	11.21%
22.5°-accuracy	98.91%	2.79%	95.38%	1.96%	97.18%	4.47%	98.11%	5.20%

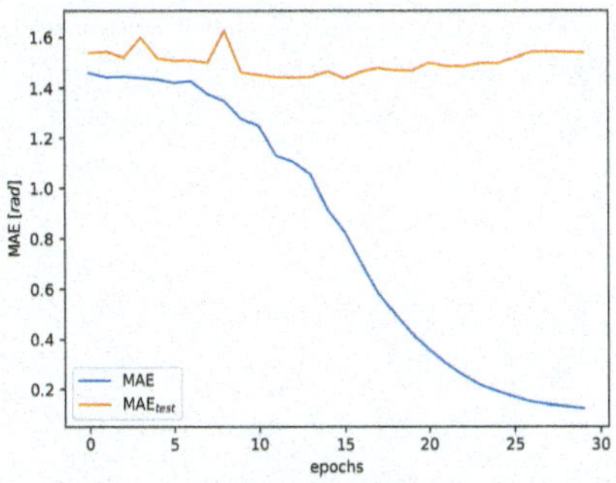

(a) AM parts dataset

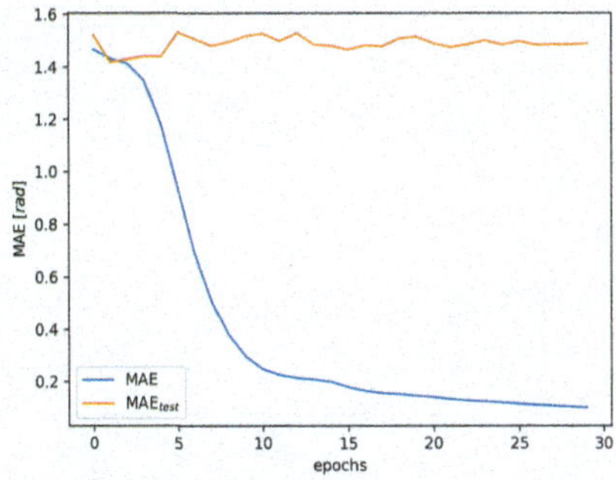

(b) Chairs dataset

Fig. 2 MAE trend during epochs in regression task—*Approach 1*

Table 3 Variation of regression metrics in AM parts dataset, changing the number of convolution layers

	Approach 1		Approach 2	
	2 layers	3 layers	2 layers	3 layers
MAE	88.33°	87.64°	87.59°	88.35°
SD	102.39°	101.84°	102.19°	102.54°
45°-accuracy	6.93%	7.32%	7.14%	7.23%
22.5°-accuracy	1.39%	2.09%	1.96%	2.00%

of overfitting. Modifying the number of convolutional layers did not lead to any significant improvement, as reported in Table 3.

5.2 *Classification Task*

The performance of the classification task has been evaluated using an *accuracy* metric. The results obtained are significantly better with respect to the ones of regression task, even if the model trained through the FDM dataset still shows overfitting even using only one convolutional layer (Fig. 3a). In fact, from the results in Table 4, the value of the accuracy for the validation set is significantly lower with respect to the one reached in the training set. However, promising results have been achieved using the chairs dataset, displaying a final accuracy higher than 98% as shown in Fig. 3b, where it is possible to observe the trend over the epochs showing an increase until epoch 13 of accuracy evaluated over validation set.

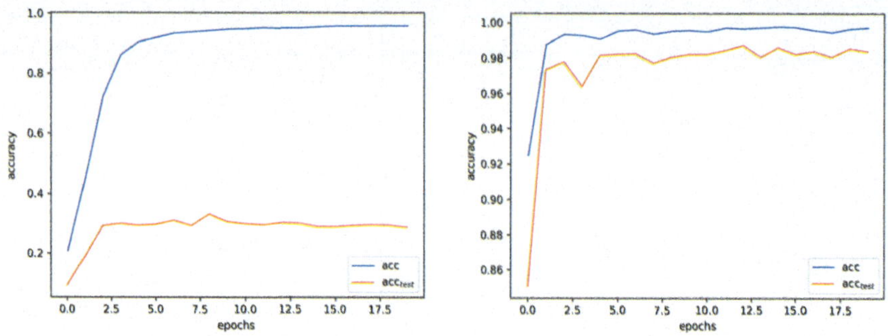

(a) AM parts dataset

(b) Chairs dataset

Fig. 3 Accuracy trend during epochs in classification task

Table 4 Accuracy for classification model

	AM parts dataset (%)	Chairs dataset (%)
Training set	95.94	99.66
Validation set	29.12	98.45

5.3 Analysis of the Results

The low performance of the results obtained through the regression model can be explained by considering the high number of data required to properly train ML algorithms in general, and in particular NNs. A continuous angular prediction requires a very large number of data, and even reducing the shape variability amongst the examples in the dataset through chairs experiment has shown no significant improvements in the final results. On the other side, the effects of a reduced geometrical shape variety inside the dataset showed great results in the classification task, together with the increased number of examples. Still, some components have been mistakenly predicted by the CNN, some examples are shown in Fig. 4. It is not possible, due to the complexity of the network, to clearly understand what led to a wrong prediction, however, some hypothesis can be made: the components involved in wrong predictions often present features (seat, back, legs, armrests) that can be mistaken by the network and misinterpreted due to their resemblance to other chairs details.

To further analyze the generalization capacities of the model trained with the *chairs dataset*, we tested the predictions of the model when applied to objects not only not present in the original dataset, but with a geometrical shape progressively different from the one of a chair. This allowed testing further the robustness of the model towards geometrical variations, an aspect of fundamental importance for applications involving high shape variety. The new components are obtained starting from a generic chair and progressively adding or subtracting details, then each object is rotated around all the 16 angular pairs and the number of correct predictions is listed in Fig. 5. It is interesting to observe that when the element strongly differs

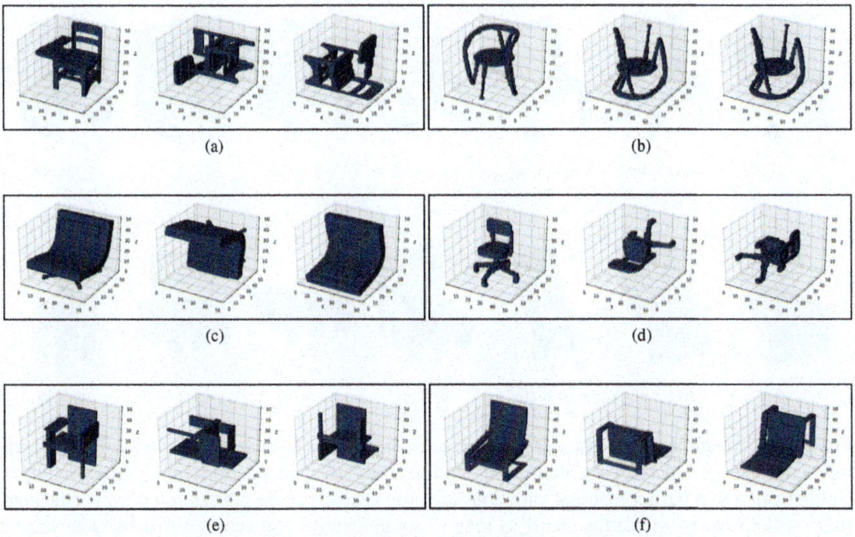

Fig. 4 Some examples of incorrect predictions. The first image represent the chair in reference position. The second image represents the chair as it is fed to the CNN. The third picture represents the chair rotated using the predicted rotation angles. In a correct prediction first and third images coincide. Each chair is represented in a volume of 50 × 50 × 50 voxel

the accuracy is reduced. This is the case in Fig. 5l. The back is absent and the legs are numerous. These differences result in low accuracy. On the other hand, if the same elements on the back are added in the traditional configuration Fig. 5k, or on the central part of the base Fig. 5m, two different results in terms of accuracy are obtained. The results show good accuracy as long as those feature characteristics of a chair are present, such as legs, armrests, seats, and backs. Finally, to test model accuracy also on a real mechanical component, a cycloidal drive output shaft (Fig. 6) has been chosen due to the similarity of pins with the legs of a chair. It is clear that the mechanical component strongly differs from the chair. Despite the axial symmetry and presence of similar features, the absence of a clear substitute for the back of a chair produces only faulty predictions. As shown in the example in Fig. 5l, the back is a crucial feature of chairs, which is present in most of the models of the dataset. The back is manually added to verify this evidence. Figure 6b and c show the sweeping increase in accuracy compared with Fig. 6a. The choice of the dataset affects the accuracy suggesting the adoption of multiple datasets for a specific component or the implementation of alternative strategies for solving the problems encountered (e.g. the addition of the back on the mechanical component). In conclusion, the model shows the capacity of correctly orientating components which can be quite different from chairs, but still keep geometrical features in key positions to allow a correct evaluation. Nevertheless, this results show the promising application of CNN to predict part orientation. It is interesting to observe that despite several works adopting the use of CNN in AM they are commonly adopted in material design,

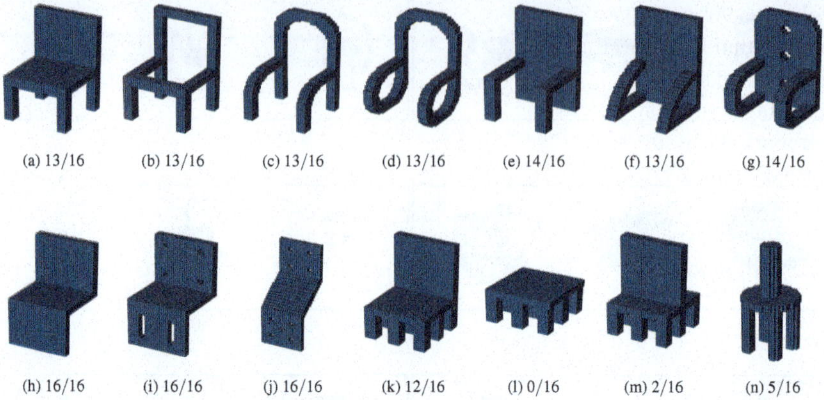

Fig. 5 Test components generated from chairs to evaluate the robustness of the model. The ratio represent the number of correct orientation prediction over the 16 angle rotation combination employed. Going from **a** to **n** the geometrical variations are progressively more substantial, however the algorithm still correctly orients the chairs as long as some features that can be ascribed as back, seat, armrest or legs are present

Fig. 6 Cycloidal drive output shaft test component. The ratio in case **a** is formulated taking into account axial symmetry of the component. Placing an extrusion resembling the back of a chair increase drastically the accuracy

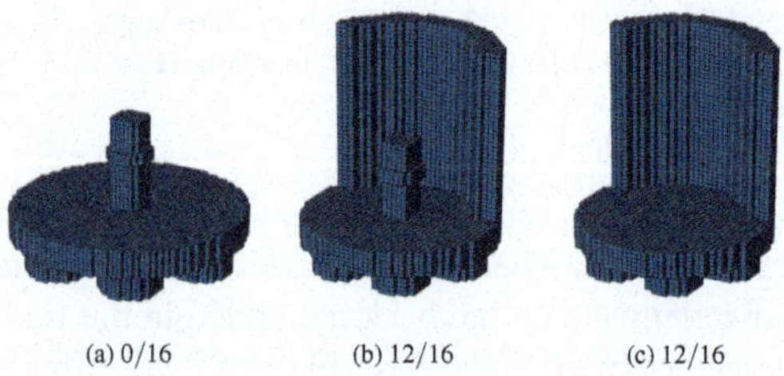

defect detection and quality assessment, and a direct comparison is not possible in terms of outcomes [16–19, 37].

6 Conclusions

The importance of part orientation in AM has been proved to be crucial due to its impact on manufactured components in terms of mechanical properties, affecting anisotropy, tolerances and aesthetic quality, due to staircase effect. However, the positioning procedure on the building platform is still conducted mostly manually, due to the absence of reliable alternative solutions. In this perspective, the use of ML techniques is a challange. This work proposed a new approach for part orientation in AM based on CNN. The results obtained show the possibility of creating a model able to provide the rotation angles to bring an object to a reference position chosen during the creation of the dataset. This was obtained through an orientation classification task involving a 90° rotation span for two rotations around the x and y axes. Furthermore, the algorithm has proved to be robust towards discrete variations from the shape of the objects in the training dataset. At this stage of the work, it was possible to obtain interesting results only by reducing the geometrical variety to one object shape, but other works have suggested the possibility of implementing a multi-task algorithm to improve the performances by adding a subsidiary task [32], making the orientation algorithm object recognition-boosted. Furthermore, the necessity for a larger dataset of components specifically designed for AM production has emerged to improve the capacity of predicting the orientation for different shape classes.

References

1. Mohsen Attaran. The rise of 3-D printing: The advantages of additive manufacturing over traditional manufacturing. *Business Horizons*, 60 (5): 677–688, 2017. ISSN 00076813. https://doi.org/10.1016/j.bushor.2017.05.011.
2. Wei Gao, Yunbo Zhang, Devarajan Ramanujan, Karthik Ramani, Yong Chen, Christopher B. Williams, Charlie C.L. Wang, Yung C. Shin, Song Zhang, and Pablo D. Zavattieri. The status, challenges, and future of additive manufacturing in engineering. *CAD Computer Aided Design*, 69: 65–89, 2015. ISSN 00104485. https://doi.org/10.1016/j.cad.2015.04.001.
3. Praveena B.A, Lokesh N, Abdulrajak Buradi, Santhosh N, Praveena B L, and Vignesh R. A comprehensive review of emerging additive manufacturing (3d printing technology): Methods, materials, applications, challenges, trends and future potential. *Materials Today: Proceedings*, 52: 1309–1313, 2022. ISSN 2214-7853. https://doi.org/10.1016/j.matpr.2021.11.059, https://www.sciencedirect.com/science/article/pii/S2214785321070632. International Conference on Smart and Sustainable Developments in Materials, Manufacturing and Energy Engineering.
4. Ramakrishna Arni and S. K. Gupta. Manufacturability analysis of flatness tolerances in solid freeform fabrication. *Journal of Mechanical Design, Transactions of the ASME*, 123 (1): 148–156, 2001. ISSN 10500472. https://doi.org/10.1115/1.1326439.
5. Ratnadeep Paul and Sam Anand. Optimal part orientation in Rapid Manufacturing process for achieving geometric tolerances. *Journal of Manufacturing Systems*, 30 (4): 214–222, 2011. ISSN 02786125. https://doi.org/10.1016/j.jmsy.2011.07.010.
6. Paramita Das, Ramya Chandran, Rutuja Samant, and Sam Anand. Optimum Part Build Orientation in Additive Manufacturing for Minimizing Part Errors and Support Structures. *Procedia Manufacturing*, 1: 343–354, 2015. ISSN 23519789. https://doi.org/10.1016/j.promfg.2015.09.041.
7. S. H. Masood, W. Rattanawong, and P. Iovenitti. A generic algorithm for a best part orientation system for complex parts in rapid prototyping. *Journal of Materials Processing Technology*, 139 (1–3 SPEC): 110–116, 2003. ISSN 09240136. https://doi.org/10.1016/S0924-0136(03)00190-0.
8. D. T. Pham, S. S. Dimov, and R. S. Gault. Part orientation in stereolithography. *International Journal of Advanced Manufacturing Technology*, 15 (9): 674–682, 1999. ISSN 02683768. https://doi.org/10.1007/s001700050118.
9. Caterina Casavola, Alberto Cazzato, Vincenzo Moramarco, and Carmine Pappalettere. Orthotropic mechanical properties of fused deposition modelling parts described by classical laminate theory. *Materials and Design*, 90: 453–458, 2016. ISSN 18734197. https://doi.org/10.1016/j.matdes.2015.11.009.
10. M. Somireddy and A. Czekanski. Anisotropic material behavior of 3D printed composite structures - Material extrusion additive manufacturing. *Materials and Design*, 195: 108953, 2020. ISSN 18734197. https://doi.org/10.1016/j.matdes.2020.108953.

11. G Meneghetti, D Rigon, D Cozzi, W Waldhauser, and M Dabalà. Influence of fatigue of maraging steel specimens produced Thermo-mechanical modeling a high pressure turbine blade of an gas turbine engine. *Procedia Structural Integrity*, 7: 149–157, 2017. ISSN 2452-3216. https://doi.org/10.1016/j.prostr.2017.11.072.
12. Lin Cheng and Albert To. Computer-Aided Design Part-scale build orientation optimization for minimizing residual stress and support volume for metal additive manufacturing : Theory. *Computer-Aided Design*, 113: 1–23, 2019. ISSN 0010-4485. https://doi.org/10.1016/j.cad.2019.03.004.
13. Xinbo Qi, Guofeng Chen, Yong Li, Xuan Cheng, and Changpeng Li. Applying Neural-Network-Based Machine Learning to Additive Manufacturing: Current Applications, Challenges, and Future Perspectives. *Engineering*, 5 (4): 721–729, 2019.
14. C. Wang, X. P. Tan, S. B. Tor, and C. S. Lim. Machine learning in additive manufacturing: State-of-the-art and perspectives. *Additive Manufacturing*, 36 (January): 101538, 2020a. ISSN 22148604. https://doi.org/10.1016/j.addma.2020.101538.
15. Jimeng Yang, Yi Chen, Weidong Huang, and Yun Li. Survey on artificial intelligence for additive manufacturing. *ICAC 2017-2017 23rd IEEE International Conference on Automation and Computing: Addressing Global Challenges through Automation and Computing*, (September), 2017. https://doi.org/10.23919/IConAC.2017.8082053.
16. Zeqing Jin, Zhizhou Zhang, Kahraman Demir, and Grace X. Gu. Machine learning for advanced additive manufacturing. *Matter*, 3 (5): 1541–1556, 2020. ISSN 2590-2385. https://doi.org/10.1016/j.matt.2020.08.023, https://www.sciencedirect.com/science/article/pii/S2590238520304501.
17. C. Wang, X.P. Tan, S.B. Tor, and C.S. Lim. Machine learning in additive manufacturing: State-of-the-art and perspectives. *Additive Manufacturing*, 36: 101538, 2020b. ISSN 2214-8604. https://doi.org/10.1016/j.addma.2020.101538, https://www.sciencedirect.com/science/article/pii/S2214860420309106.
18. Guo Dong Goh, Swee Leong Sing, and Wai Yee Yeong. A review on machine learning in 3d printing: applications, potential, and challenges. *Artificial Intelligence Review*, 54 (1): 63–94, 2021.
19. Jian Qin, Fu Hu, Ying Liu, Paul Witherell, Charlie C.L. Wang, David W. Rosen, Timothy W. Simpson, Yan Lu, and Qian Tang. Research and application of machine learning for additive manufacturing. *Additive Manufacturing*, 52: 102691, 2022. ISSN 2214-8604. https://doi.org/10.1016/j.addma.2022.102691, https://www.sciencedirect.com/science/article/pii/S2214860422000963.
20. Sachin Kumar, T Gopi, N Harikeerthana, Munish Kumar Gupta, Vidit Gaur, Grzegorz M Krolczyk, and ChuanSong Wu. Machine learning techniques in additive manufacturing: a state of the art review on design, processes and production control. *Journal of Intelligent Manufacturing*, 34 (1): 21–55, 2023.
21. Zhirong Wu, Shuran Song, Aditya Khosla, Fisher Yu, Linguang Zhang, Xiaoou Tang, and Jianxiong Xiao. 3D ShapeNets: A deep representation for volumetric shapes. *Proceedings of the IEEE Computer Society Conference on Computer Vision and Pattern Recognition*, 07-12-June: 1912–1920, 2015. ISSN 10636919. https://doi.org/10.1109/CVPR.2015.7298801.
22. Glen Williams, Nicholas A. Meisel, Timothy W. Simpson, and Christopher McComb. Design Repository Effectiveness for 3D Convolutional Neural Networks: Application to Additive Manufacturing. *Journal of Mechanical Design*, 141 (11), 09 2019, 111701. ISSN 1050-0472. https://doi.org/10.1115/1.4044199.
23. Ian Goodfellow, Yoshua Bengio, and Aaron Courville. *Deep learning*. MIT press, 2016.
24. Bastian Leutenecker-Twelsiek, Christoph Klahn, and Mirko Meboldt. Considering Part Orientation in Design for Additive Manufacturing. *Procedia CIRP*, 50: 408–413, 2016. ISSN 22128271. https://doi.org/10.1016/j.procir.2016.05.016.
25. Marijn P. Zwier and Wessel W. Wits. Design for Additive Manufacturing: Automated Build Orientation Selection and Optimization. *Procedia CIRP*, 55: 128–133, 2016. ISSN 22128271. https://doi.org/10.1016/j.procir.2016.08.040.

26. Amar M. Phatak and S. S. Pande. Optimum part orientation in Rapid Prototyping using genetic algorithm. *Journal of Manufacturing Systems*, 31 (4): 395–402, 2012. ISSN 02786125. https://doi.org/10.1016/j.jmsy.2012.07.001.
27. Kristian Hildebrand, Bernd Bickel, and Marc Alexa. Orthogonal slicing for additive manufacturing. *Computers and Graphics (Pergamon)*, 37 (6): 669–675, 2013. ISSN 00978493. https://doi.org/10.1016/j.cag.2013.05.011.
28. i. lke Demir, Daniel G. Aliaga, and Bedrich Benes. Near-convex decomposition and layering for efficient 3D printing. *Additive Manufacturing*, 21 (February 2017): 383–394, 2018. ISSN 22148604. https://doi.org/10.1016/j.addma.2018.03.008.
29. Kota Hara, Raviteja Vemulapalli, and Rama Chellappa. Designing Deep Convolutional Neural Networks for Continuous Object Orientation Estimation. 2017.
30. Rohan Ghosh, Abhishek Mishra, Garrick Orchard, and Nitish V. Thakor. Real-time object recognition and orientation estimation using an event-based camera and CNN. *IEEE 2014 Biomedical Circuits and Systems Conference, BioCAS 2014 - Proceedings*, pages 544–547, 2014. https://doi.org/10.1109/BioCAS.2014.6981783.
31. Daniel Maturana and Sebastian Scherer. VoxNet: A 3D Convolutional Neural Network for real-time object recognition. *IEEE International Conference on Intelligent Robots and Systems*, 2015-Decem: 922–928, 2015. ISSN 21530866. https://doi.org/10.1109/IROS.2015.7353481.
32. Nima Sedaghat, Mohammadreza Zolfaghari, Ehsan Amiri, and Thomas Brox. Orientation-boosted Voxel nets for 3D object recognition. *British Machine Vision Conference 2017, BMVC 2017*, pages 1–18, 2017. https://doi.org/10.5244/c.31.97.
33. Aliakbar Eranpurwala, Seyedeh Elaheh Ghiasian, and Kemper Lewis. Predicting build orientation of additively manufactured parts with mechanical machining features using deep learning. *Proceedings of the ASME Design Engineering Technical Conference*, 11A-2020 (May), 2020. https://doi.org/10.1115/DETC2020-22043.
34. Dawson-Haggerty et al. trimesh. https://trimsh.org/.
35. Sandeep Patil and B. Ravi. Voxel-based representation, display and thickness analysis of intricate shapes. *Proceedings—Ninth International Conference on Computer Aided Design and Computer Graphics, CAD/CG 2005*, 2005: 415–420, 2005. https://doi.org/10.1109/CAD-CG.2005.86.
36. Abadi Martin, Agarwal Ashish, Barham Paul, Brevdo Eugene, Chen Zhifeng, Citro Craig, Corrado Greg, S., Davis Andy, Dean Jeffrey, Devin Matthieu, Ghemawat Sanjay, Goodfellow Ian, Harp Andrew, Irving Geoffrey, Isard Michael, Yangqing Jia, Jozefowicz Rafal, Kaiser Lukasz, Kudlur Manjunath, Levenberg Josh, Mane Dandelion, Monga Rajat, Moore Sherry, Murray Derek, Olah Chris, Schuster Mike, Shlens Jonathon, Steiner Benoit, Sutskever Ilya, Talwar Kunal, Tucker Paul, Vanhoucke Vincent, Vasudevan Vijay, Viegas Fernanda, Vinyals Oriol, Warden Pete, Wattenberg Martin, Wicke Martin, Yu Yuan, and Zheng Xiaoqiang. Tensorflow: Large-scale machine learning on heterogeneous systems, 2015. https://www.tensorflow.org/.
37. Mahsa Valizadeh and Sarah Jeannette Wolff. Convolutional neural network applications in additive manufacturing: A review. *Advances in Industrial and Manufacturing Engineering*, 4: 100072, 2022. ISSN 2666-9129. https://doi.org/10.1016/j.aime.2022.100072, https://www.sciencedirect.com/science/article/pii/S2666912922000046.

CI in Recognition and Processing

SINATRA: A Music Genre Classifier Based on Clustering and Graph Analysis

Fernando Terroso-Saenz, Jesús Soto, and Andres Muñoz

Abstract At the dawn of a new era of intelligent applications in the music sector, the automatic genre-based classification of music tracks is a paramount task for the development of different services, such as music recommenders. In that sense, current solutions to uncover the genre of a song usually follow a multi-class approach revealing only a single genre per target song. However, songs do not usually belong to a single music genre but a mixture of them. In this context, the present work introduces SINATRA, a novel multi-label classifier of music genres of songs. By following an iterative procedure that continuously reduces the dimensional space of the genres, SINATRA is able to tag a song with multiple and complementary genres. The aforementioned dimensionality reduction is done by computing a graph comprising the co-occurrences of genres in songs. The evaluation results shows that SINATRA achieve an accuracy score above 0.5 given genre space covering more than 2,000 music genres.

Keywords Music · Genre classifier · Multi-label · Graph analysis · Clustering · Deep learning

1 Introduction

Music has always been strongly related to the evolution of the human race. The capability of composing, playing and enjoying music is one of the reasons that make us truly humans. In that sense, it is known the importance of music in ancient

---

F. Terroso-Saenz (✉) · J. Soto
Catholic University of Murcia (UCAM), Murcia, Spain
e-mail: fterroso@ucam.edu

J. Soto
e-mail: jsoto@ucam.edu

A. Muñoz
University of Cadiz, Cadiz, Spain
e-mail: andres.munoz@uca.es

K. Daimi et al. (eds.), *Cutting Edge Applications of Computational Intelligence Tools and Techniques*, Studies in Computational Intelligence 1118,
https://doi.org/10.1007/978-3-031-44127-1_9

civilizations such as Greece, Babylonia and Egypt as several tables of music and instruments have been found in many buried cities [1]. This importance has been transmitted through different regions and civilizations thought human history and now music is one of the most important cultural expression that better defines the identity of a human community [2].

In this context, the digital era has brought the opportunity to listening to any kind of music anywhere and anytime, regardless of its origin, through many different streaming services like Spotify, Youtube or Apple Music. As a matter of fact, Spotify offers instant access to more than 70 million songs[1] whereas Apple Music claims that its catalogue comprises 90 millions tracks.[2]

This calls for intelligent solutions able to catalogue such large number of music tracks in an automatic manner. In that sense, the genres associated to a song (e.g. pop, rock, folk and so forth) are one of the most meaningful tags that we can obtain. They usually are a key piece of information to develop, for example, music recommenders to end-users [3–5]. In the current literature, it is possible to find two different approaches to infer the genre of a song based on the considered features.

On the one hand, a large number of works rely on the song's audio signal and its directly associated features, like spectograms, to perform the genre classification [6, 7]. Despite their high accuracy, these approaches sometimes suffer from limitations to be fully deployed in certain scenarios. This is because the access to the raw audio signal of a song is usually quite restricted due to regulatory and copyright policies. As a matter of fact, the Spotify Developer API[3] only gets access to a 30-second sample of a song as MP3 file. This time length might not be enough to fully capture the characteristics of a song and identify its associated genre.

On the other hand, some works have focused on using the song's metadata as input features to infer their genre [8, 9]. Such metadata usually defines several aspects of a song related its context (e.g. liveness), mood (e.g. danceability) or audio features (e.g. instrumentalness or speechiness). Despite being slightly less accurate than the approaches based on raw audio signals, these type of solutions are more feasible to be deployed at large scale. This is because a song's metadata is usually more easily available than the raw audio signal on, for instance, the Spotify Developer API.[4]

However, existing solutions based on songs' metadata usually suffer from two important limitations in operational terms,

- First of all, they usually limit themselves to provide a single genre label to a song. However, songs frequently combine different musical styles and, thus, providing a unique label would not not be accurate enough to fully characterize a song. As a matter of fact, the 65% of the songs in the LFM-2b dataset [10], a well-known feed for musical analysis, are tagged with more than one music genre. However,

[1] https://newsroom.spotify.com/company-info/.
[2] https://www.apple.com/apple-music/.
[3] https://developer.spotify.com/documentation/web-api/reference/get-track.
[4] https://developer.spotify.com/documentation/web-api/reference/get-several-audio-features.

the development of multi-label classifiers has not fully explored in the music genre classification domain.

- Furthermore, the number of possible output genres is also rather limited, focusing on the most popular and general music genres (e.g. pop, rock, rap, latin, electronic, classic and jazz). This might provide a too coarse-grained tagging of the songs in certain context where, for example, is is necessary to distinguish between pop and j-pop songs.

In this context, the present work introduces, SINATRA, a muSIc geNre clAssifier based on clusTering and gRaph Analysis. SINATRA is a novel mechanism to perform a multi-label classification of the music genres of a song based on its associated metadata. Unlike existing works, the proposed solution is able to firstly classify a song among 48 different *core genres* and then incrementally perform a more detailed tagging up to 2,100 music styles.

To do so, SINATRA has been designed to follow an incremental process to obtain each of the genres of a song. In this manner, a recursive mechanism *refines* the classification process to predict a new genre based on the previously inferred tags for a target song. Thus, it makes use of well-known machine-learning tools like the k-Nearest Neighbours (kNN) clustering algorithm and the Random Forest (RF) classifier. While the kNN instance is in charge of providing an initial classification of a target song among the 147 core genres, different instances of RF provide the incremental tagging inferring multiple labels for a song.

A major contribution of this work is that it takes into account the frequencies of the music genres when they occur together in a song in order to generate the multi-label classification. This is modeled as a graph where nodes are the music genres and the links are labelled with the frequency of co-occurrence between pairs of genres. Also, this graph is used to define the *core genres* mentioned above as its *communities*. To the best of the authors' knowledge, this is the first time that this correlation among genres is introduced in the processing pipeline of a music-genre classifier.

All in all, the salient contributions of the present proposal are twofold, (1) the development of a fine-grained multi-label music genre classifier based on the tracks' metadata and (2) the usage of a graph modelling the correlation of music genres as a driving factor in the inference process.

The remainder of the chapter is structured as follows. Section 2 gives an overview about existing trends and techniques for music genre classification. Then, Sect. 3 describes the whole pipeline of SINATRA and Sect. 4 puts forward its evaluation. Lastly, Sect. 5 summarizes the main conclusions and potential future research lines motivated by this work.

2 Related Work

The two main approaches in the literature regarding the automatic classification of songs' musical genres are characterized by the type of data used for such a task. On the one hand, there is a line focused on the analysis of the different characteristics of the audio signal. On the other hand, the study of the songs' metadata, without taking into account the audio signal, has also proven to be a viable alternative for the classification of the musical genre. In this section we review the most relevant recent works in both lines.

2.1 Genre Classification Based on Song's Audio Signals

The authors in [6] leverage the MEL feature (a 2D representation of the audio signal) to convert audio files into spectograms that are then analyzed through a computer vision model. A spectogram representation carries rich information about the different frequency bands, which could be useful in identifying distinct patterns of music genres. Thus, this model utilizes a transfer learning approach using pre-trained Convolutional Neural Networks (CNNs) to learn the underlying features of such spectrograms. The model is tested using the GZTAN audio dataset (converted into MEL spectograms), containing a total of 10 genres, with 100 songs for each genre. The Resnet34 model shows the the best average performance with an accuracy of 79%, obtaining an almost perfect classification for the 'jazz' genre. A similar work is found in [11]. Here, the authors use a combination of wavelet and spectrogram analysis coupled with Deep Learning for classifying the song's genre. To improve the accuracy of the classification provided by the spectogram analysis, they employ a discrete wavelet transform to decompose each audio signal into multiple scales and subbands, and then calculate statistical features for each subband. These features are then passed to a CNN for genre classification along with the spectogram results. The proposal is evaluated using the GTZAN dataset again along with the Ballroom dataset, which contains 8 music genres for 698 audio files. The results show an accuracy of 81% for the GTZAN dataset and 71% for the Ballroom dataset, respectively. The authors also conducted an ablation study to analyze the impact of the individual components of their proposed method. They found that wavelet features alone had lower performance for genre classification than spectrogram features alone. However, combining the two types of features using a hybrid approach resulted in the highest accuracy rates. They also found that the specific wavelet used in their method (symlet-5) was critical for achieving the highest performance.

The work in [7] extends the idea of using spectograms by using the audio's auditory image, spectral and acoustic features. In particular in this work, it is noteworthy of the use of the auditory image, which is a representation of the spectral and temporal information of an audio signal that is similar to how the human ear processes sound. The authors propose the extraction of the relevant data related to the song's genre from three aforementioned features to then fuse the individual genre information predicted for each one. The classifiers used for this task are Support Vector Machines (SVM), K-Nearest Neighbor (KNN) and Sparse Representation-based Classification (SRC). This proposal is tested using four datasets, including GTZAN, with a maximum of 10 genres for the classification task. The results show that the SVM classifier usually offers the best accuracy for acoustic and spectral features, and the final model considering the combination of all the features show an accuracy between 65 and 92% depending on the dataset evaluated. Finally, the authors in [12] aim to improve the efficiency and accuracy of genre classification by proposing a holistic approach that combines both Transfer and Deep Learning techniques. Thus, the authors propose five new methods based on these techniques, namely, a Stacked Denoising Autoencoder (SDA) classifier, the Riemannian Alliance based Tangent Space Mapping (RA-TSM) transfer learning technique, a Transfer Support Vector Machine (TSVM) algorithm, and a BAG deep learning model consisting in a Bidirectional Long Short-Term Memory (BiLSTM) combined with an Attention model with Graphical Convolution Network (GCN). All these models are fed with the spectograms and pitch features, among others, from the raw audio files. All these classifiers are evaluated with three datasets, namely GTZAN, ISMIR 2004 and MagnaTagATune. The best classifier was the BAG deep learning model, which outperformed all the other proposed models in a range of 1% to 10% and producing the highest classification accuracy of 93.51% for the GTZAN dataset and 92.49% for the ISMIR 2004 dataset.

2.2 *Genre Classification Based on Song's Metadata*

Several works have proposed the use of ML techniques to leverage the song's metadata in the genre classification task. Hence, in [13] the performance of SVM, KNN and Naive Bayes was evaluated. The authors utilized a Spotify music dataset with over 200,000 songs, extracting 13 metadata features, with acousticness, instrumentalness and popularity as the top three relevant ones. Although this dataset covered 26 genres, due to the high complexity generated for processing all the features for all the genres, the number of genres was finally reduced to a range of 5 to 8 for the study. The results showed that the SVM classifier outperfomed KNN and Naive Bayes for all the genre classification experiments. The best result was obtained when classifying songs into 5 possible genres with an accuracy of 80%. The work in [14] proposed the analysis of 11 musical traits such as popularity, acousticness, danceability and track duration, among others, to categorize the songs' genre. The authors evaluated a music dataset from Kaggle with more than 50000 songs homogeneously grouped in ten genres. They trained and compared several ML models using the 11 features, being Random Forest the most accurate and efficient model with an astonishing accuracy of 99.6% on the test data. Popli et al. [9] focused on the classification of the sub-genres of electronic dance music by using metadata from a Spotify dataset. Although there are 166 sub-genres within this category, the authors selected five for their task: House, Drum and Bass, Techno, Hardstyle and Trap. They analyzed over 15 metadata features of 34,500 songs, including danceability, energy, valence, tempo and duration, among others. For this task, the ML models employed were logistic regression, K-nearest neighbours, SVM and Random Forest. The results showed an accuracy ranging between 83.3% and 91.3% for these sub-genres, with some difficulties in differentiating between the pairs House-Techno and Hardstyle-Trap.

The differences of our proposal with respect to these related works are twofold. First, the SINATRA framework is a multi-label classifier, whereas the related works only focused on single-label classification. Second, the range of available music genres is substantially wider than that of other works, thanks to the recursive classification mechanism implemented in SINATRA.

3 Description of the SINATRA Framework

In this section we describe in detail the inner architecture of SINATRA. In that sense, Fig. 1 provides an overview of its general pipeline regarding its training and its functionally when it is deployed as a service. As we can see, during the training stage (left side of the figure), the system takes the general dataset $\mathcal{D}$ and compose the genre graph, $\mathcal{G}$ and finds its nodes' communities as core genres (step 1). In step 2, a new version of the dataset is created by including the core genres associated to each song ($\mathcal{D}_{CG}$). In step 3, a kNN instance (CG-kNN) is fed with $\mathcal{D}_{CG}$. In the production stage (right side of the figure), the input song s is firstly classified by CG-KNN to obtain its CG ($\hat{cg}$) (step 4). Then, we refine $\mathcal{D}_{CG}$ by only keeping the songs labelled with the inferred CG giving rise to the $\mathcal{D}_{CG}^{\hat{cg}}$ dataset (step 5). In step 6, we train a new instance of RF with such a dataset and a fine-grained music genre is inferred $\hat{g}$. This outcome is used to refine again the dataset to only comprise songs labelled with $\hat{cg}$ and $\hat{g}$ (step 7) and then train a new RF instance to generate new genre label (step 8). In step 9, the probability distributions from the RF instance and the ones from the genre graph $\mathcal{G}$ are fused to provide the prediction $\hat{g}_i$ which is included in the outcome set $\hat{\mathcal{M}}_{s_{target}}$. Finally, steps 7, 8 and 9 are repeated until the desired number of music genres to infer is reached. In the following sections, we describe the different steps involved in these two stages in detail.

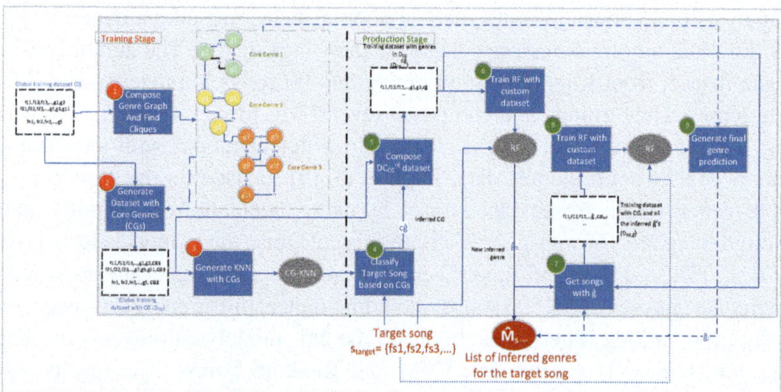

Fig. 1 Overview of the SINATRA's pipeline

3.1 Training of the Classifier

In order to train the framework, we rely on a raw training dataset, $\mathcal{D}$. As Fig. 1 shows, this dataset comprises a set of n different songs. This way, the i-th song in $s_i \in \mathcal{D}$, is defined as a vector $s_i = \langle f_{i1}, f_{i2}, \ldots, f_{ij}, g_{i1}, g_{i2}, \ldots, g_{il} \rangle$ comprising j different numerical features and labelled with l different genres. Let us call $\mathcal{F}$ the set of features of the songs and $\mathcal{M}$ the set comprising all the different music genres in $\mathcal{D}$.

Given this initial datataset, Fig. 1 shows that the first step in the training stage is the generation of an undirected graph of music genres $\mathcal{G} = \langle \mathcal{M}, \mathcal{E} \rangle$ where the music genres $\mathcal{M}$ are the nodes and $\mathcal{E}$ is the set of links connecting such nodes. In that sense, an edge $e_{u,v} = \langle u, v, t_{u,v} \rangle \in \mathcal{E}$ indicates that genres u and v have appeared $t_{u,v}$ times together in the songs in $\mathcal{D}$. In that sense, the graph is further refined by removing those edges with $t \leq \Theta_t$ to not consider irrelevant and rare connections among genres.

Once $\mathcal{G}$ is composed, we obtain its communities of nodes by means of the asynchronous label propagation algorithm [15]. In brief, this algorithm firstly initializes each node with a unique tag. Next, it endlessly sets a node's label to be the most frequent one among its neighbours. The algorithm stops when each node has the label that appears most frequently among its neighbors. Then, a community is defined as a set of nodes comprising the same label.

As a result, we obtain a set of m *core genres* $\mathcal{CG} = \langle cg_1, cg_2, \ldots, cg_m \rangle$ where cg_i corresponds to the i-th community in $\mathcal{G}$, C_i. It is important to note that, this mechanism ensures that each genre will be assigned to a single community. By means of these core genres, we are able to provide a higher-level space for the music genres definition because $|\mathcal{CG}| \leq |\mathcal{M}|$. As we will see in Sect. 3.2, this constitutes a paramount aspect to develop the incremental classification of the genres of a song followed by SINATRA.

Furthermore, as the step 3 of Fig. 1 depicts, these core genres are used to enrich the dataset $\mathcal{D}$ so that each song is now assigned with a particular CG. This gives rise to a dataset $\mathcal{D}_{CG}$ where each song $s_i \in \mathcal{D}_{CG}$ is defined as a vector $s_i = \langle f_{i1}, f_{i2}, \ldots, f_{ij}, cg_{si} \rangle$ where cg_{si} is the core genre that contains all the genres $\langle g_{i1}, g_{i2}, \ldots, g_{il} \rangle$ of s_i.

Given $\mathcal{D}_{CG}$, the third step of the proposed framework is to train an instance of K-Nearest Neighbours clustering algorithm (kNN) [16] which has been used in multiple domains [17–21]. Basically, it is a non-parametric, instance-based supervised learning algorithm where the k-nearest examples of an input record are selected and majority class majority among such neighbours makes a prediction for a the input record.

In SINATRA, we fit the kNN algorithm with $\mathcal{D}_{CG}$ to compose the CG-kNN model. As we will put forward in Sect. 4.4, a song record s_i is transformed into a vector $\Phi(s_i) = \langle c_{i1}^{\alpha_{i1}}, c_{i2}^{\alpha_{i2}}, \ldots, c_{ij}^{\alpha_{ij}} \rangle$ before being classified by CG-kNN. In this case, $c_{ij}^{\alpha_{ij}}$ is the centroid of the cluster assigned to the feature f_{ij} by means of the Fuzzy C-Mean

(FCM) clustering algorithm [22]. Given the new vector $\Phi(s_i)$, the CG-KNN instance predicts the core genre $\hat{cg}_{s_i} \in \mathcal{CG}$ of the song s_i.

3.2 Production Stage

Once the training stage has been completed with the generation of CG-KNN instance, SINATRA is ready to classify genre terms any new target song s_{target}. In that sense, the classification task performed by SINATRA can be formulated as follows:

Given the vector of features of a target song s_{target}, $\mathcal{F}_{s_{target}} = \langle f_{s1}, f_{s2}, \ldots, f_{sj} \rangle$, **Classify** s_{target} in n_g different genres by a function $\mathcal{S}$,

$$\mathcal{S}(\mathcal{F}_{s_{target}}, n_g) \rightarrow \hat{\mathcal{M}}_{s_{target}}$$

where $\hat{\mathcal{M}}_{s_{target}}$ is the set of genres tagging s_{target}. The whole procedure to perform the classification is depicted in Algorithm 1 whose key steps will be put forward next.

First of all, it is worth noticing that these tasks follow an incremental approach to eventually disclosing the music genres of a song. To begin with, the CG-KNN instance is fed with the vector of features $\mathcal{F}_{s_{target}}$ of the input song s_{target}(line 2 in Algorithm 1). As a result, the system infers the core genre $\hat{cg}$ of s_{target}. By means of this first classification, the system assumes that the set of genres that can be used to tag s_{target} are the ones included in gender community of $\hat{cg}$, $\mathcal{M}_{\hat{cg}}$. Therefore, the system focuses at this point to infer which genres in $\mathcal{M}_{\hat{cg}}$ can be accurately assigned to s_{target}.

To do so, we filter $\mathcal{D}_{CG}$ by only keeping the songs that complain with 2 conditions, (1) their core-genre is $\hat{cg}$ and (2) they are only labelled with a single fine-grained genre $g \in \mathcal{M}_{\hat{cg}}$ (line 3 in Algorithm 1). This gives rise to the dataset $\mathcal{D}_{CG}^{\hat{cg}}$ that comprises the songs labelled with a single genre which also is included as part of $\hat{cg}$. Next, this dataset fits a Random Forest (RF) classifier (line 4). This algorithm has widely and successfully used in many different domains and scopes [23]. In brief, RF is mainly based on ensembles of decision trees using if-then rules that sample the input space. In it important to remark that this RF instance is trained by only using a subset of all the training data, so that it focuses on distinguishing the differences among the genres in $\mathcal{M}_{\hat{cg}}$. As a result, this RF infers the first genre $\hat{g}$ of s_{target} and it is included in the resulting set $\hat{M}_{s_{target}}$ (lines 5–6).

At this point, SINATRA starts an iterative process until it labels the input song with n_g genres (lines 7–19 in Algorithm 1). In that sense, the system refines $\mathcal{D}_{CG}$ by keeping the songs with 3 conditions, (1) their core-genre is $\hat{cg}$, (2) they are labelled with 2 genres and (3) one of them is $\hat{g}$. This gives rise to a dataset $\mathcal{D}_{CG}^{\hat{cg}}$. (line 9) This dataset is used to fed a new instance of RF (line 10). In this case, the model returns the probability distribution of each genre $g \in \mathcal{M}_{\hat{cg}}$ to be the correct tag for s_{target}, $\mathcal{P}_{RF}$ (line 10). As lines 12-18 show, the system also extract another probability distribution from the sub-graph $\mathcal{G}_{\hat{g}} \subseteq \mathcal{G}$ that only comprises the nodes

Algorithm 1: Pseudo-code of SINATRA'S classification procedure.

Input: The target song s_{target}, the kNN for core genres CG-kNN, the dataset with core genres $\mathcal{D}_{CG}$, graph of genre co-occurrences $\mathcal{G}$, and the number of genres to label the song n_g

Output: n_g-dimensional set $\hat{\mathcal{M}}_{s_{target}}$ with the inferred genres for s_{target}

$\hat{\mathcal{M}}_{s_{target}} \leftarrow \emptyset$

`/* 1. Get the core genre of the input song. */`

$\hat{cg} \leftarrow$ CG-KNN($\mathcal{F}_{s_{target}}$)

`/* 2. Filter the dataset to keep only songs with` $\hat{cg}$. `*/`

$\mathcal{D}_{CG}^{\hat{cg}} \leftarrow filter\_data(\mathcal{D}_{CG}, \hat{cg})$

`/* 3. Train RF with filtered dataset. */`

$RF \leftarrow generate\_model(\mathcal{D}_{CG}^{\hat{cg}})$

`/* 4. Generate first-level genre classification. */`

$\hat{g} \leftarrow RF(\mathcal{F}_{s_{target}})$

$\hat{\mathcal{M}}_{s_{target}} \leftarrow \hat{\mathcal{M}}_{s_{target}} \cup \hat{g}_0$

while $|\hat{\mathcal{M}}_{s_{target}}| < n_g$ **do**

- `/* We account for the number of genres already inferred. */`
- $i \leftarrow |\hat{\mathcal{M}}_{s_{target}}| + 1$
- `/* 5.Filter dataset to keep` $\hat{cg}$ `and` $\hat{\mathcal{M}}_{s_{target}}$ `genres. */`
- $\mathcal{D}_{\hat{cg},\mathcal{M}_{s_{target}}} \leftarrow filter\_data(\mathcal{D}_{CG}, \hat{cg}, \mathcal{M}_{s_{target}}, i)$
- `/* 6. Train RF with filtered dataset. */`
- $RF_i \leftarrow generate\_model(\mathcal{D}_{\hat{cg},\mathcal{M}_{s_{target}}})$
- `/* 7. Generate new-level genre classification with RF. */`
- $\mathcal{P}_{RF}^i \leftarrow RF_i(\mathcal{F}_{s_{target}})$
- `/* 8. Filter graph to keep only edges and nodes linked with` $\hat{g}$. `*/`
- $\mathcal{G}_{\hat{g}} \leftarrow filter\_graph(\mathcal{G}, \hat{g})$
- `/* 9. Compute probability distribution of the graph based on edges weights. */`
- $\mathcal{P}_{\mathcal{G}}^i \leftarrow \emptyset$
- **for** ***each*** $e_{(v,\hat{g})} \in \mathcal{G}_{\hat{g}}.\mathcal{E}$ **do**
  - $\rho_v = \frac{(e_{v,\hat{g}}).t}{\sum_{u \in \mathcal{G}_{\hat{g}}} e_{(u,\hat{g})}.t}$
  - $\mathcal{P}_{\mathcal{G}}^i \leftarrow \mathcal{P}_{\mathcal{G}}^i \cup \rho_v$
- `/* 10. Compute final probability distribution genres based on RF and graph outcomes. */`
- $\mathcal{P}^i \leftarrow \mathcal{P}_{RF}^i \times \mathcal{P}_{\mathcal{G}}^i$
- `/* 11. We keep the genres with maximum probability as the i-th inferred genre. */`
- $\hat{g} \leftarrow \mathcal{P}^i.max$
- $\hat{\mathcal{M}}_{s_{target}} \leftarrow \hat{\mathcal{M}}_{s_{target}} \cup \hat{g}$

return $\hat{\mathcal{M}}_{s_{target}}$

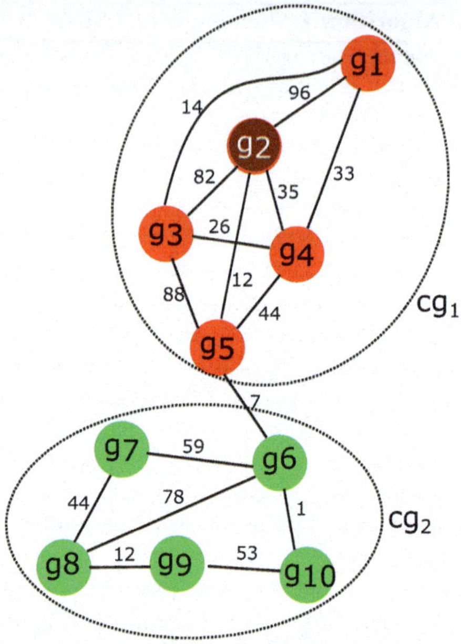

Fig. 2 Example of genre graph $\mathcal{G}$ comprising 2 different core genres, one with genres from g_1 to g_5 and a second one comprising genres from g_6 to g_{10}. The edges are labelled with the frequency of occurrence between pairs of genres

and edges directly connected to the music genre $\hat{g}$. This way, we extract a probability distribution $\mathcal{P}_{\mathcal{G}}$ where the probability of a genre v is computed as the rate of its edge's weight with $\hat{g}$ ($e_{(v,\hat{g})}.t$) with respect overall sum of edges' weights connected to $\hat{g}$ (line 15 of Algorithm 1).

At this point, we have a probability distribution of genres based on the features of the target song ($\mathcal{P}^i_{RF}$) and another based on their co-occurrence ($\mathcal{P}^i_{\mathcal{G}}$). Then, both distributions are joined giving rise to the final genre distribution $\mathcal{P}^i$ (line 17). Among all the genres in this distribution, the one with maximum probability is selected as the new inferred genre, $\hat{g}$, and included in the outcome set of genres $\hat{\mathcal{M}}_{s_{target}}$ (lines 18 and 19). Finally, this set is returned as the provided classification when the required number of genres is reached.

For the sake of clarity, Fig. 2 shows an example setting to compute this graph-based probability distribution extraction. Giving this figure, let us assume that CG-KNN infers cg_1 as $\hat{cg}$ (line 2 of Algorithm 1) and then RF infers g_2 as the first $\hat{g}$ (line 5). Consequently, the dataset $\mathcal{D}_{\hat{cg},\mathcal{M}_{s_{target}}}$ only comprises the songs of the training set $\mathcal{D}$ with core genre cg_1 and including g_2 as one of its labelled genres $\langle g_1, g_3, g_4, g_5\rangle$. Then, the first instance of RF in the processing loop (RF$_1$, line 10) generates the following distribution of genres $\mathcal{P}^1_{RF} = \langle g_1 : 0.32, g_3 : 0.09, g_4 : 0.42, g_5 : 0.17\rangle$. Given graph structure in Fig. 2, its probability distribution is $\mathcal{P}^1_{\mathcal{G}} = \langle g_1 : 0.43, g_3 : 0.36, g_4 : 0.16, g_5 : 0.05\rangle$. In this case, the sum of the weights of all the edges of g_2 is 225 (82+12+35+96), hence the probability of g_1 is computed as $\frac{96}{225} \approx 0.43$ as the edge e_{g_1,g_2} has a weight of 96 (which would be the number of times that g_1 and g_2 have appeared together as genres of a song). The resulting joined distribution would be $\mathcal{P}^i = \langle g_1 : 0.14, g_3 : 0.03, g_4 : 0.06, g_5 : 0.01\rangle$ as, for instance $0.32 \times 0.43 \approx 0.14$ in the case of g_1. Finally, g_1 is set to $\hat{g}$ and included to the outcome set, $\hat{\mathcal{M}}_{S_{target}} = \langle g_2, g_1\rangle$. In case the required number of genres n_g were set to 2, that set would be the final classification of the mechanism. Otherwise, the inference process is repeated but in this case $\mathcal{D}_{\hat{cg},\mathcal{M}_{S_{target}}}$ would only comprise the songs of the training set $\mathcal{D}$ with core genre cg_1 and labelled with g_1 *and* g_2.

4 Evaluation of SINATRA

In this section we describe the evaluation of the SINATRA framework by firstly describing the dataset used and then discussing the obtained results.

4.1 Dataset Description

The evaluation of SINATRA has relied on two different raw music feeds. On the one hand, we made use of the *Top 200* daily rankings released by Spotify, which contain the 200 most played songs in 69 different countries per day.[5] By means of an ad-hoc crawler, daily rankings were extracted for each single country for a 2-year period from 2020-07-17 to 2022-10-11. As a result, 14,816 unique songs were extracted from 2,569 artists. On the other hand, we have also processed the Last-FM 2b (LFM-2b) dataset [10]. Among other features, this second dataset comprises 2,378,113 tracks included in the Spotify catalogue from 266,479 artists covering a 15-year time period (2005/02/18-2020/03/20) in Last.fm, a very well-known online music service.

For each song in these datasets, we extracted 10 features from the Spotify Developer Platform (SDP).[6] These features are related to the song's audio properties (loudness, speechiness and instrumentalness), context (liveness and acousticness) and mood (danceability, valence, energy, and tempo), along with its release date as the tenth feature.[7] These features constitute the $\mathcal{F}$ set of the framework (Sect. 3.1). The reason of including the release date in the feature set is because some studies point out that genres evolve through time and *static* classifiers that do not consider the temporal dimension of the tracks may suffer from serious inaccuracies [24].

Eventually, we fused both datasets by removing duplicates giving rise to a global dataset comprising 1,354,932 tracks from 259,698 unique artists covering a 17-year period from 2005-02-18 to to 2022-10-11. Furthermore, it included 2,166 music genres labeling such tracks. These genres composed the $\mathcal{M}$ set of SINATRA.

---

[5] https://charts.spotify.com/charts/overview/global.

[6] https://developer.spotify.com/discover/.

[7] We assumed that the release date of a song was the one of the album on which it appeared.

4.2 Exploratory Analysis

Regarding the exploratory analysis of the aforementioned dataset, Fig. 3 shows the distribution of songs based on their associated number of genres. To see the suitability of developing multi-label classifiers for music genres, we would like to remark that 926,085 songs were labeled with 2, 3 or 4 genres (475,765 + 289,339 + 160,984) versus the 600,985 songs with a unique genre.

Concerning the distribution of the genres, Fig. 4 shows the number of times each genre has labelled a song in the dataset. Unsurprisingly, the 2 most frequent genres are pop and rock and among the top-10 we can find well-known genres as rap, hip-hop, dance or folk. These genres have been commonly included as the target tags of many genre classifiers [8, 9, 25]. However, there are others like alternative, punk or house that are not usually taken into account by such solutions. This calls for alternative methods that *enlarge* the dimensionality of the target genres.

Concerning the relationships among genres, Fig. 5a shows the graph $\mathcal{G}$ defining the relationships among genres as it is put forward in Sect. 3.1. For its generation we set Θ_t to 12 to remove the irrelevant edges. Despite the fact that the density of the graph is very low, 0.0085, we can observe a dense cloud of genres in its center. In that sense, Fig. 5b provides more details of this cloud by showing the most frequent

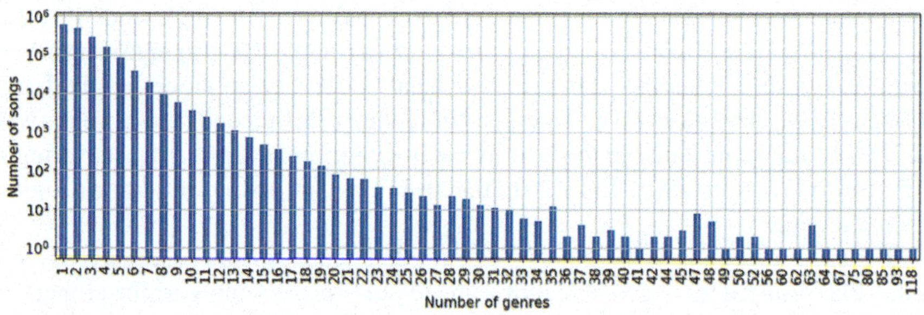

Fig. 3 Distribution of the songs in the dataset based on their associated number of genres

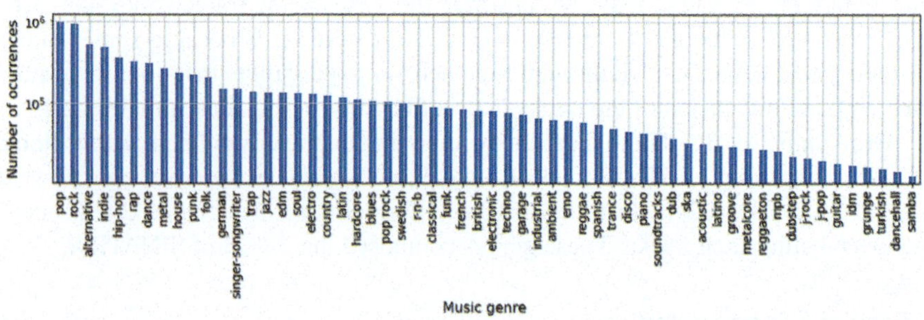

Fig. 4 Number of songs associated to each music genre. For the sake of clarity, we have limited this plot to genres with a frequency above 10,000 songs

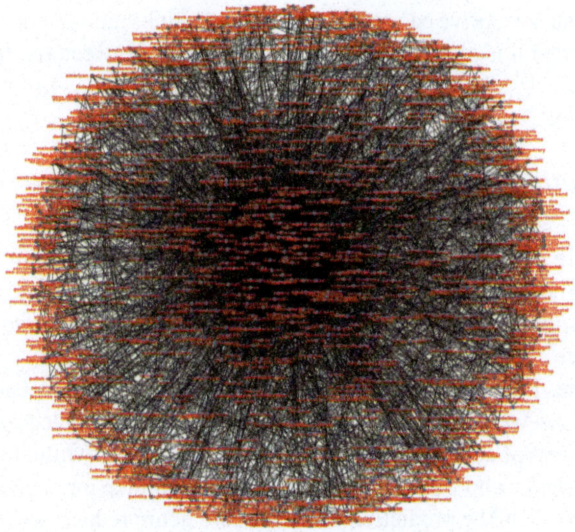

(a) Global graph $\mathcal{G}$.

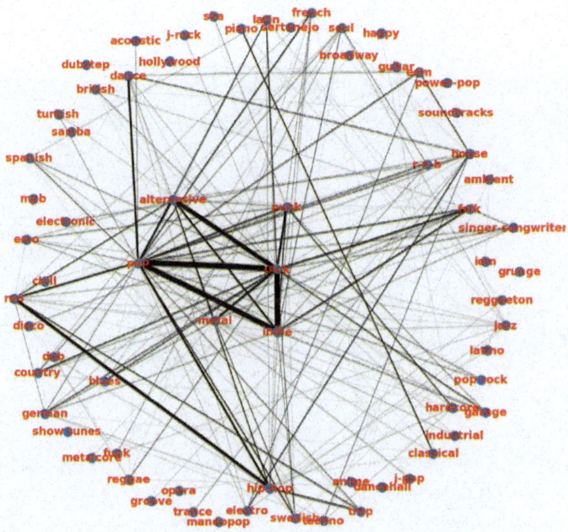

(b) Sub-graph only comprising the most frequent genres.

Fig. 5 Analysis of the co-ocurrences of genres. The nodes are the subset of most frequent music genres and the wide of each edge is proportional to the frequency of each pair of genre genres of the dataset. As we can see, there are strong links among pop, indie, rock and alternative genres. It is also remarkable the strong link between the rap and hip-hop genres.

4.3 Generation of the Core Genres

Given the genre graph described above, we extracted 147 different core genres as Fig. 6 shows where each genre is coloured according to its core genre. As we can see, the genres comprising the dense network in the center of the graph belong to the same core genre.

In that sense, Fig. 7 shows the number of genres included in each top-30 core genre by size. As we can see, the distribution is quite imbalanced as cg_0 is much larger than the rest of core genres by comprising 311 genres while, for example, cg_2 comprises 60 genres and cg_3, 53. However, we should take into account that, in the worst-cased scenario, the prediction of cg_1 as core genre reduces the dimensionality of the output space of the classifier from 2,166 genres to 311. This would be helpful in the generation of RF instances during the classification pipeline.

For the sake of completeness, Fig. 8 shows the latent graph of the first 4 core genres. As we can see, g_1 comprises most of the *mainstream* genres like pop, rock

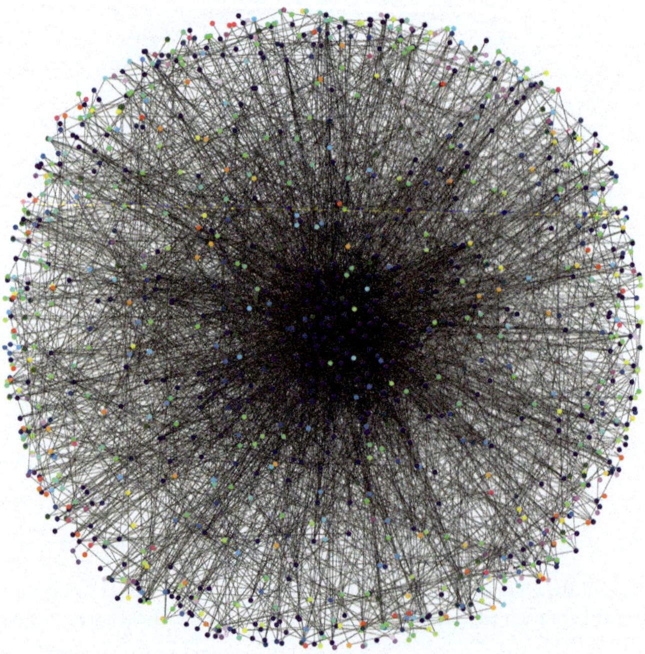

Fig. 6 Genre graph $\mathcal{G}$ where each node is coloured according to its own core genre

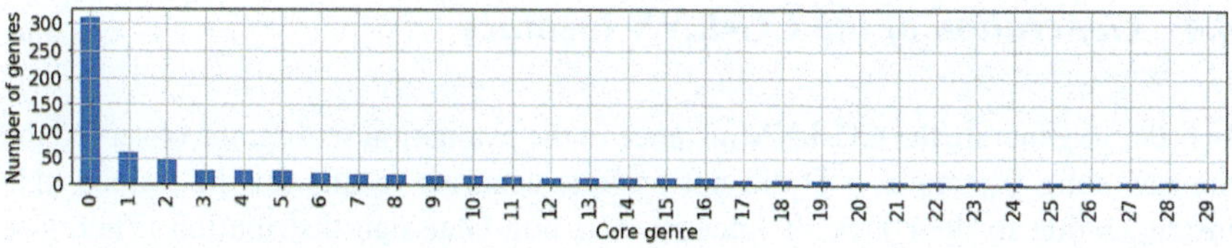

Fig. 7 Number of genres associated to the first 30 core genres ordered by size

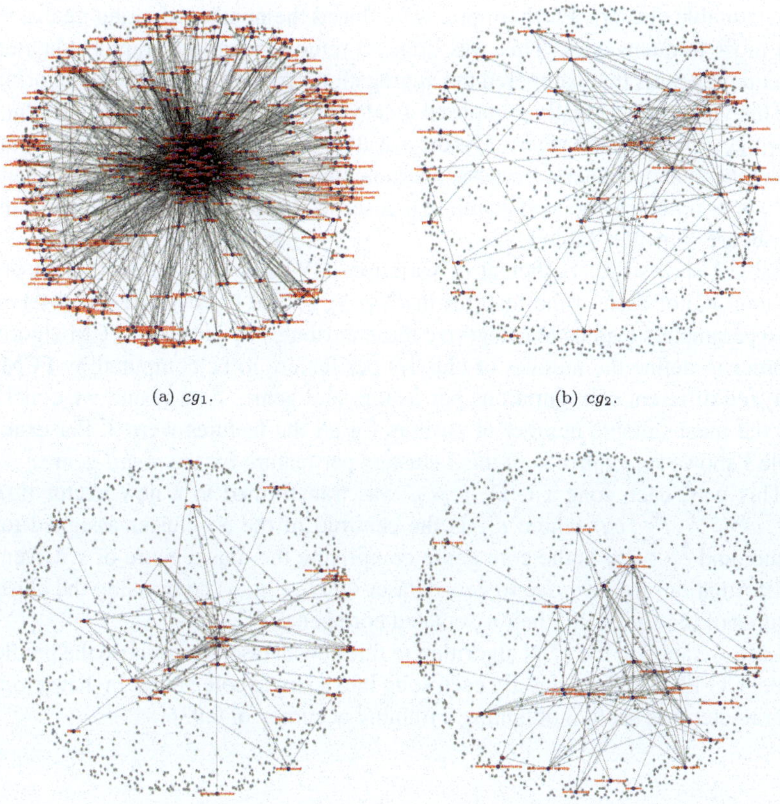

(a) cg_1. (b) cg_2.

(c) cg_3. (d) cg_4.

Fig. 8 Representation of the 4 largest core genres identified in graph $\mathcal{G}$

or rap showing a quite dense community. Concerning g_2 it mainly comprises very specific genres like neo-kraut, modern swing or motown. In the case of g_3, its most representative genre is electronica, and this causes this core genre to be biased towards such a type of music by including other genres such as trip-hop, microhouse or big-beat. Finally, c_4 is mainly related to religious music with genres like worship, christian-music or messianic-praise.

4.4 Generation of the CG-KNN Instance

In order to generate the CG-kNN instance in the evaluation setting, we observed that the raw song vectors $s_i \in \mathcal{D}$ did not provide accurate results for the trained kNN instances due to their lack of linearity and non-Gaussian distribution. Therefore, we followed the approach described in [19]. As a result, we eventually kept with the energy, loudness, speechiness, acousticness, liveness, instrumentalness, valence, danceability and tempo features to carry out the classification as they gave rise to more suitable clusters. Furthermore, we reduced the number of target genres to the 400 most frequent ones. This was done to remove from the training dataset the infrequent genres that hampered the aggregation capability of the kNN algorithm.

This way, we eventually composed a dataset $\mathcal{D}_{CG-kNN} = \{\langle f_1, f_2, \ldots, f_{10}, g\rangle\}$, $|\mathcal{D}_{CG-kNN}| = 486,256$ where f_j is the j-th attribute in the list {key, energy, loudness, speechiness, acousticness, liveness, instrumentalness, valence, danceability, tempo} and g is its unique associated genre. In that sense, the 20 most frequent genres in this dataset are shown in Table 1.

Table 2 shows the coefficient of variation (CV) and quartile coefficient of dispersion(QC) of each of the features in $\mathcal{D}_{CG-kNN}$. The high variability observed in these parameters lead us to categorize these attributes by means of FCM algorithm. In order to define the number of clusters per feature to be computed by FCM, we analyzed different configurations per feature and genre. As a result, we concluded that the most suitable number of clusters for all the features were 3. For example, Table 3 shows the centroids of the 3 clusters per feature for the *metal* genre.

This way, each song $s \in \mathcal{D}_{CG-kNN}$ was transformed to a new vector $\Phi(s) = \langle c_1^{\alpha_1}, c_2^{\alpha_2}, \ldots, c_j^{\alpha_j}, cg\rangle$ where $c_j^{\alpha_j}$ is the centroid of the α_j cluster assigned to j-th feature and $cg \in \mathcal{CG}$ is the core genre comprising the music genre of s. Given this transformation, we give rise to new dataset $\Phi(\mathcal{D}_{CG-kNN})$ comprising the clustered features of the songs and their associated core genre.

Last, we applied the kNN algorithm to this new dataset to compose the predictive model CG-kNN able to classify each song into a single core genre. In that sense, the k parameter was set to 3 obtaining a training accuracy of 0.936.

Table 1 Most frequent genres in the dataset $\mathcal{D}_{CG-kNN}$

Genre	Frequency	Genre	Frequency	Genre	Frequency	Genre	Frequency
Metal	63974	Ambient	14418	Folk	9068	Electronic	5229
Indie	43377	Punk	11099	Metalcore	8928	k-pop	4773
Pop	34685	Hardcore	10091	House	8815	Electro	4770
Rock	31872	Hip-hop	10003	Funk	5907	Alternative	4331
Jazz	16157	Soundtracks	9252	Trance	5787	Rap	3367

Table 2 Parameters computed for the tracks' features in $\mathcal{D}_{CG-kNN}$

	Mean	SD	CV	QC[a]
Key	5.28838183	3.57625009	0.6762466	0.6363636
Energy	0.66049992	0.25506911	0.3861758	0.2860278
Loudness	−8.40717255	4.91206237	0.5842704	0.3232573
Speechiness	0.08719028	0.09907875	1.1363508	0.4308176
Acousticness	0.25336094	0.32227873	1.2720143	0.9864500
Liveness	0.20701897	0.17675012	0.8537870	0.4668471
Instrumentalness	0.26564973	0.36615754	1.3783471	0.9999853
Valence	0.41206921	0.25560388	0.6202936	0.5131086
Danceability	0.51641888	0.18787191	0.3637975	0.2591171

[a] The quartile coefficient of dispersion (QC) is a descriptive statistic which measures dispersion and is used to make comparisons within and between data sets. Since it is based on quantile information, it is less sensitive to outliers than measures such as the coefficient of variation

Table 3 Feature centroids of the 3 FCM clusters for the *metal* genre

Feature	Cluster 1	Cluster 2	Cluster 3
Key	1.263	5.901	9.728
Energy	0.725	0.874	0.970
Loudness	−9.337	−5.995	−3.703
Speechiness	0.057	0.122	0.222
Acousticness	0.003	0.230	0.698
Liveness	0.117	0.330	0.707
Instrumentalness	0.031	0.468	0.846
Valence	0.091	0.290	0.546
Danceability	0.210	0.366	0.514
Tempo	104.704	123.143	142.818

4.5 Evaluation Metric

In order to evaluate our framework, we needed to take into account that we were not dealing with fixed number of tags for each song, as we have seen in Fig. 4. Hence, we defined the accuracy of classification provided by SINATRA for a song s, acc_s, given its true set of genres $\mathcal{M}_s$ as follows,

$$acc_s = \frac{|\hat{\mathcal{M}}_s \cap \mathcal{M}_s|}{|\mathcal{M}_s|}$$

where we basically compute the overlap level between the classification of SINATRA ($\hat{\mathcal{M}}_s$) and the ground truth. This way, the global accuracy (ACC) of SINATRA given an evaluation dataset $\mathcal{D}_{eval}$ is computed as,

Table 4 Key parameters of the performed evaluation

Element	Parameter	Value
Strategy	Train-test split	0.90–0.10
	Evaluation approach	Cross-validation
	Num. of splits	10
	Repetitions	5
RF	Num. estimators	100
	Min. num. samples to split	2

$$ACC = \frac{\sum_{s \in \mathcal{D}_{eval}} acc_s}{|\mathcal{D}_{eval}|}$$

4.6 Evaluation Parameters

Table 4 shows the key parameters of the configuration applied to evaluate SINATRA. Given the cross-validation approach, in each iteration we used as training set 1,219,439 records to evaluate 135,493 songs.

4.7 Result Discussion

Table 5 shows the ACC score obtained by SINATRA when it came to predict songs comprising between 1 and 9 genres. For comparison purposes, we have included the results of an alternative version of the framework where the probabilities of the graph only relies on the RF outcomes (we removed lines 12–17 of Algorithm 1 so that $\mathcal{P}^i = \mathcal{P}^i_{RF}$). Besides, the results of a RF for multi-label classification has been also used ($RF_{multilabel}$). For this model, we followed the *label power set* policy, a well-known approach for multi-label classification [26]. Basically, this policy transforms a multi-label problem to a multi-class problem with 1 multi-class classifier trained on all unique label combinations found in the training data.

From this table we can see that SINATRA achieved an ACC score above 0.46 in all the configurations with a mean score of 0.5064. Moreover, these results also show the clear positive impact of considering the correlations among genres. For example, the ACC score of SINATRA was 0.58 in terms of predicting songs with a single genre and the alternative version only achieved and ACC score of 0.26. At the same time, the accuracy of SINATRA to classify songs with 2 genres was 0.45 and the no-graph alternative achieve an ACC of 0.34. It is true that the difference between the full version of SINATRA and the no-graph one becomes smaller as the number of genres to predict becomes larger. This is because as we incrase the genres to predict,

Table 5 ACC score based on the number of genres of the target song. The best score per configuration is shown in bold

Number of genres	SINATRA	SINATRA (no-graph)	$RF_{multilabel}$
1	**0.5818 (± 0.3858)**	0.2665 (± 0.4421)	0.1923 (± 0.2498)
2	**0.4901 (± 0.3618)**	0.3428 (± 0.3745)	0.1312 (± 0.2193)
3	**0.4950 (± 0.3035)**	0.4199 (± 0.3433)	0.2066 (± 0.2298)
4	**0.4916 (± 0.2553)**	0.4344 (± 0.3062)	0.2550 (± 0.2524)
5	**0.4867 (± 0.2886)**	0.4786 (± 0.3275)	0.2760 (± 0.2429)
6	0.4722 (± 0.2456)	**0.5542 (± 0.3022)**	0.2833 (± 0.2362)
7	0.4464 (± 0.2194)	**0.5714 (± 0.2359)**	0.4228 (± 0.2927)
8	**0.6500 (± 0.3297)**	0.5866 (± 0.2278)	0.3804 (± 0.2344)
9	0.4444 (± 0.1543)	**0.5563 (± 0.1855)**	0.3751 (± 0.1396)
Mean	**0.5064**	0.4678	0.2589

the classification algorithm must execute more loops and refine more and more the training dataset, so the RF instances must deal with a smaller dimensionality space and, thus, they become more accurate reducing the positive effect of the graph.

Furthermore, Table 6 shows the ACC score based on the *classification deep* of SINATRA, that is, the number of iterations of the classification loop. For example, the ACC of the framework when it predicted the second genre of a song, that is, in the second iteration of the classification loop (lines 7–20 of Algorithm 1) was 0.4490. In this case, we can see that the accuracy of the proposal steadily decreases as we went deep in the classification.For example, the ACC to reduce from 0.5256 in the first iteration (level) was 0.2187 in the sixth one. This decrement is due to the fact that the classification error of an iteration *propagates* to the new ones. In comparison terms, we can see that the full version of SINATRA achieved a much higher ACC score in the first two iterations with respect the no-graph alternative. Nevertheless, this alternative version achieved slightly higher ACC values for almost the rest of levels. Again, this due to the fact that propagation error from previous layers hampers the classification of a layer as the *filtered* graph $\mathcal{G}_{\hat{g}}$ used to compose the probability distribution of genres $\mathcal{P}^i_{\mathcal{G}}$ (lines 12–16 of Algorithm 1) focuses on a wrong part of the global graph $\mathcal{G}$. Nonetheless, the average ACC of the full version of the framework (0.2902) was slightly higher than the one operating without genre graphs (0.2860).

In order to analyze the ACC results per genre, Fig. 9 shows this score based on the frequency of each genre. As we can see, the genres with the highest ACC (pop, rock, indie, metal and rap) were also some of the most frequent ones. Actually, the Pearson Correlation between the ACC of a genre and its frequency was 0.4252 showing a slight positive correlation. Regarding the more unusual genres, we can see that SINATRA also achieved quite high scores in some of them such as, samba (0.62), dum (0.60) or electro (0.76). This type of genres are usually secondary ones used to label a song with more than one genre.

Table 6 ACC score based on the classification level of the proposal

Classification level	SINATRA	SINATRA (no-graph)
1	**0.5256 (± 0.3858)**	0.2976 (± 0.4572)
2	**0.4490 (± 0.4973)**	0.4244 (± 0.4942)
3	0.3588 (± 0.4796)	**0.4214 (± 0.4937)**
4	0.2682 (± 0.4430)	**0.3982 (± 0.4895)**
5	0.2385 (± 0.4262)	**0.3366 (± 0.4752)**
6	0.2187 (± 0.4133)	**0.2450 (± 0.4301)**
7	**0.2962 (± 0.4566)**	0.1666 (± 0.3726)
8	0.1428 (± 0.3499)	**0.1739 (± 0.3790)**
9	**0.1142 (± 0.3219)**	0.1111 (± 0.3142)
Mean	**0.2902**	0.2860

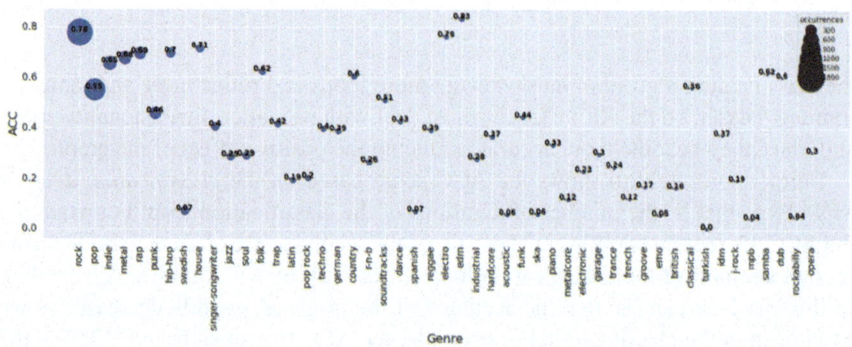

Fig. 9 ACC score per genre. Each bubble is labelled with the ACC score of the genre and its size is positively correlated with the number of occurrences of the genre in the evaluation dataset

Figure 10 shows the average execution time required by SINATRA to classify a song based on its number of genres. These times have been obtained by executing SINATRA in a server with 187GB, 64 CPUs Intel Xeon Gold 6226R CPU @ 2.90GHZ and Linux Ubuntu 20.04.4 LTS as operating system. As we can see, this time did not increased exponentially wit respect the number of genres to predict, increasing from an average execution time of 3.76 milliseconds (ms) to predict songs with a single genre to 6.82 ms for songs with 9 genres.

Last, Fig. 11 shows the average importance of the input features computed by the different instances of RF. This relevance is computed as the mean and standard deviation of accumulation of the impurity decrease within each tree for each feature. As we can see, the year, acousticness, loudness and danceability were the three most meaningful features to classify the songs. On the contrarly, the mode and key features were much less relevant.

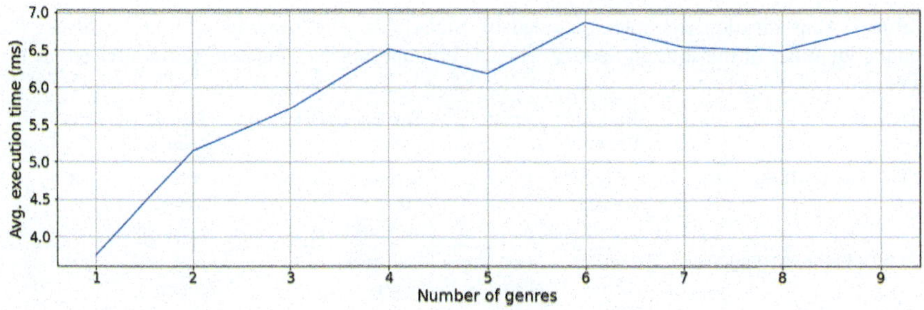

Fig. 10 Average execution time required by SINATRA to classify a song based on its total number of genres

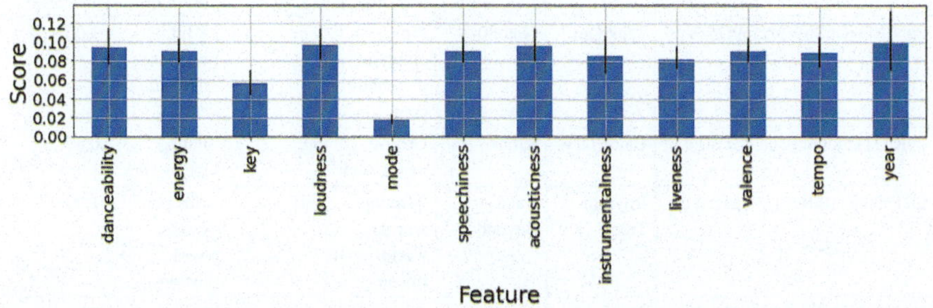

Fig. 11 Average importance scores of the featuers according to the RF instances created during ehe evaluation

4.8 Classification Examples

For the sake of completeness, Table 7 shows some classification examples provide by the framework. As we can see, the mechanism was able to correctly predict songs covering a different range of genres from *indie rock* ones such as *Love for Granted* of the popular French band Phoenix to video-game based ones like *Tänk Om* of *Goto80* an independent electronic artist.

Furthermore, the system was also able to predict *local* genres such as *german* for the song *Eternity: 666 Weeks Beyond Eternity* of Freedom Call or *afropop* in the case of the song *Rhythm Tree* . Regarding the number of genres, we can see examples of songs comprising a limited number of genres like *Black Days* with 3 genres to others with 5 or 6 genres like *Welcome To the World* or *19 Million Ac.*

5 Conclusion and Future Work

Music genre recommendation is gaining attention from both the academia and the music industry. From a research viewpoint, understanding how to effectively recommend music genres could provide insights into how people consume and interact

Table 7 Classification examples provided by SINATRA. The *Track id.* columns indicates the unique identifier of the song in Spotify. The CG column indicates the *core genre* inferred for the song

Track id.	Track Name	Artist	True Labels	CG	Pred. Genres	acc_s
3Yrk1Ytp3Vg1IRshQud90x	Love for Granted	Phoenix	Alternative, dance, indie, rock	cg_1	Alternative, dance, rock, indie	1.00
2YWDM0rElh3YIqFtjrhTvd	Can I Forgive Him	Paul Simon	Folk, rock, singer-songwriter	cg_1	Folk, rock, singer-songwriter	1.00
0puFmcIvB53ZUtvVP78o0e	Welcome To The World	T.I., Kanye West, Kid Cudi	Dance, hip-hop, hip-hop, pop, rap	cg_1	Pop, rap, electro, house, hip-hop	0.60
2qNiXQgChS6W5A5p9G5zVp	Tänk Om	Goto80	8-bit, chiptune, nintendo-core	cg_1	8-bit, chiptune, nintendo-core	1.00
2yRtQTkhvwnP836V42zWTd	Black Days	Klone	Groove, metal, rock	cg_1	Punk, metal, rock	0.60
0ETYvknTR8Reb1MCp4BWct	Rhythm Tree	Baka Beyond	Afropop, world, world fusion	cg_1	Afropop, world, world fusion	1.00
0oeOUA6H4BmR11kIiNmeNd	Balance (Joe Mason Remix)	The Him, Oktavian, Joe Mason	Edm, electro, house, pop	cg_1	Electro, alternative, edm, house	0.75
7eEkIGFuI4XTzFIamCBOhm	This Is Who We Are	Run Kid Run	Punk, rock	cg_1	Punk, rock	1.0
7mEk5bQPqfR22BULHoaD1D	19 Million Ac	The Spits	Garage, hardcore, indie, pop, punk, rock	cg_1	Alternative, country, indie, garage, punk, rock	0.66
0NwbMiopi2MAEJarvmSYN6	Eternity: 666 Weeks Beyond Eternity	Freedom Call	german, metal	cg_1	German, metal	1.00

with music, with implications in areas such as music psychology and music therapy. Regarding the music industry, genre recommendation systems may help music streaming platforms improve their user experience. This increases the likelihood of users returning to the platform, potentially leading to increased revenue by securing subscriptions from these users.

In this context, the automatic classification of music genres of a song catalogue is a paramount task. To this end, in this work we have tackled the music genre classification problem by developing SINATRA, a framework that combines Random Forests (RF) and knowledge graphs to achieve a multi-label music genre classification system. As input data, SINATRA relies on the songs' metadata, which have been previously proved as a compelling alternative to the use of more complex techniques such as the analysis of raw audio signals. In particular, 10 metadata features have been included in SINATRA grouped into three groups, namely song's context (e.g., liveness), mood (e.g., danceability), and audio features (e.g., instrumentalness or speechiness).

The classification process in SINATRA follows two stages. First, a training phase is performed to generate an undirected graph of related music genres based on their co-occurrences, thus creating a set of *core genres*. These core genres are passed to a kNN instance to predict the core genres that are likely to be associated to each particular song. After the initial classification with kNN, SINATRA applies multiple instances of RF, with each instance tagging the songs with different sub-genres of the core genres. Besides, the proposed framework leverages the genre graph and the output of the RF models to refine the classification results. The graph is used to take into account the correlations among sub-genres, allowing SINATRA to produce a set of complementary genres that accurately describe the input song. Finally, this process is repeated recursively until the desired number of sub-genres is detected.

SINATRA has been evaluated by a dataset with almost 1,350,000 songs' metadata extracted from Spotify and Last-FM, including 2,166 different music genres. For this dataset, SINATRA has been asked to predict up to 9 relevant genres of each song. Moreover, in order to compare the accuracy of the proposal, two different configurations of SINATRA has been tested, namely the full version and a second version without the use of the knowledge graph in the second stage. The results shows that the full version outperforms the no-graph version for all the experiments predicting between 1 and 9 genres, with an average ACC score of 0.5064. These results shows that SINATRA is a promising framework to classify songs with respect to other works that only consider a rather reduced initial set of genres (usually up to 10 versus the 2,166 genres included in SINATRA) and only produce a single-label classification.

Future work will focus on further refining the graph-based inference step in SINATRA. This could involve exploring different graph algorithms and metrics for identifying genre communities and relationships, as well as considering the incorporation of additional information sources, such as artist or album data, to enrich the graph structure. Another potential direction for future work is the extension of SINATRA to other music-related tasks beyond genre classification. For example, the framework could be used to identify potential collaboration among artists or to recommend music for different activities. Finally, the integration of the SINATRA as part of a portal or service for music recommendation is also foreseen.

Acknowledgements Financial support for this research has been provided under grant PID2020-112827GB-I00 funded by MCIN/AEI/10.13039/501100011033. It is also partially granted by the "EMERGIA" programme, funded by the Junta de Andalucía through the grant EMC21_004171.

References

1. Franklin, J.C.: Ancient Greek Music and the Near East. A Companion to Ancient Greek and Roman Music, 229–241 (2020)
2. Terroso-Saenz, F., Soto, J., Muñoz, A.: Evolution of global music trends: An exploratory and predictive approach based on spotify data. Entertainment Computing 44, 100536 (2023)
3. Elbir, A., Aydin, N.: Music genre classification and music recommendation by using deep learning. Electronics Letters 56(12), 627–629 (2020)
4. Gunawan, A.A., Suhartono, D., et al.: Music recommender system based on genre using convolutional recurrent neural networks. Procedia Computer Science 157, 99–109 (2019)
5. Singh, J., Bohat, V.K.: Neural network model for recommending music based on music genres. In: 2021 International Conference on Computer Communication and Informatics (ICCCI), pp. 1–6 (2021). IEEE
6. Mehta, J., Gandhi, D., Thakur, G., Kanani, P.: Music genre classification using transfer learning on log-based MEL spectrogram. In: 2021 5th International Conference on Computing Methodologies and Communication (ICCMC), pp. 1101–1107 (2021). IEEE
7. Cai, X., Zhang, H.: Music genre classification based on auditory image, spectral and acoustic features. Multimedia Systems 28(3), 779–791 (2022)
8. Jiang, Y., Jin, X.: Using k-means clustering to classify protest songs based on conceptual and descriptive audio features. In: Culture and Computing: 10th International Conference, C&C 2022, Held as Part of the 24th HCI International Conference, HCII 2022, Virtual Event, June 26–July 1, 2022, Proceedings, pp. 291–304 (2022). Springer
9. Popli, C., Pai, A., Thoday, V., Tiwari, M.: Electronic Dance Music Sub-genre Classification Using Machine Learning. In: Artificial Intelligence and Sustainable Computing: Proceedings of ICSISCET 2021, pp. 321–331. Springer (2022)
10. Schedl, M., Brandl, S., Lesota, O., Parada-Cabaleiro, E., Penz, D., Rekabsaz, N.: LFM-2b: A Dataset of Enriched Music Listening Events for Recommender Systems Research and Fairness Analysis. In: Proceedings of the 2022 Conference on Human Information Interaction and Retrieval. CHIIR '22, pp. 337–341. Association for Computing Machinery, New York, NY, USA (2022). DOI https://doi.org/10.1145/3498366.3505791
11. Jena, K.K., Bhoi, S.K., Mohapatra, S., Bakshi, S.: A hybrid deep learning approach for classification of music genres using wavelet and spectrogram analysis. Neural Computing and Applications, 1–26 (2023)
12. Prabhakar, S.K., Lee, S.-W.: Holistic Approaches to Music Genre Classification using Efficient Transfer and Deep Learning Techniques. Expert Systems with Applications 211, 118636 (2023) DOI https://doi.org/10.1016/j.eswa.2022.118636
13. Ignatius Moses Setiadi, D.R., Satriya Rahardwika, D., Rachmawanto, E.H., Atika Sari, C., Irawan, C., Kusumaningrum, D.P., Nuri, Trusthi, S.L.: Comparison of SVM, KNN, and NB Classifier for Genre Music Classification based on Metadata. In: 2020 International Seminar on Application for Technology of Information and Communication (iSemantic), pp. 12–16 (2020). DOI https://doi.org/10.1109/iSemantic50169.2020.9234199
14. Singhal, R., Srivatsan, S., Panda, P.: Classification of Music Genres using Feature Selection and Hyperparameter Tuning. Journal of Artificial Intelligence and Capsule Networks 4(3), 167–178 (2022)
15. Raghavan, U.N., Albert, R., Kumara, S.: Near linear time algorithm to detect community structures in large-scale networks. Physical review E 76(3), 036106 (2007)
16. Cover, T., Hart, P.: Nearest neighbor pattern classification. IEEE transactions on information theory 13(1), 21–27 (1967)
17. Abdulameer, A.S., Tiun, S., Sani, N.S., Ayob, M., Taha, A.Y.: Enhanced clustering models with wiki-based k-nearest neighbors-based representation for web search result clustering. Journal of King Saud University - Computer and Information Sciences 34(3), 840–850 (2022) DOI https://doi.org/10.1016/j.jksuci.2020.02.003
18. Cheng, D., Huang, J., Zhang, S., Wu, Q.: A robust method based on locality sensitive hashing for K-nearest neighbors searching. Wireless Networks, 1–14 (2022)

19. Elshenawy, L.M., Chakour, C., Mahmoud, T.A.: Fault detection and diagnosis strategy based on k-nearest neighbors and fuzzy C-means clustering algorithm for industrial processes. Journal of the Franklin Institute 359(13), 7115–7139 (2022) DOI https://doi.org/10.1016/j.jfranklin.2022.06.022
20. Tsai, C.-F., Eberle, W., Chu, C.-Y.: Genetic algorithms in feature and instance selection. Knowledge-Based Systems 39, 240–247 (2013) DOI https://doi.org/10.1016/j.knosys.2012.11.005
21. Uddin, S., Haque, I., Lu, H., Moni, M.A., Gide, E.: Comparative performance analysis of K-nearest neighbour (KNN) algorithm and its different variants for disease prediction. Scientific Reports 12(1), 1–11 (2022)
22. Bezdek, J.C., Ehrlich, R., Full, W.: Fcm: The fuzzy c-means clustering algorithm. Computers & geosciences 10(2-3), 191–203 (1984)
23. Genuer, R., Poggi, J.-M., Tuleau-Malot, C., Villa-Vialaneix, N.: Random forests for big data. Big Data Research 9, 28–46 (2017) DOI https://doi.org/10.1016/j.bdr.2017.07.003
24. Nie, K.: Inaccurate Prediction or Genre Evolution? Rethinking Genre Classification. In: Ismir 2022 Hybrid Conference (2022)
25. Zhang, R., Zhou, X., Song, J.: Music and musician influence, similarity measure, and music genre division based on social network analysis. In: 2nd International Conference on Artificial Intelligence, Automation, and High-Performance Computing (AIAHPC 2022), vol. 12348, pp. 95–104 (2022). SPIE
26. Read, J., Puurula, A., Bifet, A.: Multi-label classification with meta-labels. In: 2014 IEEE International Conference on Data Mining, pp. 941–946 (2014). IEEE

Towards an Enhanced and Lightweight Face Authentication System

Ying Zhang, Roger Zimmermann, Zhiwen Yu, and Bin Guo

Abstract Face recognition is one of the most well-adopted ways to verify someone's identity, which however might be spoofed by presenting for example a fake image or video. Therefore, it is essential to include additional face liveness detection for a safer application. Among the existing solutions, it is common to plug in a separate model for face liveness detection. Therefore, the current authentication platform will provide two models in order to provide a safe authentication. However, in many practical scenarios, the platform (e.g., IoT devices) has limited resources in terms of computation power and storage, and this may prevent the two-models from being deployed successfully. Observed that both recognition and liveness detection work on the same face image, we believe it is possible to integrate two functions into a unified model, which will reduce the computational workload and storage requirements. To achieve this, we explore two works with different model designs, research focuses, and potential solutions. In the first work, we try to enhance a usual face recognition model with additional task capability without any additional storage cost. Concretely, we first analyze the two task's relationship, and by a mathematics formulation, we insert the observed dual-task relationship to a novel deep model with distance-ranking feature. The training of the model focuses on the feature-learning and it does not directly use the task ground truth labels, which makes the model has a good generalization capability on new data. We have conducted experiments on a benchmark dataset and the results show that our average performance has a minimal 15% improvement compared to the baselines. In the second work, we adopt the classic multi-task learning model to combine the two tasks. Rather than using a deep multi-task model, we compress the original deep model to a lightweight version.

Y. Zhang (✉) · Z. Yu · B. Guo
Northwestern Polytechnical University, Xian, China
e-mail: izhangying@nwpu.edu.cn

Z. Yu
e-mail: zhiwenyu@nwpu.edu.cn

B. Guo
e-mail: guob@nwpu.edu.cn

R. Zimmermann
National University of Singapore, Singapore, Singapore
e-mail: dcsrz@nus.edu.sg

K. Daimi et al. (eds.), *Cutting Edge Applications of Computational Intelligence Tools and Techniques*, Studies in Computational Intelligence 1118,
https://doi.org/10.1007/978-3-031-44127-1_10

Additionally, in order to compensate the performance degradation due to compression, a multi-teacher assisted knowledge distillation is applied where a good balance between accuracy and model size is achieved.

Keywords Face recognition · Face liveness detection · Biometric authentication · Lightweight modeling · Forensics

1 Introduction

Biometric authentication is a trending technique in computer science for access control or user identification thanks to the distinctive feature of these biometric identifiers. Popular biometrics include, but not limited to face, fingerprint, iris and voice. Among these biometrics, face might be the most popular one due to its good performance [8], contact-free and user-friendy process [24], and less dependence on the hardware resources [17]. The global facial recognition market size is estimated to reach USD 8.5 billion in the next five years.

However, majority of the current face recognition systems might be cheated by the face spoofing attacks [11] by presenting a fake copy of the photo, a video, or a even 3D mask of a targeted person. The creation of such spoofing is not hard as the face information could be easily found on many social platforms. In many critial applications, it is necessary and important to provide additional safeguard to the face verification system and face liveness detection is one of the most popular techniques for this purpose. Specifically, face liveness detection will determine if a face image is live or fake. In order to achieve this detection, there are usually two categories of solutions. The first category will be the hardware-based methods where additional hardware is required. For example, the recent iPhone and Microsoft Surface start to adopt three-dimensional camera to do the detection if the fake image comes from e.g. an image. Although the performance is good, the high cost make these hardware-based solutions not widely-deployed [20]. The second category is the software-based solutions where a liveness detection algorithms is carefully designed to distinguish the fake face copies from the live ones and these solutions are more compatible with ordinary 2D camera platforms(e.g., [1, 12, 15, 22]) (Fig. 1).

For the above software-based solutions, they are usually implemented as a separate model from the current face recognition module. Figure 2a shows the model settings. Under such a setting, the liveness detection model and the face recognition model will be trained independently from each other. Under many practical scenarios such as in the IoT platforms, the computation resource, energy and storage space are usually very limited, all of which will prevent the two-model settings from a successful deployment. The load can increase further if some deep models with millions of parameters are used.

Therefore, it is necessary to explore a model that can achieve both face recognition and face liveness detection without requiring too many extra resources, i.e., to obtain **a lightweight and enhanced face authentication system**. To this end, this

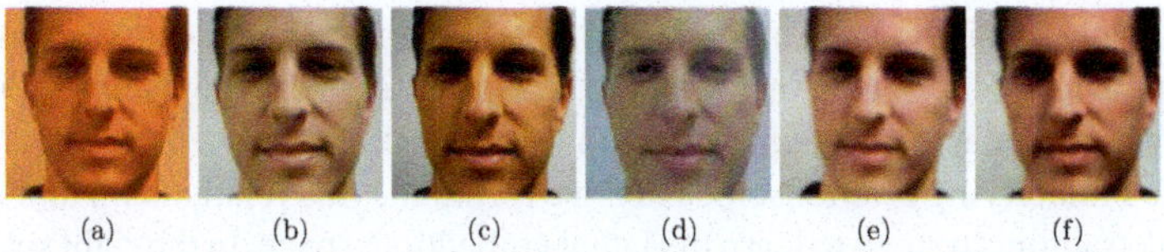

Fig. 1 Some example from the REPLAY [5] dataset of the same person. The first column (**a**) is the live face. The second column (**b**) is a fake face captured from a high-definition photo. The column (**c**) is a fake face from high-definition video frame. The column (**d**) is a fake face from a print photo. The column (**e**) is a fake face from a mobile phone's photo. The last column (**f**) is a fake face from a mobile video

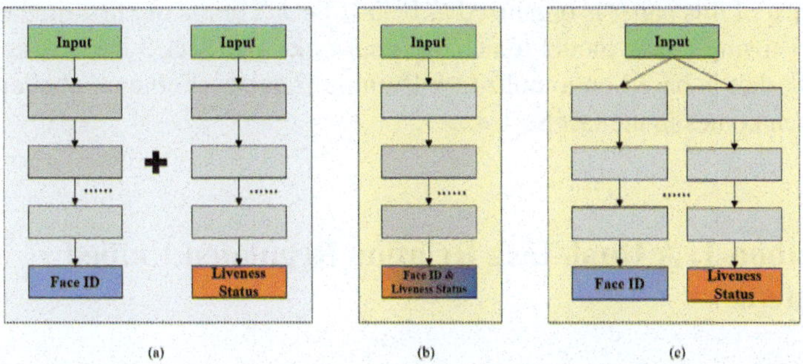

Fig. 2 The model difference between **a** the conventional work and **b–c** our two potential lightweight and unified solutions where (**b**) is based on a single neural network while (**c**) is based on the multi-task learning framework

work introduces our recent two works [4, 25] with different model designs, research focuses and solutions. In the first work, we do not modify the usual face recognition model but enable it to distinguish the liveness status. Concretely, we first analyze the two task's relationship, and by a mathematics formulation, we insert the observed dual-task relationship to a novel deep model with a distance-ranking feature. The training of the model focuses on the feature-learning and it does not directly use the task ground truth labels, which makes the model has a good generalization capability on new data. In the second work, we adopt the classic multi-task learning model to combine the two face-related tasks together. However, rather than directly using a deep multi-task model, we compress the model to a lightweight version. Considering the performance degradation due to such compression, a multi-teacher assisted knowledge distillation is further applied where a good balance between accuracy and model storage is achieved.

Here we clarify the model differences among the traditional solution and our two works in Fig. 2. The left sub-figure illustrates the traditional method where two models exist for face recognition and liveness detection, respectively. The input to both models is the same face image and the output is either the face ID or the face liveness status. Here we want to remind the audience that the design and training of these two models are totally separately. The middle sub-figure shows our first attempt, where a single and fully shared model is used to complete two tasks at the same time. The core of this method is to learn a single embedding that could well distinguishes the faces that appear in both tasks. To this end, we first investigate the two task's underlying relationship. Then we formulate this relationship and design a dual-task training strategy to learn the latent embedding space. In the right sub-figure, our second work integrates the two tasks by a multi-task learning model, where a task-shared segment and two task-specific branches exist. For the two branches, one branch is designed for face recognition and the other is for face liveness detection. We will introduce the details of each of these two works in the following parts of this paper.

The rest of this paper is organized as below: Sect. 2 briefs our first method which leverages a single-task model to achieve dual-tasks and Sect. 3 introduces the second work that is based on a multi-task learning framework. Future challenges and conclusion comes in the last Sect. 4.

2 Method 1: A Dual-Task Relation Regulated Unified System

2.1 Background

We firstly brief the overall workflow of a face authentication system. An authentication system usually has two stages, a registration stage and an authentication stage. The registration, as shown in Fig. 3, will be triggered when a client uses the system for the first time. Usually, the system will take a photo of the client and this photo is stored in a database for future identify pairing. When the registered client reuses the system, the face authentication stage will start by comparing the current input face to the stored copy. Our first solution in this section will combine the two tasks in just one go.

2.2 Formulation of the Relationship Between Two-Tasks

To achieve the above goal, we first introduce a few definitions. In the face authentication stage, a new face image will be imported into the system, which is expected to be recognized as a previously registered person. We name this to-be-recognized person as the *referred person* and the corresponding face as the *referred face*. Thus, a candidate's face image could fall into one of the following face-type categories:

Definition 1: Positive face is a face image that is captured alive from its referred person.

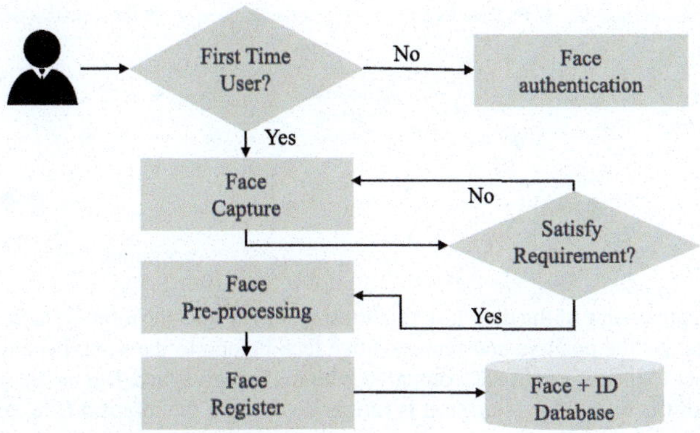

Fig. 3 The overview of the face/identity registration stage

Definition 2: Semi-positive face is a face image from its referred person but from a fake resource.

Definition 3: Negative face is a face image from another person other than the referred one and it is from either a live or a fake resource.

In the rest of this paper, without explicitly indicating, we will use the term *face* to represent a face image for simplicity. Mathematically, the referred face is denoted as a, the positive face is denoted as p, semi-positive face is denoted by s and negative face is denoted by n. By analyzing the data, the objective of the work is to learn a latent feature space which can reflect the following distance-ranking relationship:

$$D(g(a), g(p)) < D(g(a), g(s)) < D(g(a), g(n)) \tag{1}$$

Here $g(x)$ is the feature embedding of the face image x. $D(x, y)$ is the distance between x and y. and this work uses the L2 norm function.

Figure 4 visualized the above relationship and the intuition can be better explained with the following division:

$$\begin{cases} D(g(a), g(p)) < D(g(a), g(s)) \\ D(g(a), g(k)) < D(g(a), g(n)) \end{cases} \tag{2}$$

In the above equation pairs, the $k = \{p, s\}$ is a superset of positive faces and the semi-positive faces. Since the semi-positive is a fake version of the referred person, it is further away from the referred face, compared to the positive face. This relation is shown in the first Eq. 2a. On the other hand, as the positive and semi-positive face belong to the same person, we expect both of them to be closer to the referred face, compared with the negative face and this derives the second equation.

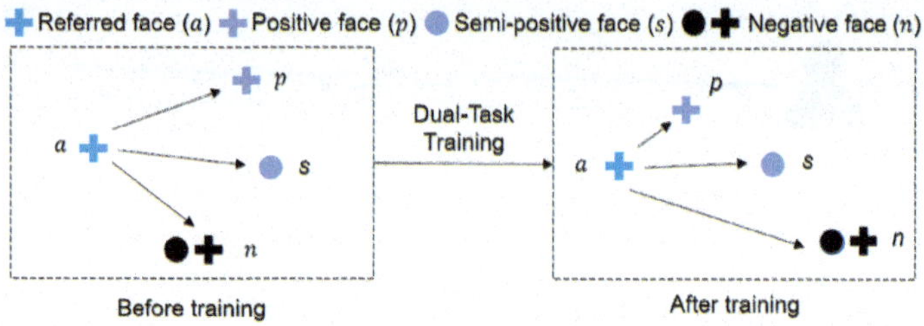

Fig. 4 A conceptual view of the dual-task relationship. Other than the referred face, there are three other face-types. As the positive and semi-positive face belong to the same person, so we expect them to be closer to the referred face, compared with the negative face. But as the semi-positive is a fake version of the referred person, so it is further away from the referred face, compared to the positive face

2.3 Design of Loss and Training Strategy

Once the objective is set, we will design proper losses and training strategy in this section. Denote a face image as x and it is associated with $y_{liveness}$ – a liveness label, and also $y_{identity}$ – a person's identity label. The liveness status label is either fake or live and therefore it is a binary value $y_{liveness} = \{0, 1\}$, if a face is live then the value goes as one and if a face is fake then the value is zero. The symbol $y_{identity}$ is a label to identify a person. Accordingly, the Eq. 2a is converted to Eq. 3.

$$\begin{aligned} D(g(x_1), g(x_2)) + margin_1 &< D(g(x_1), g(x_3)) \\ \text{where } margin_1 > 0,\ y_{liveness}(x_1) &= y_{liveness}(x_2) \\ y_{liveness}(x_1) &\neq y_{liveness}(x_3) \\ y_{identity}(x_1) = y_{identity}(x_2) &= y_{identity}(x_3) \end{aligned} \tag{3}$$

where $margin_1$ is a distance-control parameter to ensure the two classes should not overlap too much. This equation indicates that for the same person, the live faces should be well separately from the fake faces.

The Eq. 2b is converted in a similar manner to Eq. 4:

$$\begin{aligned} D(g(x_1'), g(x_2')) + margin_2 &< D(g(x_1'), g(x_3')) \\ \text{where } margin_2 > margin_1,\ y_{identity}(x_1') &= y_{identity}(x_2') \\ y_{identity}(x_1') &\neq y_{identity}(x_3') \end{aligned} \tag{4}$$

In order to obtain the relationship above, we should conduct training on different triplets. Let us use the symbol $L = \{(x_1, x_2, x_3)\}$ to denote the face-triplets in Eq. 3 and use the symbol $P = \{(x_1', x_2', x_3')\}$ to denote the face-triplets in Eq. 4. Accordingly, two triplet losses can be designed to regulate the liveness detection training as well as the recognition training, and their formulation is represented as below:

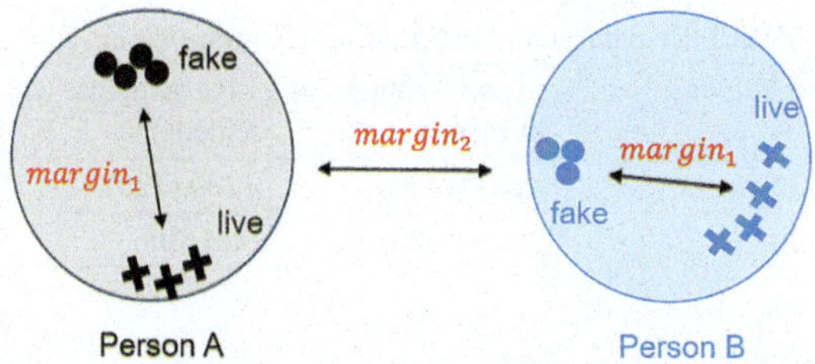

Fig. 5 We include two margin parameters to formulate the face recognition and liveness detection relationship. One parameter is set between Person A and Person B ($margin_2$), the other parameter is set between the fake and live embeddings of the same person ($margin_1$)

$$L_{liveness}((x1, x2, x3) \in L) = \max(0, D(g(x_1), g(x_2)) \\ +margin_1 - D(g(x_1), g(x_3))) \tag{5}$$

$$L_{identity}((x_1', x_2', x_3') \in P) = \max(0, D(g(x_1'), g(x_2')) \\ +margin_2 - D(g(x_1'), g(x_3'))) \tag{6}$$

Eventually, the dual-task loss is expressed in Eq. 7.

$$L_{dual-task} = L_{liveness} + L_{identity} \tag{7}$$

The above training idea is illustrated by Fig. 5 and we expect that at one hand, faces for each person will be grouped together. On the other hand, the containing fake image are still separable from the live images.

Training strategy. An intuitive way to conduct the above training is to use all valid triplet data, that are L and P. But in order to achieve a faster training convergence, we choose the batch-hard strategy [16] which discards the triplets that easily meet the loss requirement and this strategy shows to be very effective in various applications.

Backbone model. We adopt the InceptionNet based FaceNet [19] as the backbone model so that the feature mapping function g is determined. We chose the FaceNet because it is one of the most-adopted face recognition models. Yet, this model could be updated to any other state-of-the-art models as the key techniques do not rely on the specific model structure.

Once we have finished the above dual-task training and given a face image, we could use the well-trained model to extract the universal descriptor which could indicate the face identity and liveness status at the same time. Following some common solutions, we use thresholding strategy as in the previous work [19]. Specifically, we can verify the authenticity of x based on its referred face a through the following comparison:

Table 1 Statistics of REPLAY dataset and the training-test non-overlap split

	# Person	Total # Images (# Videos)	Live # Images (# Videos)	Fake # Images (# Videos)
Training set	30	2400 (720)	1200 (120)	1200 (600)
Test set	20	1600 (480)	800(80)	800 (400)

$$\text{facetype}(x|a) = \begin{cases} p, & \text{if } D(g(x), g(a)) < t_1 \\ s, & \text{if } t_1 < D(g(x), g(a)) < t_2 \\ n, & \text{otherwise} \end{cases} \tag{8}$$

Here, t_1 and t_2 denotes two boundary parameters and their concrete values could be determined empirically (Table 1).

2.4 *Experiments and Discussion*

Dataset We evaluate this method on benchmark dataset for face liveness detection—**REPLAY** [5] which is a video dataset that captures human faces and also contains the identifies for the captured person. To ensure the generalization ability, we require a person-wise split between training and testing data. In total, the number of videos is 1300 in REPLAY. There are two types of spoofing in the fake videos—photo attacks and video attacks, which include fifty persons in various lighting or environment conditions. Each person is associated with six live videos and twenty fake videos. Specifically, we use two live videos for registration. The readers could refer to the Table 1 to check the data statistics.

Face image preparation The videos are preprocessed by the following before importing them to the authentication system. **(1) Frame Extraction.** Our method studies face liveness detection and face recognition for single face image, so the first step is to extract frames from the videos. It is known that nearby videos frames usually share similar contents and in order to reduce redundancy, we apply a uniform sampling from all video frames for the later evaluation. And because the subject might step in or leave the field of view of the camera, the face sometimes does not appear at the video beginning or the video ending part and we drop the first and last 20% of the frames of a video while keep the middle 60% as the sample images. We also consider the fake-live class balance problems in the sampling processing as there are different numbers of live videos and fake videos. Therefore, different selection rates are used to sample these two types of videos. Based on the above process, we finally obtain two image datasets and their containing live and fake samples are similar in terms of the amount. The statistic summary is also included in Table 1. **(2) Face Extraction.** Secondly, we crop out only the face regions using the well-known MTCNN tool [23].

Note that we only use the standard RCB color images for evaluation and do not use the depth information in REPLAY.

Evaluation Metrics For either face recognition or liveness detection task, our evaluation is carried out in a pair-wise manner. Concretely, for each task, we denote a face-pair as (i, j) and construct two groups of face-pairs [19]. The first group, P_{same}, contains all pairs of the same label: $P_{same} = \{(i, j)|y_i = y_j\}$; The second group, P_{diff}, contains all pairs of different labels: $P_{diff} = \{(i, j)|y_i \neq y_j\}$. 1) For the task of face liveness detection, the *same* means that both faces in a given pair are live copies, and *different* means that one face is live and the other face is fake. Note that the fake–fake pairs are not included because any registered face stored in system is ensured to be the live ones. Therefore, for any real prediction case, we should have at least one live face in each pair. 2) For the face recognition task, *same* means the faces belong to the same person, otherwise, the label will be *different*. Following the prior works, the widely used biometric evaluation metric HTER - half total error rate [18] will be used for our evaluation. The calculation of HTER is shown in Eq. 9:

$$HTER(t) = \frac{FAR(t) + FRR(t)}{2} \tag{9}$$

Here, FAR denotes the false acceptance rate, FRR denotes the false rejection rate. Their calculation is expressed in the following two equations.

$$FAR(t) = \frac{\{(i, j) \in P_{same}, D(g(i), g(j)) > t\}}{|P_{same}|} \tag{10}$$

$$FRR(t) = \frac{\{(i, j) \in P_{diff}, D(g(i), g(j)) \leq t\}}{|P_{diff}|} \tag{11}$$

Baselines We named our proposed method FaceLivePlus and compared to a few baseline methods:

- **FaceNet/P** [19] is a classic network for face recognition and it achieves very excellent performance. Its accuracy will be the best performance our model aims to get for recognition.
- **FaceNet/L** was first proposed for face recognition task. To make it also applicable to the face liveness detection task, we re-train the network with the objective is the liveness classification and triplet loss is used for the model training.
- **LiveFace** [21] is one of the most recent works that also try to integrate the two tasks of face liveness detection and face recognition into a multi-task learning system [21]. In this work, the model has a shared segment of three convolutional blocks and two branches. The original training of this method requires the labels. As our dataset is non-overlap in terms of the persons between the training and testing and this might be unfair for LiveFace method, so we take LiveFace as a feature extractor. We take the fine-tuned features from either branch and evaluate the performance on the corresponding task.

- **MTL + FaceNet**. During our experiments, we found that the network of LiveFace is very shallow so that the extractor feature does not work very well. So we further replaced the shallow shareable network to a deep version to ensure a good fine-tuned feature could be extracted.
- **Proposed, Separate** is another version of our proposed FaceLivePlus. The structure of this model is the same as FaceLivePlus, however, each training triplets of L may include the faces of different persons.

Quantitative Performance We summarize the quantitative results in the table below. Compared to the four baselines, our method has a clear improvement of around 15%. We also observed the following:

- As expected, FaceNet/P and FaceNet/L has the best performance on the recognition and liveness detection task, respectively.
- Our method works well on both tasks in a balanced manner.
- The LiveFace and MTL+FaceNet two solutions do not balance the two tasks well. A plausible reason is the task complexity level varies greatly and thus a good fitting on both tasks around the similar timing is hard to be obtained [9].

To show the comparison and difference more intuitively, we visualize the two-task performance in Fig. 6 (Table 2).

Feature Visualization To understand why the proposed model works better, we extract their TSNE features and show the 2D TSNE in the Fig. 7. For each person, we also annotated his/her fake center as well as the live center. E.g., 123F is the center of all the fake faces of the person whose identity is 123. Similarly, 123L is the center of all the live faces of the person whose identity is 123. From the plots we have a few observations:

- FaceNet/P can cluster the persons correctly but for the same person, the fake faces are heavily overlapped with the live ones.
- FaceNet/L can well separate the live and fake. But they mix the different persons messily.

Table 2 Half total error rate comparison on REPLAY dataset where a lower value is better

Model	Target	REPLAY		
		Face Liveness Detection	Face recognition	Average
FaceNet/L	FL	**0.0000**	0.5000	0.2500
FaceNet/P	FR	0.4025	**0.0019**	0.2022
LiveFace	Both	0.4550	0.4349	0.4450
MTL + FaceNet	Both	0.3900	0.0275	0.2088
Proposed, Separate	Both	***0.0100***	0.3321	0.1711
FaceLivePlus	Both	0.0805	***0.0264***	**0.0535 (>15% ↑)**

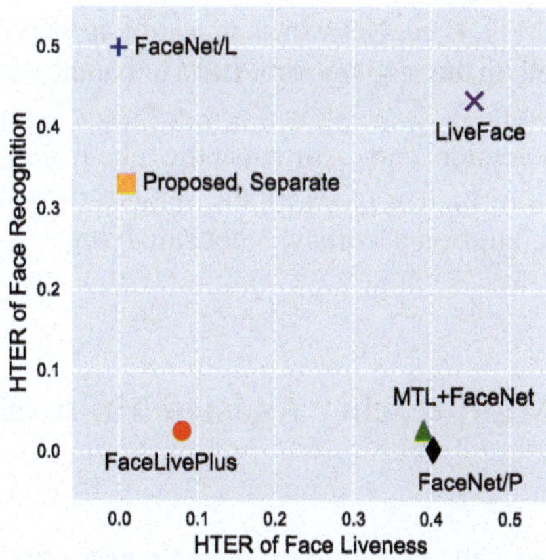

Fig. 6 An intuitive visualization of the HTER results. The (0,0) coordinate is the best performance. We can see our proposed method performs well on both tasks most balancely

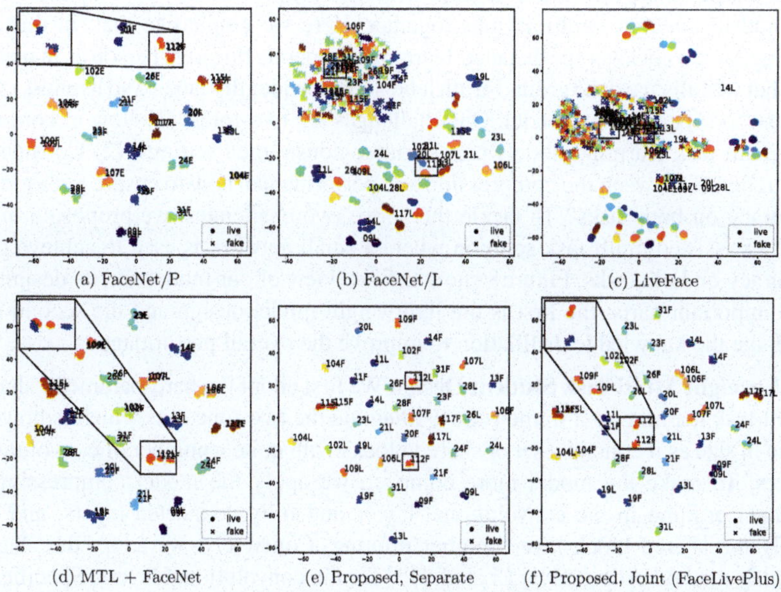

Fig. 7 Visualization of the TSNE features from baselines and our model on REPLAY dataset. Different colors represents different persons and different symbols represent different liveness status. The text labels the person identify and also the liveness status (live or fake). We highlight the person with identify 112 in the black box(s). We also zoom-in the plots for the person 112 in the left-upper corners in some subplots

- It is surprising that MTL+FaceNet works in a similar way of FaceNet/P and they both do not work well on the recognition. And a unbalance source allocation might be the potential reason.
- Our models, in both version, can distinguish the fake from live. But the fake faces are very far away from its live faces by the version (Proposed,Separate) ad this explains why the recognition accuracy is not satisfying.

3 Method 2: A Multi-teacher Assisted Multi-task Learning Framework

In last section, we discuss the possibility if a single network can be used to achieve both liveness detection and face recognition tasks and the methodology is mainly from a feature-representation aspects. In this section, we change the model backbone from a single-task network to a multi-task network, which is a popular and classic method to integrate multiple jobs together. Here we emphasize that, in this new setting, we do not leverage distance-learning to separate different person as in method 1 2, but we fully used the ground-truth labels (person identity labels and liveness status labels) to supervise our model. Our challenges are two-folds, (1) How to compress the multi-task learning models in a resource-constrained settings. (2) Considering the trade-off between the compression and performance, how to ensure a reasonable accuracy on both tasks? To tackle the above two challenges, we propose a multi-teacher assisted multi-task solution where a small network is used to achieve good accuracy on both tasks. Figure 8 show the overview of our model and its design has two important parts: the first is the lightweight model design and the second is to leverage the knowledge distillation to improve the overall performance.

Lightweight Multi-task Student Design We first adopt the hard-parameter sharing multi-task learning as a framework to integrate the two functions which is however large in size as it contain a set of shared blocks with quite a number of convolutional layers. To make the model more compact, we apply the model compressing by channel pruning in the convolutional layers and fully connected layers, and also apply the shared block layer number pruning. Concretely, we adopt the channel pruning method proposed by Li et al. [13] for the convolutional layers. Specifically, during the training process, a weight will be determined for each convolutional layer channel and the channels with relatively smaller weight values will be discarded. After this pruning process, the total number of parameters of our multi-task model is smaller than one million and its training is optimized by Eq. (12).

$$\mathcal{L}_{mtl} = \alpha\mathcal{L}_{liveness} + \beta\mathcal{L}_{recognition} \tag{12}$$

where $\mathcal{L}_{liveness}$ and $\mathcal{L}_{recognition}$ is the loss for liveness detection task and the recognition task, respectively. α and β are weight factors. For simplicity, both face liveness detection single-task branch and the face recognition branch share the same structure

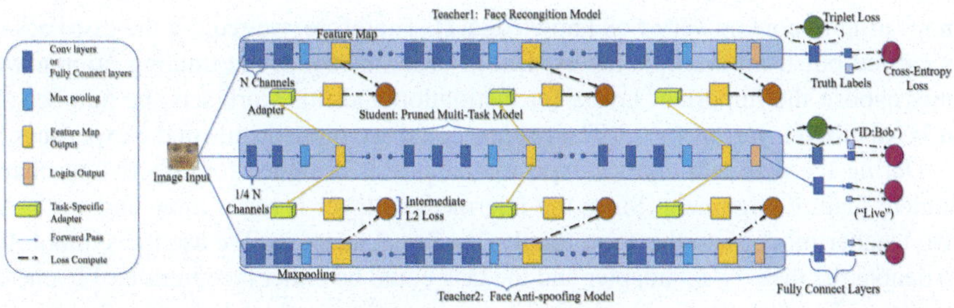

Fig. 8 An overview of our proposed method which includes three models in the training. The upper and lower part are two single-task models, which are usually based on deep neural network and work for recognition and liveness detection, respectively. Due to their powerful performance, we name them as the teacher models. The middle part is our proposed lightweight model and we call it a student model as it will grasp the knowledge from two teacher models (knowledge distillation) for satisfying performance. During the distillation, we jointly leverage multiple layers' features via multiple task-specific adaptors

except the final classification category setting part. Cross-entropy loss is used for the liveness detection as follows.

$$\mathcal{L}_{liveness} = \mathcal{L}^{ce}\left(\varphi\left(\cdot, \vartheta_s^{\mathcal{T}}\right) \circ \phi\left(x, \theta_s\right), y_{\mathcal{T}}\right) \tag{13}$$

where x is the input image, $\mathcal{T}$ is the task and $y_{\mathcal{T}}, \mathcal{T} = FL$ is the ground truth labels. Note that x is in a form of triplet (a, p, n) where a denotes an anchor, p denotes the positive and n denotes the negative. $\phi(\cdot, \theta_s)$ is a function to convert the image to a feature from the shared part of the network. The symbol $\varphi\left(\cdot, \vartheta_s^{\mathcal{T}}\right)$ denotes the predictor for a specific task and its input is the shared embedding, say ϕ and the output is the task $\mathcal{T}$'s logits.

For the face recognition, we adopt FaceNet [19] and obtain a 128-dimension feature and the recognition loss is formulated as follows:

$$\mathcal{L}_{recognition} = \sum_{a,p,n \epsilon \mathcal{D}} \left[max\left(\|a_{tri} - p_{tri}\|^2 - \|a_{tri} - n_{tri}\|^2 + \gamma, 0\right)\right] + \mathcal{L}^{ce} \tag{14}$$

where the first part is the triplet loss, γ is a threshold set as 0.2, and the second part $\mathcal{L}^{ce}$ is the cross entropy loss. a_{tri}, p_{tri} *and* n_{tri} are three extracted 128-dimensional features from the anchor image a, positive image p and negative image n, respectively.

Multi-Teacher Knowledge Distillation Based on the above, the obtained lightweight multi-task student network will benefit its deployment in resource-limited platform. However, a trade-off would be the performance drop compared to the original full-sized teacher model. So this section will introduce a knowledge distillation technique to provide the precision compensation. For the visual tasks, the intermediate feature maps usually contain very rich knowledge[2]. Therefore, we require the distillation is carried out from multiple layers. But in order to reduce the training efforts, we only choose the important layers for distillation and this work sets the number of to-be-distilled layers as $I = 3$. The readers could see more details in the experiment.

During the distillation, a task-specific adapter is designed for effective teacher-student feature alignment. Since the intermediate layer is set as three and we have two teacher models, so there are totally $2 \times 3 = 6$ adapters. We use the symbol A_τ^i to denote the task τ's i^{th} adapter and the task could be either recognition of liveness detection. For the adapter, we design it as a linear layer which has a $3 \times 3 \times C_{in} \times C_{out}$ convolution. For the output of the student encoder, say $\phi\left(x, \theta_s^i\right)$, the adapter will map it to the same size as for the corresponding task: $\mathbb{R}^{C_{si} \times H_{si} \times W_{si}} \Rightarrow \mathbb{R}^{C_{ti} \times H_{ti} \times W_{ti}}$ where the symbol C denotes the number of the channel. The symbol H and W denotes feature map's height and width, respectively. Therefore, the distillation loss is represented by Eq. (15):

$$\mathcal{L}_{kd} = \sum_{\tau}^{T=2} \sum_{i}^{I=3} \lambda_\tau^i \mathcal{L}^d\left(A_\tau^i\left(\phi\left(x, \theta_s^i\right)\right), \phi\left(x, \delta_\tau^i\right)\right) \tag{15}$$

In the above equation, the symbol $\phi\left(x, \delta_\tau^i\right)$ denotes the i_{th} feature map from τ_{th} teacher, δ_τ^i is the i_{th} shared layer parameters of teacher τ. $\mathcal{L}^d$ is a distance to measure the L2 normalized feature maps' difference, λ_τ^i is the parameter. The distance is calculated as in Eq. 16:

$$\mathcal{L}^d\left(f_1, f_2\right) = \left\| \frac{f_1}{\|f_1\|_2} - \frac{f_2}{\|f_2\|_2} \right\|_2^2 \tag{16}$$

where $f^i, f^j \in \mathbb{R}^{C \times H \times W}$ and C, H, W are channels' number, height and width of the feature maps by the feature encoders $A_\tau^i\left(\phi\left(x, \theta_s^i\right)\right)$ and $\phi\left(x, \delta_\tau^i\right)$. Based on the above, the model's final loss is represented by the following equation:

$$\mathcal{L} = \mathcal{L}_{mtl} + \mathcal{L}_{kd} \tag{17}$$

3.1 Experiments and Discussion

Dataset Similar to Method 1, we also adopt **REPLAY** [6] as the evaluation dataset.
Baselines We named of method as MMAKD and we carried out the comparison with a few baselines including:

- **STL** is the model designed for a single task (either recognition or liveness detection). We choose the classic VGG16 as the backbone . The two STL models are also served as the teacher models.

Table 3 Performance comparison on REPLAY dataset

Type	Model	Distillation	Parameters	FR accuracy	FL accuracy	Average accuracy	Time
STL	FR	–	29.55M	0.941	–	0.966	4.8/5.1ms
	FL	–	29.55M	–	0.990		
1/8 MTL	MTL	–	0.25M	0.865/0.819	0.883/0.955	0.874/0.887	2.0/2.1ms
	LKD	Yes	0.25M	0.888/0.812	0.878/0.927	0.883/0.870	
	BAM	Yes	0.25M	0.885/0.814	0.869/0.918	0.877/0.866	
	KR	Yes	1.99M	0.891/0.835	0.873/0.908	0.882/0.872	
	CKD	Yes	25.48M	0.916/0.867	0.930/0.984	0.923/0.926	
	MKD	Yes	0.85M	0.919/0.856	0.909/0.992	0.914/0.924	
	Ours	Yes	0.99M	0.928/0.882	0.933/0.990	0.931/0.936	

- $\frac{1}{k}$ **MTL**: our lightweight multi-task model and the fraction $\frac{1}{k}$ means we prune off 1-$\frac{1}{k}$ channels of the shared blocks in the multi-task model.
- We also compare with five classic and recent knowledge distillation methods including **LKD** [10], **BAM** [7],**MKD** [14] ,**KR** [3] and **CKD** [2].

Metrics For the liveness detection task, we choose the widely used metrics top-1 accuracy. For the recognition task, we evaluate the triplet accuracy.

Quantitative Performance The accuracy performance is compared in Table 3. From the statistics, our proposed method works best on both tasks at the same time compared to the rest multi-task methods. The smallest gap to the two teacher models indicates that our distillation method from multiple layers (both deep and shallow features) is effective.

We further compare the inference speed, compared with the original single task models, we reduce the time by around 56% on the cost of an average 2.1% accuracy drop. However, we also found that the speed-up does not linearly related with parameter amount. A possible reason is that, different operators in a neural network have different resource requirements. For example, Relu activation functions are memory-intensive operators but Conv and FC functions are computationally intensive operators. Since the inference time might depends more on the computation quantity, some parameter pruning might not help with the inference speed-up, which is left for future work.

Feature Map Visualization We further visualize the feature maps [26] to understand the distillation process in Fig. 9. Each row shows different methods. These different patterns might explain which areas might trigger the detection in either task.

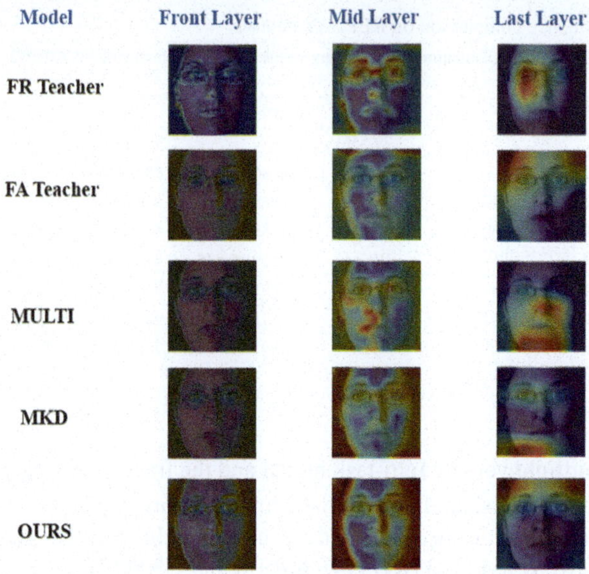

Fig. 9 We visualize the feature maps of multiple layers that locate at the front, in the middle and at the ending part. The visualisation is implemented by a well-adopted technique, CAM [26] . The heatmap represent the focus region of each model and a red part means that region is of more importance to the decision. From the visualization, our model in the last row well includes more key regions from both teachers and this explain its out-performance

4 Conclusion

This work explores the possibility to embed an extra face anti-spoofing ability to the existing face recognition system without requiring too much other resources. We introduce our two recent works with different model architectures, research focuses and solutions. Firstly, we discuss a case where a usual single-task neural network is used to achieve both tasks in one-go. We analyze the two task's relationship, and by a mathematics formulation, we insert the observed dual-task relationship to a novel deep model with distance-ranking feature. Secondly, we discuss a case where a classic multi-task model is used for dual-task integration. To suit such a model to resource-constraint devices, we leverage model compression and knowledge distillation techniques to achieve a good balance between the model size and the dual-task performance. Both works are working towards the building of a lightweight and enhanced face authentication system and intial experiments indicate the effectiveness of either method. Considering that Method 1 is from a feature learning aspect (internal knowledge) while Method 2 discusses from a knowledge distillation aspect (external knowledge), so it would be interesting to further explore how to integrate these two works in the future.

References

1. Bao, S.H., Li, M., Qian, W.H., Su, Z.: Secure face authentication with liveness detection for mobile (Jun 20 2017), uS Patent 9,684,779
2. Chen, D., Mei, J.P., Zhang, Y., Wang, C., Wang, Z., Feng, Y., Chen, C.: Cross-layer distillation with semantic calibration. In: Proceedings of the AAAI Conference on Artificial Intelligence. vol. 35, pp. 7028–7036 (2021)
3. Chen, P., Liu, S., Zhao, H., Jia, J.: Distilling knowledge via knowledge review. In: Proceedings of the IEEE/CVF Conference on Computer Vision and Pattern Recognition. pp. 5008–5017 (2021)
4. Cheng, T., Zhang, Y., Yin, Y., Zimmermann, R., Yu, Z., Guo, B.: A multi-teacher assisted knowledge distillation approach for enhanced face image authentication. In: 2023 International Conference on Multimedia Retrieval (2023)
5. Chingovska, I., Anjos, A., Marcel, S.: On the effectiveness of local binary patterns in face anti-spoofing (2012)
6. Chingovska, I., Anjos, A., Marcel, S.: On the effectiveness of local binary patterns in face anti-spoofing. In: BIOSIG. pp. 1–7 (2012)
7. Clark, K., Luong, M.T., Khandelwal, U., Manning, C.D., Le, Q.: Bam! born-again multi-task networks for natural language understanding. In: Proceedings of the 57th Annual Meeting of the Association for Computational Linguistics. pp. 5931–5937 (2019)
8. Ge, S., Zhao, S., Gao, X., Li, J.: Fewer-shots and lower-resolutions: Towards ultrafast face recognition in the wild. In: ACM MM. pp. 229–237 (2019)
9. Guo, M., Haque, A., Huang, D.A., Yeung, S., Fei-Fei, L.: Dynamic task prioritization for multitask learning. In: ECCV (2018)
10. Hinton, G., Vinyals, O., Dean, J.: Distilling the knowledge in a neural network. stat **1050**, 9 (2015)
11. Kuang, H., Ji, R., Liu, H., Zhang, S., Sun, X., Huang, F., Zhang, B.: Multi-modal multi-layer fusion network with average binary center loss for face anti-spoofing. In: ACM MM. pp. 48–56 (2019)
12. Lee, C.E., Zheng, L., Zhang, Y., Thing, V.L., Chu, Y.Y.: Towards building a remote anti-spoofing face authentication system. In: TENCON 2018-2018 IEEE Region 10 Conference. pp. 0321–0326. IEEE (2018)
13. Li, H., Kadav, A., Durdanovic, I., Samet, H., Graf, H.P.: Pruning filters for efficient convnets. In: International Conference on Learning Representations
14. Li, W.H., Bilen, H.: Knowledge distillation for multi-task learning. In: Computer Vision–ECCV 2020 Workshops: Glasgow, UK, August 23–28, 2020, Proceedings, Part VI 16. pp. 163–176. Springer (2020)
15. Li, Y., Li, Y., Xu, K., Yan, Q., Deng, R.H.: Empirical study of face authentication systems under osnfd attacks. IEEE Transactions on Dependable and Secure Computing **15**(2), 231–245 (2016)
16. Moindrot, O.: Triplet Loss and Online Triplet Mining in TensorFlow. https://omoindrot.github.io/triplet-loss##batch-hard-strategy (2018)
17. Pan, G., Sun, L., Wu, Z., Lao, S.: Eyeblink-based anti-spoofing in face recognition from a generic webcamera. In: ICCV. pp. 1–8. IEEE (2007)
18. Poh, N., Bengio, S.: Database, protocols and tools for evaluating score-level fusion algorithms in biometric authentication. Pattern Recognition **39**(2), 223–233 (2006)
19. Schroff, F., Kalenichenko, D., Philbin, J.: Facenet: A unified embedding for face recognition and clustering. In: CVPR. pp. 815–823 (2015)
20. Yang, Y., Sun, J., Guo, L.: Personaia: a lightweight implicit authentication system based on customized user behavior selection. IEEE Transactions on Dependable and Secure Computing **16**(1), 113–126 (2016)
21. Ying, X., Li, X., Chuah, M.C.: Liveface: A multi-task cnn for fast face-authentication. In: ICMLA. pp. 955–960 (2018)

22. Yoo, B., Youngjun, K., Kim, J., Jinwoo, S., Changkyo, L., Choi, C.K., JaeJoon, H.: Liveness test method and apparatus (Jan 31 2019), uS Patent App. 16/148,587
23. Zhang, K., Zhang, Z., Li, Z., Qiao, Y.: Joint face detection and alignment using multitask cascaded convolutional networks. IEEE Signal Processing Letters **23**(10), 1499–1503 (2016)
24. Zhang, Jian, Y.K.H.Z.Y., Xu, Y.: A collaborative linear discriminative representation classification method for face recognition. In: International Conference on Artificial Intelligence and Software Engineering (2014)
25. Zhang, Y., Zheng, L., Thing, V.L., Zimmermann, R., Guo, B., Yu, Z.: Faceliveplus: A unified system for face liveness detection and face verification. In: 2023 International Conference on Multimedia Retrieval (2023)
26. Zhou, B., Khosla, A., Lapedriza, A., Oliva, A., Torralba, A.: Learning deep features for discriminative localization. In: Proceedings of the IEEE conference on computer vision and pattern recognition. pp. 2921–2929 (2016)

CI in Finance, Business, Economics and Education

Conceptual Intelligence, Digital Transformation, and Leadership Skills: Key Concepts for Modern Business Success

Nadire Cavus and Seyedali Aghamiri

Abstract One of the primary characteristics of a leader in the digital era is conceptual intelligence. It requires creative thinking, opportunities and obstacles identification, and developing new solutions that fuel corporate development and success. Conceptual intelligence is vital for leaders in the digital transformation context to anticipate the forthcoming of their firm's environment. Digital leaders with great conceptual intelligence grasp the most important business aspects. They can connect the dots between different parts of their business, such as technology, marketing, finance, competitors, customers, and human resources. They can discover improvement areas and formulate methods to improve their operations, cost savings, and customer experience. Conceptual intelligence helps leaders to adopt a long-term perspective and forecast upcoming trends and opportunities by thinking outside the box and using data-driven decision-making in their business. Leaders must prioritize ongoing learning for themselves and their employees, which can lead to growth. Digital leaders can achieve business success in the rapidly shifting digital world by incorporating digital transformation, conceptual intelligence, and leadership skills such as strategic thinking, emotional intelligence, customer centricity, digital literacy, agility, and data-driven decision-making.

Keywords Digital transformation · Conceptual intelligence · Leadership skills · Business success · Modern business

N. Cavus (✉) · S. Aghamiri
Department of Computer Information Systems, Near East University, Nicosia, Cyprus
e-mail: nadire.cavus@neu.edu.tr

Computer Information Systems and Technology Centre, Ankara, Turkey

S. Aghamiri
e-mail: 20210614@std.neu.edu.tr

K. Daimi et al. (eds.), *Cutting Edge Applications of Computational Intelligence Tools and Techniques*, Studies in Computational Intelligence 1118,
https://doi.org/10.1007/978-3-031-44127-1_11

1 Introduction

The world is fast changing, and the rise of digital technology is reshaping many aspects of our life. Individuals and organizations must adapt to these changes and seize the opportunities afforded by digital transformation in this scenery. Organizations of all sizes must have the essential abilities and expertise to lead and manage this shift successfully. Conceptual intelligence is more important than ever in the digital era. Leaders must be able to think critically, connect ideas and concepts, and discover new solutions to complicated challenges as the speed of technological change accelerates; this requires not only conceptual intelligence but also requires other essential skills for today's leaders. Conceptual intelligence equips leaders to understand the complexity of digital transformation and find new possibilities for innovation in both their organization's digital transformation and their product or services as well. Developing conceptual intelligence takes a mixture of cognitive abilities, emotional intelligence, and creativity. Leaders must be able to evaluate complex information, combine different perspectives, and effectively see through complex concepts. To shape successful, cooperative teams, they must also recognize and control their own emotions, which means emotional intelligence is necessary for a digital leader. High Emotional intelligence enables the leader to listen and understand others' feelings and perspectives correctly actively. In a world where communications and networking are super-powers for digital leaders, training an incredibly high emotional intelligence can be a winning card for any leader. According to [1], emotional intelligence is highly trainable and is not only an inborn talent, but it demands high effort to get to high emotional intelligence. Lastly, leaders should possess the ability to think outside the box and provide innovative solutions to organizational pain points, which requires the willingness to take risks and be ready to experiment.

1.1 Digital Transformation

The incorporation of digital technology into various divisions of an organization and its procedures in a way that offers added value to customers and eliminates the need for paperwork and complicated bureaucracy is called digital transformation. This transition requires a strategic strategy considering that leverages technology to stimulate innovation and improve efficiency. Digital transformation may be complex since it dictates considerable changes in corporate culture, procedures, tools, and business models. However, it provides valuable opportunities for development and innovation within an organization, enabling the organizations to utilize technology for generating new services and products that solve customers' pain points. To triumph in digital transformation, organizations must have a clear vision and plan, strong leadership, innovative culture, and a proper digital transformation model or framework that matches their organizational strategy [2]. They must be willing to invest in new technology and procedures, experiment, and take risks. They must also be able to articulate the benefits of digital transformation to all stakeholders and generate consensus on the objectives.

1.2 Conceptual Intelligence

The capacity to grasp complicated concepts, evaluate information, and discover answers to difficult situations is conceptual intelligence. Since firms face complex issues and unpredictable surroundings, this form of intelligence has become increasingly vital in today's corporate world. In this introduction, an outline of the relevance of conceptual intelligence in business and its many facets will be presented. Abstract thinking is another name for conceptual intelligence, the capacity for original thought and problem-solving. Comprehending and handling abstract ideas and concepts requires applying complex cognitive processes, including conceptualization, abstraction, and critical thinking. The conceptual and general intelligence relationship has received considerable attention in psychology and cognitive science. According to [3], Fig. 1, Practical Intelligence can be compared to "street smarts." Being practical means a person can find the solutions which may be employed in everyday life by using information based on personal experiences. Practical intelligence seems to be distinct from the traditional understanding of Intelligence; in a way that entities who gain a high score in practical intelligence may or may not possess comparable marks in creative and analytical intelligence. Analytical intelligence is narrowly associated with academic problem-solving and computational intelligence, and Creative Intelligence is known for creating or visualizing a solution to a problem or condition. The definition of human intelligence is an intentional adaptation to, a selection from, and structuring of real-world surroundings vital to one's existence. Therefore, Sternberg defined *intelligence* as the capacity of an individual to adapt to environmental changes throughout their lifetime. He claimed that the ability to apply abstract reasoning to real-world issues should utilize to gauge one's intelligence. The idea of multiple intelligences, which [4] also advanced, lists logical-mathematical intelligence as one of its eight forms of intelligence. Moreover, studies have shown that conceptual intelligence significantly indicates academic achievement and employment performance. For instance, those with high conceptual intelligence are more prone to succeed in leadership or management positions and flourish in subjects like science, technology, engineering, and mathematics (STEM) [5, 6]. The importance of this factor in predicting academic and professional success highlights the need for more study and the creation of techniques to enhance this intelligence component.

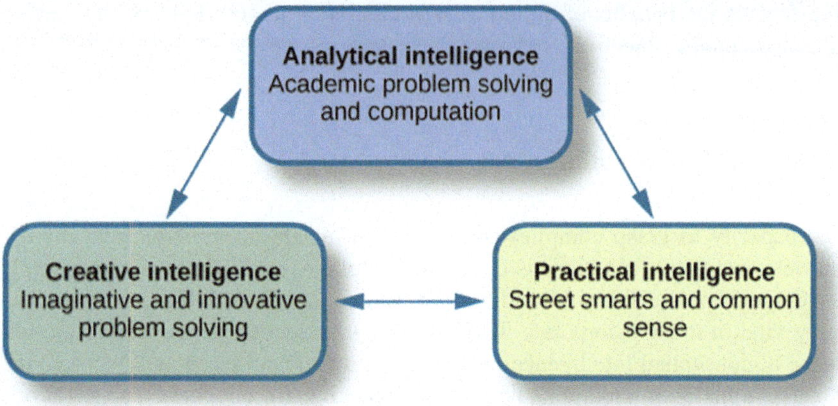

Fig. 1 The capacity to break down complicated situations into smaller components, evaluate each item, and uncover patterns and links between them is referred to as analytical intelligence (www.courses.lumenlearning.com)

1.3 Leadership Skills for Digital Transformation

Leadership skills are vital for success in the digital era. Leaders must have various skills to drive digital change, including strategic thinking, communication, cooperation, and the capacity to inspire and encourage others. Leaders must create and formulate a clear vision, have a robust digital transformation strategy, engage and motivate stakeholders, and successfully interact with team members and customers. Effective communication is crucial for digital transformation because leaders must be able to convey the benefits and risks of novel technologies while also developing a shared knowledge of the digital transformation's goals and objectives. Collaboration is another vital skill since driving transformation and innovation in the context of digital transformation certainly requires collaborating within multiple roles. Finally, leaders must inspire and encourage their employees, and even their customers, nurture an innovative environment and follow the path of continuous improvement and innovation, and empower team members to try new things and take risks without the fear of being penalized.

Digital leaders are people who have the abilities and expertise necessary to guide their businesses through the process of digital transformation. They understand digital technology's potential and how it can be leveraged for business success. They can inspire their employees to embrace change and create fresh and creative procedures, novel perspectives, and approaches toward old problems while managing the barriers of digital transformation. A digital leader is someone who has technical capabilities, strong digital literacy, and other skills. They must be able to comprehend and use new technology while also acquiring a complete understanding of their company's business strategy and the market they are involved in. Digital leaders must foster constant innovation and growth, encourage experimentation and risk-taking, and

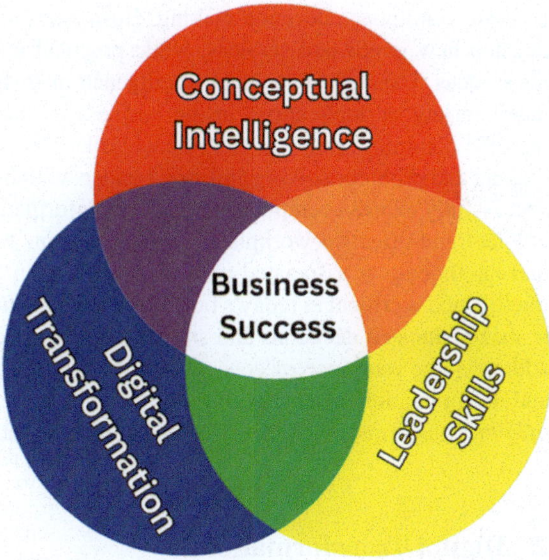

Fig. 2 Business success concept

guarantee that failure is accepted as a means to learn and improve. Figure 2 shows that business success is at the center of the three concepts, and digital leaders can achieve business success by utilizing conceptual intelligence, digital transformation, and leadership skills for digital transformation.

Businesses desire individuals who can evaluate complex data, identify patterns and correlations between diverse concepts, and offer creative solutions to issues in today's highly competitive industries. Businesses that lack these characteristics risk falling behind their competitors and missing out on new opportunities. Decision-making requires a high level of abstract understanding since organizations oppose progressively complicated and vigorous situations; having people who can evaluate information and make educated decisions is vital for them. Leaders with high conceptual intelligence can help firms identify risks and opportunities and develop effective strategies to mitigate upcoming risks. Conceptual intelligence is also essential in problem-solving. Business challenges are becoming increasingly multidimensional, requiring personnel to study and comprehend different aspects of the situation and use the available tools to overcome hardships and threats.

Another issue is that digital leaders with high conceptual intelligence can break complex issues into smaller components, evaluate each part, and discover novel solutions. The capacity to break down complicated situations into smaller components, evaluate each item, and uncover patterns and links between them is referred to as analytical intelligence. Creative intelligence is the capacity to produce novel ideas and solutions to challenges. Practical intelligence is the capacity to apply information and abilities to real-world circumstances successfully. These characteristics of conceptual intelligence are significant in the context of business. Analytical intelligence may assist firms in identifying trends and patterns in data, which can then be used to influence decision-making. Businesses may use creative intelligence to develop new goods and services, while practical intelligence can help implement these ideas. Uses of Conceptual Intelligence in Business: Conceptual intelligence has a variety of applications in the business world. Conceptual intelligence may increase communication skills in the workplace in addition to improving decision-making and problem-solving abilities. Leaders with high conceptual intelligence can successfully express complicated ideas and information to others, resulting in more efficient and effective workplace communication. According to research, conceptual intelligence is associated with work success in various industries. Conceptual intelligence is critical in today's business environment because it enables companies to make educated decisions and stay ahead of their competition. This introduction addresses the relevance of conceptual intelligence in business, its many elements, and its applications. The demand for people with great conceptual intelligence will only grow as firms expand and become more complicated.

2 Digital Transformation

Integrating digital technology into every area of the business, fundamentally changing the way that the organization works, and delivering value to our customers is digital transformation. It is also a cultural shift that pushes organizations to constantly challenge the status quo and have the will to experiment and embrace failure as a learning point to become better. Another function of digital transformation is to use digital technologies to streamline organizational operations, improve the customer experience, and generate new streamlines of income. It forces organizations to be agile and react to market changes, requiring an in-depth awareness of the industry and the capacity of the organization to be innovative in such environments. Conceptual intelligence is understanding, analyzing, learning, and using abstract concepts. It involves seeing patterns and trends, linking the information to form a novel and meaningful conclusion, and producing new ideas and solutions. Digital transformation can be utilized to find new digital transformation possibilities and build innovative solutions that exploit digital technologies to achieve business success. For example, a company that uses data-driven decision-making to predict the market trends is more likely to be ahead of its competitors in building new products and services that fit the changing demands of its customers.

Several factors can motivate a company to board on digital transformation. But by far, the most probable explanation is that they have no choice; it is a matter of survival. A business must be able to adjust quickly to interruptions in the supply chain, time-to-market challenges, and fast-changing customer expectations. Businesses may acquire and analyze massive volumes of data using machine learning tools and techniques and data science analytical technologies to find previously unseen patterns and trends, which can provide insightful information on customer behavior, market trends, and business operations and can lead to informed strategic decision-making and create extra potency towards innovation. Digital transformation has recently become a major focus for enterprises across sectors. The use of digital technologies in organizational procedures has the potential to boost efficiency, revenue, and customer experience.

Nevertheless, executives face considerable difficulties in implementing digital transformation, including change management, integrating new technologies such as artificial intelligence, machine learning, cloud computing, and augmented reality, and also dealing with cultural disagreements within their organization. Employees that lack the digital knowledge to deal with the novel technologies and used to do their tasks in an old fashion now see themselves in a different situation that demands learning new skills and changes in daily operations. Effective leadership must address these obstacles to achieve successful digital transformation. The tremendous benefits provided by digital transformation show its high impact on all areas of the business. According to a survey conducted by Capgemini [7], firms that thrive at digital transformation beat their rivals regarding revenue growth, profitability, and market valuation.

On the other hand, digital transformation is challenging. According to [8], two of the top five critical digital transformation challenges include organizational silos, which can be described as a necessity for supplementary skills, comprehension, and knowledge, Organizational Silos in decision-making and strategy were cited as a high-priority problem by 51% of respondents. Organizational silos are a massive problem because they influence all facets of a successful digital transformation, including strategy, finance, and execution. Governments, departments, and commercial sectors all have these silos, each needing a specific type of intervention. The following top key digital transformation challenge is the digital skill gap; the fourth most frequent obstacle to leaders' successful digital transformation is a need for appropriate depth or breadth of digital capabilities across the firm; for digital transformation initiatives to be successful, core specialized capabilities in fields like enterprise architecture, cybersecurity, cloud, analytics, and digital experience design are essential. Developing digital dexterity throughout the company is crucial to raising change preparedness. Leaders must have the necessary skills and expertise to overcome these difficulties and drive effective digital transformation programs. Also, the crucial leadership abilities needed for effective digital transformation will be looked into. The value of strategic thinking, customer centricity, data governance and privacy, employee empowerment, change management, innovation, strategic alliances, and continuous development will be discussed, and relying on insights from academic literature, industry reports, and case studies to give practical counsel to executives wishing to drive digital transformation efforts in their businesses.

There is a discussion about the significance of strategic thinking in digital transformation and how to establish a digital transformation plan that matches corporate goals and objectives. After that, the crucial role of customer-centricity in digital transformation and how leaders may establish customer-centric digital transformation programs will be looked into. The significance of data governance and privacy in digital transformation and suggest how to manage data ethically and openly, and the discussion of the crucial role of employee empowerment in driving successful digital transformation programs and techniques for cultivating an innovation and experimentation culture will be examined. Lastly, the significance of change management, strategic relationships, continuous improvement in digital transformation, and practical advice on approaching these essential issues will be explored. Ultimately, this chapter will give an in-depth look at the fundamental leadership abilities necessary for effective digital transformation. It will give executives practical direction for driving digital transformation efforts in their businesses and highlight the significant obstacles and possibilities connected with digital transformation. Leaders can ensure the success of digital transformation programs and generate sustainable development and innovation in their businesses by concentrating on four crucial leadership competencies. In today's corporate context, the necessity of digital transformation cannot be emphasized [9]. Incorporating digital technology into an organization's activities can result in significant benefits such as higher operational efficiency, improved customer experience, more revenue, and new growth prospects. Here are numerous benefits of digital transformation and how they may help organizations succeed.

2.1 Improved Operational Efficiency

Improved operational efficiency is a tremendous benefit of digital transformation. Organizations can automate repetitive procedures and workflows and minimize the time and effort needed to perform tasks by integrating digital technologies into business processes [10]. This saves resources, time, and money and increases accuracy while lowering the risk of errors in the whole body of the organization. Using the digital form of supply chain management systems, for instance, can help businesses in inventory tracking, monitoring performance, and optimizing logistics, resulting in increased efficiency and cost savings [11]. Adopting computerized inventory tracking systems in a manufacturing organization illustrates increased operational efficiency through digital transformation. Employees used to have to manually count the number of completed goods and raw materials in the warehouse, which was time-consuming and prone to faults. Organizations can now automatically track the number and position of their goods in real-time thanks to deploying a digital inventory management system, enabling quicker and more precise decision-making as well. It will also prevent stock problems by leveraging the data produced by the digital inventory system. Overall, this is how the organization benefits from increased productivity and cost savings.

2.2 Enhanced Customer Experience

Digital transformation has the potential to improve the customer experience significantly. By using digital technologies, organizations are able to deliver tailored and responsive services that fit customers' needs. Chatbots and AI-powered virtual assistants can help businesses by providing 24/7 customer support, responding promptly to customer inquiries or complaints, and providing tailored suggestions based on customer's profile and data. A video or music streaming service implementation of tailored suggestions is a practical illustration of how using artificial intelligence as part of the digital transformation can improve the user experience. The video streaming service can make new recommendations that are catered to a customer's interests by looking at their watching history, clicks, wait time of each video item, preferences, and search queries. Offering a more individualized and relevant selection of movies and TV series may enhance the customer experience by making it more straightforward for users to find new material they would want to watch. This can lead to an increase in customer satisfaction, loyalty, and retention. Another example is how a furniture company utilizes augmented reality (AR) technology to create a better customer experience. Customers can see how furniture will look on their property before making a purchase by using their smartphone software application encompassing AR technology; not only may it improve the purchase experience for customers by making it more immersive and exciting, but it also facilitates the ability for consumers to imagine how a product will fit in their environment by shape, color, and size. It lowers the likelihood of product returns, which leads to less overhead costs such as transportation and labor. These businesses can set themselves apart from their competitors, create a stronger relationship with their clients, and eventually boost their revenue and profit by using AR to offer tailored and exciting experiences.

2.3 *Increased Revenue*

Digital transformation can deliver considerable revenue growth by helping businesses to uncover new business opportunities and income sources; integration of digital technologies, for instance, can help firms to develop new products and services, enter new markets, increase market share, and explore new distribution channels [12]. Organizations are also finding new customer segmentation categories and establishing focused marketing and sales strategies to boost revenue growth by employing data analytics and insights. By using data analytics to comprehend customer behavior and strong knowledge of their preferences, a media company can deliver targeted ads to specific audiences, maximizing the effectiveness of its advertising campaigns and generating higher revenue from ad sales, this is an example of increased revenue with the use of digital advertising.

2.4 *New Growth Opportunities*

Fast innovation, growth, progress, and disruption define this phenomenon. Any organization that wants to thrive must be prepared to change with the digital world. Implementing new technology, purchasing tools, or modernizing current systems are only a portion of the digital transformation process. The digital transformation plan helps leaders address issues for their organization, including the current level of digitalization, the desired future, and the path to get there. Organizations must have three essential competencies linked to awareness, educated decision-making, and swift execution if they are to be safeguarded from digital disruption. Many firms across industrial industries now prioritize developing and implementing a digital transformation plan [13].

3 Leadership Skills

While digital transformation has numerous advantages for businesses, it also has significant drawbacks. Leaders are needed to lead successful digital transformation by providing a clear and understandable vision for the whole organization to take steps toward that unique vision. Leading digital transformation, on the other hand, demands a diverse set of skills and competencies from traditional leadership roles [14]. Here are some of the essential leadership qualities required for effective digital transformation in Fig. 3 that will be discussed in more detail.

Fig. 3 Digital leader skills

3.1 *Visionary Leadership*

Leaders who can provide a clear and compelling vision of how the organization can generate innovation and sustainable growth. Digital leaders must be able to convey the value proposition to stakeholders. They must also get the vision to inspire and motivate employees to embrace change and take ownership of the transformation process. A visionary leader should be able to create the why, the what, and the how for the employees, as it is crucial for the whole organization's members to fall on the same page while moving towards the vision. A leader who has a clear vision for the future of their firm, sector, or organization is an example of visionary leadership. This leader can inspire and encourage their team to work toward a single objective by effectively communicating their vision. Digital leaders constantly seek new methods to innovate, remain ahead of the curve, and quickly take chances or make sensible decisions. Furthermore, leaders can foresee future trends and modify their plans accordingly, guaranteeing that their organization is ready for any unforeseen obstacles.

3.2 *Change Management*

Digital transformation frequently necessitates large organizational structures, processes, and cultural changes. Leaders must manage change successfully by anticipating and managing opposition, including people in the transformation process, and explaining the advantages and dangers of digital transformation [15]. They must also be able to foster a culture of constant learning and experimentation to support ongoing digital transformation activities. As an illustration, a business could deploy a new software system to simplify its processes. Due to this transformation, employees may need to learn new skills, modify their work processes, and adopt new behaviors. Employees must be informed of the new system's advantages and given training and assistance to help them adjust. Any opposition to the change must be addressed as part of an effective change management strategy. In order to effectively utilize the capabilities of the new software, the organization may also need to restructure teams or processes; in this case, a change management strategy would help assure a seamless transition.

3.3 *Digital Literacy*

Leaders must know digital technologies and their potential uses in corporate situations. They must be able to analyze the impact of emerging technologies on their businesses and discover ways to exploit them to generate innovation and growth. They must also be able to successfully interact with technology specialists and stakeholders to ensure that digital transformation projects are aligned with business goals and objectives. According to [8], digital skills are one of the top reasons for digital transformation failures. One example is a business chief executive officer (CEO) who must be able to use digital technologies to enhance customer experiences, expedite communication, and improve operations. This can entail using data analytics technologies, interacting on digital platforms, and using project management software to guide strategic decision-making. Leaders can stay ahead of the curve and remain competitive in the digital age by improving their digital literacy abilities.

3.4 Data-Driven Decision-Making

Large volumes of data are generated due to digital transformation, which may be leveraged to develop insights and influence decision-making. Leaders must be able to make educated judgments regarding digital transformation programs by leveraging data analytics and insights [15]. They must also be capable of establishing data governance frameworks and ensuring that data is utilized ethically and safely to support digital transformation activities. For instance, a retail business may utilize sales data to decide which items to offer and how much to charge. The business may make educated decisions that help them stay competitive and profitable by examining sales patterns and consumer behavior. Another illustration may be a healthcare facility that analyzes patient data to pinpoint problem areas and enhance treatment regimens. The company can make informed decisions that enhance patient outcomes and cut expenses using data analytics. Instead of depending exclusively on instinct or prior experience, data-driven decision-making enables companies to make better-educated judgments supported by empirical evidence in both scenarios.

3.5 Collaborative Leadership

Cross-functional communication and coordination across departments and business divisions are required for digital transformation. Leaders must form successful teams, develop cooperation, and promote information sharing and communication throughout the business. They must also be able to define team members' roles and duties and offer them the tools and support they require to succeed. In order to solve a social issue, for instance, a non-profit organization could bring together stakeholders from several sectors. The collaborative leader would be responsible for guiding fruitful dialogues, spotting areas of consensus, and creating a shared vision for change. The stakeholders may have significant and long-lasting influence by cooperating and utilizing their distinct views and resources. Thanks to collaborative leadership, teams can cooperate more successfully to accomplish shared objectives in each situation.

3.6 Agility and Innovation

Leaders shown in Fig. 3 that are prepared to take chances and embrace experimentation and innovation are required for digital transformation. Leaders must cultivate an agile and innovative culture by allowing leaders to take ownership of digital transformation efforts and encouraging them to take chances and learn from setbacks [14]. They must also promote quick experimentation and learning to achieve continuous digital transformation. Leaders may drive successful digital transformation programs and help their businesses enjoy the benefits of digital technology by acquiring these leadership abilities. On the other hand, developing these skills necessitates constant learning and growth and a willingness to adapt to changing business situations and technology.

Another close argument proposed by Promsri [16] addresses the abilities and competencies leaders must acquire to flourish in the digital era. Leaders must gain new skills to keep up with the rapid speed of technology development and the rising digitalization of enterprises. Also, the researcher identified six critical competencies that digital leaders must have to drive digital transformation in enterprises.

- *Strategic thinking* is the first skill. Digital leaders must provide a clear vision and plan for their organization's digital transformation. This necessitates knowledge of the business landscape and a thorough awareness of digital technologies and their potential influence on the organization. For consistency and buy-in, digital leaders must identify areas where digital technologies may provide value to the business and effectively communicate their vision and plan to their team members and stakeholders. Strategic thinking is a necessary skill for digital leaders because it helps them to discover digital innovation possibilities and build a clear vision for the organization's digital future. This ability entails thinking about the long-term consequences of digital transformation activities and making data-driven decisions corresponding to the organization's strategic goals.
- *Collaboration* across departments and stakeholders within a company is required for digital transformation, which is the second skill. For the success of digital transformation initiatives, digital leaders must be able to build effective cross-functional teams and facilitate collaboration. This necessitates excellent interpersonal skills and the capacity to develop trust and connections with team members and stakeholders. Delivering digital transformation efforts necessitates cooperation across departments and organizational levels. To guarantee that all stakeholders are aligned and working toward the same objective, digital leaders must promote a collaborative culture. Effective communication, active listening, and generating a feeling of shared ownership for digital transformation efforts are all part of these skills.
- The third highlighted skill is *agility*. When new technologies emerge and business circumstances shift, digital leaders must be able to adapt and pivot their plans swiftly. This necessitates being open to new ideas and prepared to experiment and take risks. Digital leaders must develop an innovative culture inside their firm and an attitude of continual learning and progress. Firms must be nimble and

adaptive to stay ahead of the competition in the digital world. Digital leaders must be able to respond swiftly to changes in the digital environment and adapt their digital transformation activities accordingly. This skill entails being at ease with ambiguity, taking measured chances, and being open to new experiences.

- The fourth skill mentioned is *digital literacy*. Digital leaders must be thoroughly aware of digital technologies and their potential business applications. Keeping current with evolving technology and trends necessitates constant learning and growth. Digital leaders must be able to analyze the impact of digital technology on their company and find possibilities for innovation and development. As digital technologies become complicated daily, digital leaders must deeply grasp the newest digital tools and platforms. This competence entails being current on digital trends and technology and utilizing this knowledge to drive digital transformation activities.
- The fifth essential ability for digital leaders is *customer-centricity*. Customer experience is critical for business success. Thus, digital transformation efforts should be developed with the customer in mind. Digital leaders must understand customers' requirements and preferences and use digital technology to provide tailored and engaging customer experiences by putting themselves in customers' shoes and deeply understanding their true needs, and leveraging digital technology to offer tailored and engaging experiences. This demands knowledge of client behavior and the capacity to use data to influence decision-making.
- Finally, *data-driven decision-making* is the sixth and last competence highlighted skill. Data and analytics must be used to inform digital leaders' decision-making processes. This demands a deep comprehension of data and analytics technologies, such as data science. Moreover, the leader must be able to evaluate existing data, find the data trend, and reach a conclusion based on the existing evidence.

These are valuable foundations for digital leaders to build the abilities required to flourish in the digital age. Digital leaders with these six crucial abilities can drive successful digital transformation and keep their businesses competitive in the rapidly changing markets. The findings of [2] in another study demonstrated the relevance of digital transformation models and frameworks in attaining corporate success in the digital era. Nevertheless, firms must have excellent leadership capabilities to capitalize on these benefits and successfully implement digital transformation efforts. Promsri [16] highlighted six critical abilities for digital leaders: Strategic thinking, teamwork, agility, digital literacy, customer-centricity, and data-driven decision-making. These abilities are essential for executing successful digital transformation programs and realizing the benefits mentioned by [2]. In conclusion, the skills highlighted by Promsri [16] are required for digital leaders to successfully undertake digital transformation efforts and reap the benefits identified by [2]. Businesses may develop a digital culture that supports creativity, improves the customer experience, and drives corporate success by acquiring these skills and employing digital transformation models and frameworks.

Aghamiri et al. [2]'s systematic literature evaluation gives a complete grasp of the benefits of digital transformation models and frameworks for enterprises. These benefits include greater productivity, a better customer experience, innovation, and decision-making. Businesses may attain these benefits and stay ahead in the continuously changing digital world by adopting digital transformation methods and holding essential leadership capabilities. In the digital era, digital transformation projects assist firms in overcoming difficulties and capitalizing on possibilities. However, to successfully implement these efforts, firms must have the appropriate leadership capabilities and the ability to use digital transformation models and frameworks. Businesses can develop a culture of innovation and drive business success in the digital era. Digital leaders understand how to use technology to achieve company goals and can manage the complicated and fast-changing digital ecosystem. They recognize the significance of digital transformation and can develop and implement technology strategies to drive innovation, enhance operational efficiency, and generate new value for consumers. They may also build an environment of innovation and constant learning and lead and encourage employees in the digital era. Digital leaders play a vital role in promoting organizational digital transformation and innovation. They forefront digital projects, discover new growth opportunities, and use emerging technology to provide value to all stakeholders. Nevertheless, due to the quick rate of technological growth and the rising relevance of digital transformation in today's business landscape, the needs for digital leaders have significantly altered over the past few years.

4 Advantages and Possibilities for Leaders with Excellent Digital Literacy

Being innovative is crucial for a digital leader; they must always look for newly released tools and areas that can put the automation in place; implementing innovative technologies into the procedures is a daunting task and may need many resources to take place, but it is definitely worth it because after the automation is replaced with old methods of working and old procedures, it will make everyone's job easier. Another benefit of innovation is that it incorporates data, and that data can later be utilized to make data-driven decision-making. According to research published in the MIT Sloan Management Review, companies embracing digital innovation are more likely to generate considerable revenue growth than those not [10], digital leaders play a crucial role in injecting this innovation into organizations.

According to [17], Digital literacy encompasses a wide range of sophisticated cognitive, physical, social, and emotional abilities that users require to properly participate in digital contexts. It goes beyond simply being able to utilize software or operate a digital device. The tasks necessary in this context include, for instance, "reading" instructions from the graphical user interface (GUI) displays, using digital imitation to create new, meaningful materials from old ones, making knowledge from non-linear, hypertextual map routing system, evaluating the quality and validity of information, and having an established and realistic comprehension of the "rules" that govern cyberspace. This recently developed idea of "digital literacy" may be used as a gauge of the caliber of staff in digital contexts and may provide academics and practitioners with a more effective way of interaction when creating more user-centered settings. In the research by the author, a comprehensive and sophisticated conceptual framework for digital literacy is proposed, one that considers socioemotional, photo-visual, reproduction, branching, and information literacy. The European Union Expert Group has pointed out that digital literacy is a vital life skill and that "the incapacity to access or use Information Communication Technologies (ICT) has become an effective barricade to social and personal development." [18]. For example, in New Zealand as an innovative country, "Digital Literacy is now an essential life skill and the right of every NZ citizen, addressing ICT competence within the workforce would potentially bring about a productivity gain of up to $1.7 billion per annum for New Zealand [19].

The Analysis by Bunker [19] examines case studies and worldwide research on digital literacy to identify and forecast results when used in the New Zealand setting. It summarizes academic and industry leaders' studies on digital literacy, ICT, and productivity. In addition, it has examined a variety of case studies involving ICT skills projects from throughout the globe, including those from Europe, the Middle East, Africa, South America, and Asia. These studies cover large-scale, government-sponsored programs at the national level and grassroots, neighborhood-based efforts. Although there are many studies available, only a representative sample has been included in this study since it highlights crucial elements of successful initiatives that are present in all the research. The findings largely magnify the success stated in multiple studies worldwide, emphasizing both workplace efficiency (and the value to a country as a result) and e-Inclusion—greatly raising the standard of living for those in need individually and as a part of communities.

Firms must swiftly adapt to market developments to remain competitive in today's fast-paced business world. By employing technology to develop a greater understanding of their business and uncover new chances for growth and expansion, digital leaders can help their firms remain ahead of the competition. Organizations may acquire a competitive advantage and position themselves for future success by staying up to speed with the latest trends and advancements in the digital realm. According to a McKinsey report, companies embracing digital transformation are more likely to generate better profitability than those not [20]. Moreover, organizations that prioritize employee engagement are more likely than those that need to achieve better levels of profitability and efficiency. Organizations may boost employee happiness while driving productivity and creativity by embracing technology to build a more connected and engaged workforce. Digital leaders may play an essential part in this process by utilizing technology to enhance employee communication, collaboration, and knowledge-sharing. Digital leaders may utilize technology to increase operational efficiency while decreasing expenses. By automating basic task procedures, businesses may free up personnel to focus on more challenging jobs that demand human abilities. It can lead to improved productivity and better work quality. For example, McKinsey and Company research discovered that digitizing supply chain management operations showed a 50% decline in processing time and a 90% decrease in mistakes [21].

5 Leaders; Then Versus Now

In 2010, digital literacy was frequently restricted to fundamental computer abilities and familiarity with regularly used software tools. Email communication, word processing, and spreadsheets were all anticipated skills for digital leaders. However, leaders are expected to have broader digital skills and expertise in today's digital world. The increased emphasis on new technologies, such as artificial intelligence, data analytics, and machine learning, and being able to work with cutting-edge no-code tools and software, is one of the significant contrasts between digital leaders' digital backgrounds in 2010 and 2022. These technologies were still in their immaturity stage in 2010, and digital leaders needed more experience and understanding than is now crucial. Today's digital leaders must understand how to use these technologies to create digital transformation and corporate development. They must be able to read data, comprehend complicated algorithms, and formulate methods for gaining insights and creating value. Another notable distinction between digital leaders' digital backgrounds in 2010 and 2022 is the rising importance of digital marketing and e-commerce tools. Before, digital executives were mainly in charge of internal information technology systems and infrastructure. However, since digital marketing and e-commerce have grown in popularity, digital leaders must thoroughly understand customer behavior, online sales channels, and digital marketing techniques. To improve customer satisfaction and experience from using their product and service, they must be able to generate and implement efficient online marketing campaigns, optimize e-commerce platforms, and manage company data.

Moreover, digital leaders in 2022 are anticipated to have a robust digital presence and be skilled at communicating and collaborating via digital platforms. They must be able to connect with colleagues, partners, and clients worldwide through social media, messaging applications, and video conferencing technologies. They must also communicate with the most recent collaboration and project management technologies and methods, such as agile, scrum, and their associated tools, to promote remote collaboration and increase efficiency. Despite the significant benefits that come with digital leaders' evolving digital backgrounds and digital literacy, there are also some challenges that organizations face when recruiting and developing digital leaders. One challenge is the growing demand for digital skills; another challenge is that the number of young entrepreneurs that founded startups is rising compared to 2010. With the rapid pace of technological change, there is a need for more skilled professionals with crucial digital skills and knowledge. As a result, organizations may need help finding and recruiting qualified candidates for digital leadership roles, and the summary of these differences is depicted in Fig. 4.

Another challenge is continuously updating digital leaders' skills and knowledge. With new technologies emerging, digital leaders must stay current with the latest

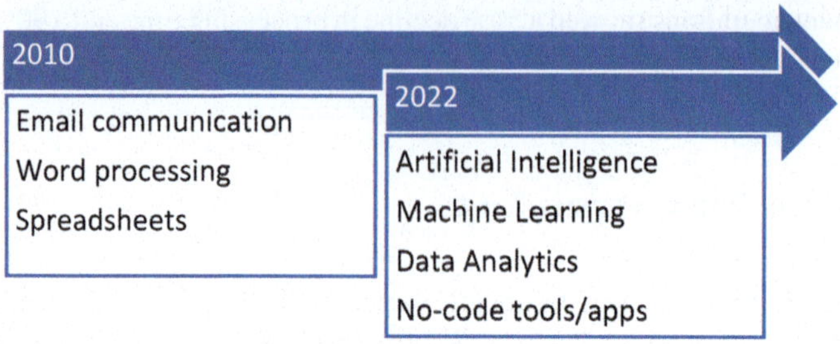

Fig. 4 Leadership digital skills in 2010 versus 2022

trends, tools, and best practices to remain effective; this requires ongoing training and development, which can be costly and time-consuming for organizations. Additionally, some employees may resist change because they need to familiarize themselves with new technology or ways of functioning. Digital leaders must be adept at managing change, overcoming resistance, and driving the organization's adoption of new technology and procedures. Notwithstanding these hurdles, having digital leaders with a solid digital background and digital literacy has considerable benefits. Improved inventiveness and agility are two advantages. Digital leaders who understand evolving technologies can spot new chances for development and innovation and rapidly react to changing market conditions. They can also establish and deploy new business models that use technology to add value to the consumer experience.

Another problem of leaders' growing digital literacy is the risk of leaning too heavily on technology and forgetting the value of human interactions and emotional intelligence. The capacity to connect with and understand employees, customers, and other stakeholders personally is critical for effective leadership and can only be replaced partially by technology. Conversely, having leaders who are well-versed in digital technology has several advantages. One significant advantage is the capacity to use technology to increase an organization's efficiency, productivity, and creativity. Leaders with digital technologies and platforms may discover new methods to optimize processes, automate mundane work, and stimulate cross-team and cross-departmental cooperation.

Furthermore, digital literacy may assist executives in staying educated and current on the newest trends and advances in their sector, as well as the larger digital world. This knowledge may help make strategic decisions and forecasting market shifts. Data and analytics experts may also get significant insights into consumer behavior, industry trends, and other crucial elements that can guide corporate strategy. It can result in more successful marketing and sales efforts, better customer service, and overall corporate performance improvements. Finally, companies face obstacles and possibilities as leaders' digital literacy grows. While there is a risk of over-reliance on technology and a potential loss of human connection and emotional intelligence, having leaders adept with digital tools and platforms has several advantages. Organizations may leverage the potential of their leaders to promote success and growth in the digital age by balancing the benefits and difficulties of digital leadership.

A new way of looking at leadership is presented by [22] to raise questions about conventional ideas and methods. According to the author, leadership is a social and relational process that develops due to interactions between leaders, followers, and the environment in which they function rather than merely a set of abilities or characteristics owned by an individual. In order to better comprehend leadership, the author suggests a new conceptual framework that considers the complexity and unpredictability of modern organizations and society, and the author examines issues, including the nature of power, the function of ethics, and the significance of context in determining leadership by utilizing concepts from philosophy, psychology, and sociology. It proposes alternative strategies emphasizing cooperation, communication, and shared accountability while critically reviewing conventional leadership ideas and practices. It also offers case studies and practical examples to show how these fresh ideas can be worked on. The findings encourage readers to reevaluate their presumptions and views regarding leading successfully in today's complicated and quickly changing world.

6 Digital Leaders with Academic Excellence Versus, Digital Leaders with Digital Hands-on Experience

A competitive and thriving business in the rapidly evolving digital market must embrace digital transformation. Businesses desire digital leaders who can help them navigate the difficulties of the digital revolution as a result. Digital leaders with both academic and hands-on experience can go into comparison because, in the digital era, these are the frontiers of leadership in digital markets.

6.1 Digital Leaders with Academic Excellence

A robust educational background in computer science, engineering, data analytics, or business administration frequently characterizes Digital Leaders with outstanding academic records and performance in digital leadership. The philosophy and guiding principles of digital technologies and trends are familiar to them. The principles and guidelines of digital technologies and trends are familiar to them. These experts can provide businesses with strategic advice and leadership based on their understanding of the digital environment. The capability of academically trained digital leaders to evaluate data and make informed judgments in light of that analysis is one of their most important benefits. They could utilize data to fuel digital transformation initiatives, always find space for improvement or at least see the potential in all the areas to be improved, and facilitate company procedures. They could also suggest how companies use digital technology to boost productivity, increase customer happiness, and streamline operations. Digital leaders with strong academic credentials can stay current with emerging trends and technologies. Digital leaders with assertive academic credentials can stay current with rising trends and technologies. Their network has a wide range of peers and experts that can provide knowledge about current and forthcoming technologies. These leaders could use their knowledge to keep their companies ahead of the curve and competitive.

6.2 Digital Leaders with Hands-on Digital Skills

Digital leaders that are hands-on are skilled at integrating and leveraging new technologies. They are knowledgeable about the techniques and instruments required for a strategy for digital transformation. These decision-makers could offer guidance on successfully and practice utilizing digital technologies. Understanding the issues and limitations associated with adopting digital technology is one of the main advantages of having digital leaders with practical digital skills. They can anticipate potential obstacles and issues and develop methods to get around them. These leaders can also suggest integrating new technologies into existing practices and processes. Digital leaders who possess practical digital skills could demonstrate how to use digital technology to overcome difficulties encountered in the real world and improve business results. They could encourage team members to adopt new techniques and technologies, boosting output, effectiveness, and innovation.

6.3 Comparing Digital Leaders with Academic Excellence and Hands-on Digital Skills

Digital leaders with academic proficiency and practical digital skills have meaningful advantages in digital transformation. However, there are some notable differences between the two groups. Strategic thinking and vision are the main advantages of having digital leaders with academic accomplishments. They provide an expansive overview of the effects of digital transformation on the business. They could also find new opportunities for innovation and growth based on their proficiency in the most recent digital trends and technologies. On the other hand, digital leaders with hands-on skills have expertise in digital technologies and, most probably, software engineering skills. They can provide suggestions on correctly carrying out a plan for digital transformation which can lead to much faster product development, especially in the IT sector. They may also foreknow issues and challenges in the future and develop proper solutions. These two categories can also be distinguished by how they approach problem-solving. Academically accomplished digital leaders approach problem-solving in a more analytical and data-driven manner. They may use data and insights to form informed decisions and identify growth opportunities. Digital leaders who use their hands only to solve problems use a more practical and hands-only approach. They could use their skills and knowledge by looking for solutions to problems that arise in the real world. When it comes to digital transformation, there is no one size fits all approach. Its unique requirements and goals determine the ideal digital leader for a firm. Conversely, businesses can profit from a mix of academically qualified and practical digital leaders. Academic excellence in digital leadership may offer strategic counsel and leadership, while practical digital competence may come from hands-on experience. Considering the company's aims and ambitions, they could create a comprehensive digital transformation strategy. Additionally, a combination of academic proficiency and practical digital skills can help create a bridge between Information technology (IT) and business divisions. Academic excellence in digital leaders might establish a common language and framework for business and IT organizations to connect effectively. Hands-on digital leaders can help align digital transformation initiatives with business goals and objectives. Additionally, comparing digital leaders' academic credentials and practical digital skills depends on the particular aims and goals of the company. Both categories have significant benefits when it comes to digital transformation. Academic excellence in digital leadership may provide tactical guidance and direction during practical digital. Academic brilliance in digital leadership might give strategic advice and leadership, while hands-on digital skills can provide practical insights and competence. Nevertheless, combining both can aid in developing a holistic digital transformation plan that addresses the objectives and goals of the organization.

7 Conclusion

Conceptual intelligence, digital transformation, and leadership skills are critical to business success. To be ahead of the competition and satisfy the changing demands of consumers and customers in today's fast-paced business market. Businesses must be able to adapt, innovate, and improve continually. Continuous improvement, or frequently analyzing and refining company processes, goods, and services, is critical to long-term success. It lets businesses streamline operations, cut expenses, increase quality, and enhance customer experience. Organizations may maintain a competitive advantage and fulfill the requirements of their consumers by constantly exploring areas for Improvement and making changes. Another critical quality for leaders in the digital era is conceptual intelligence. It entails thinking creatively, identifying opportunities and obstacles, and developing new solutions that fuel corporate development and success. Leaders with high conceptual intelligence comprehend the more significant business picture and can draw the dots between many parts of their firm. They may design plans to optimize operations, cut costs, and improve the customer experience by having a long-term view and predicting upcoming trends and opportunities. Digital transformation has become a critical driver of company success in recent years. It entails utilizing emerging technology, data analytics, and digital marketing to streamline operations, automate mundane chores, and improve customer experience. Another essential component of modern business success is leadership skills. Successful leaders must have various skills and characteristics, including communication, teamwork, flexibility, agility and innovation, networking, digital literacy, emotional intelligence, and visionary leadership. They must be able to encourage and empower their employees to accept change, learn new skills, and promote an innovative culture. Digital leaders can achieve revolutionary change and position their firms for long-term success by combining these leadership skills with digital transformation. While there has been tremendous progress in seizing these principles, there is still much to learn. Future research should concentrate on the changing nature of digital transformation and the influence of future technologies like artificial intelligence, blockchain, cloud computing, 5G (5th generation of mobile networks), 6G (6th generation of mobile networks) that can enable live operations from afar with zero latency and achieve world-changing innovations, especially in robotics and medical science, and the Internet of Things (IoT). Finally, constant improvement, conceptual intelligence, digital transformation, and leadership skills are crucial to modern business success.

To summarize, the digital era has presented significant possibilities and problems to enterprises across all industries. Leaders must have technical capabilities, business acumen, and leadership traits to flourish in this fast-paced and ever-changing world. This chapter has covered four major issues in digital transformation, conceptual intelligence, leadership skills, and leader's comparison. Conceptual intelligence is vital for success in the digital era because it allows leaders to think critically, create, and traverse complicated contexts. Digital transformation provides several potentials for development and innovation, but it needs a thoughtful strategy and strong leadership skills. Digital leaders are vital in driving digital transformation and cultivating an innovative culture inside their businesses. This chapter has created a roadmap for leaders wishing to navigate the challenges of the digital era and promote development and innovation inside their enterprises by merging these essential areas. Whether a company leader, entrepreneur, or digital professional, the insights and methods given in this chapter will assist in developing the skills and knowledge required for success in the digital world, the success of a leader of an organization in the digital era can be highly facilitated by combining conceptual intelligence, digital transformation, digital leadership, and leadership skills.

References

1. V. Mattingly, and K. Kraiger, "Can emotional intelligence be trained? A meta-analytical investigation," Human Resource Management Review, vol, 29, issue, 2, pp. 140-155, 2019.
2. S. Aghamiri, J. Karima, and N. Cavus, "Advantages of Digital Transformation Models and Frameworks for Business: A Systematic Literature Review," International Journal of Advanced Computer Science and Applications, vol, 13, issue 12, pp. 40-47, 2022.
3. R. J. Sternberg, "Beyond IQ: A triarchic theory of human intelligence," Cambridge University Press, 1985.
4. H. Gardner, "Frames of mind: The theory of multiple intelligences, Basic Books" 1983.

5. B. Erdogan, and T. N. Bauer, "Leader-member exchange (LMX) theory: The relational approach to leadership," The Oxford handbook of leadership and organizations, pp. 407–433, 2014.
6. G. Park, D. Lubinski, and C. P. Benbow, "Ability differences among people who have commensurate degrees matter for scientific creativity," Psychological Science, vol, 19, issue, 10, 2008, https://doi.org/10.1111/j.1467-9280.2008.02182.x.
7. Capgemini Consulting, "The Digital Advantage: How digital leaders outperform their peers in every industry," 2017, https://www.capgemini.com/wpcontent/uploads/2017/07/The_Digital_Advantage__How_Digital_Leaders_Outperform_their_Peers_in_Every_Industry.pdf [Retrieved: March, 2023].
8. Gartner, "5 Key Digital Transformation Challenges Government CIOs Must Tackle," 2022, https://www.gartner.com/en/articles/5-key-digital-transformation-challenges-government-cios-must-tackle, Retrieved: March, 2023].
9. N. Cavus and N. Sancar, "The Importance of Digital Signature in Sustainable Businesses: A Scale Development Study," Sustainability, vol 15, issue 6, pp. 5008, 2023. https://doi.org/ https://doi.org/10.3390/su15065008
10. G. Westerman, D. Bonnet, and A. McAfee, "Leading digital: Turning technology into business transformation," 2014.
11. J. W. Ross, and P. Weill, "Six IT Decisions Your IT People Should Refrain from Making," Harvard Business Review, 2002, https://hbr.org/2002/11/six-it-decisions-your-it-people-shouldnt-make, [Retrieved: March, 2023].
12. E. Brynjolfsson, and A. McAfee, "The Second Machine Age: Work, Progress, And Prosperity in A Time of Brilliant Technologies," W. W. Norton and Company, 2014.
13. S. Albukhitan, "Developing digital transformation strategy for manufacturing," Procedia computer science, vol, 170, pp. 664-671, 2020.
14. S. Berman and A. Marshall, "The Next Digital Transformation: From an Individual-Centered to an Everyone-to-Everyone Economy," Strategy & Leadership, vol. 42, issue 5, pp. 9–17, 2014. https://doi.org/10.1108/SL-07-2014-0048
15. I. Faisal, S. Khuram, B. Aurangzeab, and K. Jussi, "Leadership Competencies for Digital Transformation: Evidence from Multiple Cases," Advances in Human Factors, Business Management and Leadership: Proceedings of the AHFE, pp. 81–87. 2020, DOI: https://doi.org/10.1007/978-3-030-50791-6_11.
16. D. C. Promsri, "The Developing Model of Digital Leadership for a Successful Digital Transformation," GPH-International Journal of Business Management, vol. 2, issue 08, pp. 01–08, 2019. http://www.gphjournal.org/index.php/bm/article/view/249
17. Y. Eshet, "Digital literacy: A conceptual framework for survival skills in the digital era," Journal of educational multimedia and hypermedia, vol, 13, issue, 1, pp. 93-106, 2004.
18. DG Information Society and Media Group, "Digital Literacy Report: a review for the i2010," eInclusion initiative, 2008.
19. B. Bunker, "A summary of international reports, research and case studies of digital literacy," Wellington: New Zealand Computer Society Inc, 2017.
20. J. Bughin, T. Catlin, M. Hirt, and P. Willmott, "Why digital strategies fail," McKinsey Quarterly, 2018, https://www.mckinsey.com/business-functions/mckinsey-digital/our-insights/why-digital-strategies-fail, [Retrieved: March, 2023].
21. M. Lurie and L. Tegelberg, "The New Roles of Leaders in 21st Century Organizations," 2019, https://www.mckinsey.com/capabilities/people-and-organizational-performance/our-insights/the-organization-blog/the-new-roles-of-leaders-in-21st-century-organizations, [Retrieved: September, 2023].
22. D. Ladkin, "Rethinking leadership: A new look at old leadership questions, In Rethinking Leadership," Edward Elgar Publishing, 2010.
23. G. Westerman, D, Bonnet, and A. McAfee, "The Nine Elements of Digital Transformation," 2014, https://sloanreview.mit.edu/article/the-nine-elements-of-digital-transformation, [Retrieved: March, 2023].
24. PwC, "Experience is everything: get it right," 2018, https://www.pwc.com/us/en/services/consulting/library/consumer-intelligence-series/future-of-customer-experience.html, [Retrieved: March, 2023].

GEMM-SaFIN(FRIE)++: Explainable Artificial Intelligence Visualisation System with Episodic Memory

Nelson Mingwei Ko, Chen Xie, Qi Cao, and Chai Quek

Abstract Neuro-fuzzy network systems take advantage on functionalities of hybrid fuzzy logic and neural networks approaches. IF–THEN fuzzy rules allow good interpretability for human experts to understand the correlation between inputs and outputs. However, only designers know the mechanism and behavior of neuro-fuzzy systems. Details on how a neuro-fuzzy system derives and formulates the predictions or rules are unknown to users. It is due to the lack of transparency of the design and connections of neuro-fuzzy systems. We propose a novel explainable artificial intelligent (AI) visualization system for the neuro-fuzzy architecture, named general episodic memory mechanism (GEMM) *Self Adaptive Fuzzy Inference Network with Fuzzy Rule Interpolation or Extrapolation* with online learning capabilities (i.e., GEMM-SaFIN(FRIE)++). The proposed explainable AI visualization system is designed in a form of graphical user interface to assist users better understanding inner function mechanism on how rules are generated and how conclusions are drawn from the data fed into the neuro-fuzzy system. One of the challenges for neuro-fuzzy systems is making real-time predictions in the fast changing applications where data can be sparse. It may not be able to automatically detect and react to the occurrence of concept drifts, affecting the online learning capabilities. GEMM-SaFIN(FRIE)++ employs fuzzy rule interpolation and extrapolation techniques to make inference when concept drifts are detected. The GEMM mimicking human brains is employed to capture and retrieve from past events that GEMM-SaFIN(FRIE)++ learns. This is done by storing and retrieving them from an episodic memory storage during the transient event behaviors. Several experiments are conducted to evaluate the performance of the GEMM-SaFIN(FRIE)++. Firstly,

N. M. Ko · C. Xie · C. Quek
School of Computer Science and Engineering, Nanyang Technological University, Singapore, Singapore
e-mail: xiec0003@e.ntu.edu.sg

C. Quek
e-mail: ashcquek@ntu.edu.sg

Q. Cao (✉)
School of Computing Science, University of Glasgow, Glasgow, UK
e-mail: qi.cao@glasgow.ac.uk

K. Daimi et al. (eds.), *Cutting Edge Applications of Computational Intelligence Tools and Techniques*, Studies in Computational Intelligence 1118,
https://doi.org/10.1007/978-3-031-44127-1_12

three sub datasets in Nakanishi dataset including a sub dataset for daily price of stock in a stock market are utilised in the experiments. Secondly, GEMM-SaFIN(FRIE)++ is used to detect market trend reversal for two stock market indexes: S&P 500 and DJIA index, under various events through the period during COVID-19, subprime and 9–11 attack. Encouraging experiment results are observed for event detections to capture a shift in stock prices.

Keywords Neuro-fuzzy system · Inference learning visualization · Neuro-fuzzy visualization · Episodic memory · Event detection

1 Introduction

Neuro-fuzzy systems or fuzzy neural networks combine linguistic representations of fuzzy set theory and fuzzy inference with neural networks [1, 2]. It is a hybrid system combining the functionalities of fuzzy logic systems and neural network [3–5]. Neural networks are known with black-box nature, that can be resolved by combining the learning ability of neural networks with interpretability of fuzzy logic. Fuzzy logic IF–THEN rules introducing interpretability allow users to understand the reasoning process. It enables to decode the learning by the relationships between the inputs and outputs of the system.

Neural networks simulate human learning capabilities that are modelled based on neuron structures and functioning in human brains. Artificial neurons which contain information locating at various layers in neural networks are linked and interconnected [2]. Sample training input and output data are fed into neural networks in the training process. The neural networks adapt and learn information based on the training data. They are able to model complex relationships between inputs and outputs, then find patterns in data. With the training, the links of artificial neurons are connected with corresponding weights. Afterwards, the trained neural networks are ready for inferencing to work with the testing data. The prediction results at the output layer are produced accordingly.

Black-box nature of these systems allows powerful predictions, but it cannot be directly explained Prior works have been reported in the literature to provide the explainable artificial intelligent (AI) methods for transparency of machine learning models. Explainable AI systems are able to increase the transparency and trust of decision making in the human computer interaction [6]. An explainable AI decision support-system is introduced to automate the underwriting process of lend loan by a belief-rule-base approach [7]. An interpretable convolutional neural network (CNN) is presented to clarify knowledge representations and help people understand the logic inside [8]. An explainable control system is reported for deep neural networks on handling an autonomous dynamic positioning system [9]. Mendel and Bonissone discuss their research findings about the explainable AI for rule-based fuzzy systems [10]. An explainable machine learning model based on a K-means clustering and classification tree is introduced for default privacy setting prediction [11]. Cheng et al. [12] share that interactive explanations through user interfaces can improve users' comprehension on inner workings of an algorithm.

Online learning capability of neuro-fuzzy systems enables the incorporation of new data by self-organizing its rule base structure to represent the most recent knowledge. In a real-time environment, the relationships and features between input and output data of machine learning applications may change over time, that is known as concept drift [13, 14]. The detection for the concept drifts is important for neuro-fuzzy systems as it allow the system to adapt new changes and maintain system performances under such changes [15]. When neuro-fuzzy systems are adopted in fast changing scenarios such as financial applications, the changes involve sharp spikes or sudden drops in some extreme cases. It may be impossible to incorporate a complete rule base that cover all possible concept drifts in such applications. The current fuzzy rule base is considered as a sparse rule base when it cannot handle directly some dynamic scenarios in the operating environment [16]. The technique of fuzzy rule interpolation or extrapolation (FRIE) is important in neuro-fuzzy systems to address the issues of sparse rule bases. FRIE technique may still enable deriving useful conclusions, even if some observations are not covered directly by the current fuzzy rules [16, 17].

A neuro-fuzzy architecture named *Self Adaptive Fuzzy Inference Network* (SaFIN) is reported in [18], that has the flexibility to incorporate new knowledge into the knowledge base. The combination of SaFIN architecture and the FRIE technique improves the capabilities of the neuro-fuzzy system, that produces the architecture of SaFIN(FRIE). The further enhanced SaFIN(FRIE) with online learning capabilities is referred as SaFIN(FRIE)++.

One of the challenges for neuro-fuzzy systems is making real-time predictions in the financial market where data can be sparse. An evolving Mamdani Fuzzy Inference System (eMFIS) with FRIE technique has been reported which can be used in financial stock prediction [19]. A general episodic memory mechanism (GEMM) to mimic human brain is reported to detect event and predict transient behaviors using FRIE to generate suitable rules [19]. The GEMM mechanism can be adopted in the SaFIN(FRIE)++ system to form a GEMM-SaFIN(FRIE)++ architecture. When the GEMM-SaFIN(FRIE)++ system detects a concept drift, there is a possibility that an event or changes may occur in the operation environment or input dataset. Hence, episode of an event can be captured by the interpolated or extrapolated fuzzy rules. An episodic memory cache mechanism is able to store and retrieve fuzzy rules of similar events happened. The fuzzy rules stored in episodic memory will not be removed, such that they can be recalled handling similar events in the future. But fuzzy rules in the rule base of the GEMM-SaFIN(FRIE)++ system will be removed, if these rules are inactive for a long period of time. It ensures the rule base is always updated with most relevant knowledge.

In order to better understand how the neuro-fuzzy system with GEMM-SaFIN(FRIE)++ architecture perceives data, an explainable AI visualization system is proposed for visualizing, explaining and interpreting the GEMM-SaFIN(FRIE)++ architecture. It is able to greatly provide the aid for users to better understand the mechanism and inner functionality of the neuro-fuzzy systems, which are usually difficult to be achieved in the past. Such features of the proposed explainable AI visualization system bring significant conveniences for parameter tuning and new fuzzy rules being generated for the events of concept drifts. It makes the black-box features of the neuro-fuzzy systems become more transparent to users. The design and implementation of the explainable AI visualization system for the GEMM-SaFIN(FRIE)++ architecture will be described in details in this chapter.

The remaining parts of this chapter is organized as follows. Section 2 introduces the overall architecture, rule generation, and episodic memory mechanism of the GEMM-SaFIN(FRIE)++. Section 3 elaborates the design of the proposed explainable AI visualization system for the GEMM-SaFIN(FRIE)++. Experiments have been conducted in Sect. 4 where experimental results are analysed. Section 5 concludes this chapter.

2 Architecture of GEMM-SaFIN(FRIE)++

2.1 Overall Architecture

The fuzzy inference model of SaFIN [18] is adopted into the GEMM-SaFIN(FRIE)++. The computational structure, reasoning process, and self-automated rule generation process of GEMM-SaFIN(FRIE)++ are described in this section.

The SaFIN model is a five-layered neural fuzzy system [18]. Similarly, there are also five layers in GEMM-SaFIN(FRIE)++ architecture. The input layer as Layer 1 consists of input nodes, that the input vector of the neuro-fuzzy system is connected with. Antecedent nodes are at Layer 2, where the degree of similarity between input values and the membership function embedded at this layer is computed. Layer 3 is the rule layer for *IF–THEN* fuzzy rule nodes. Layer 4 consists of consequent nodes, where the consequent derivation is executed for the fuzzy rule base on the input vector. Defuzzification is performed Layer 5, as the output layer consisting of output nodes, which is connected to the output vector. The GEMM-SaFIN(FRIE)++ architecture with an example networks is shown in Fig. 1.

Layer 3 of GEMM-SaFIN++ encrypts the rule base where each rule node encodes an *IF–THEN* Mamdani-type fuzzy rule [20] shown in Eq. (1):

$$\begin{aligned} R_k\!: &\ If\ x_1\ is\ A_{j1}\ and\ x_2\ is\ A_{j2}\ and \ldots and\ x_p\ is\ A_{jp}, \\ &THEN\ y_1\ is\ C_{m1}\ and\ y_2\ is\ C_{m2}\ and\ \ldots and\ y_q\ is\ C_{mq} \end{aligned} \tag{1}$$

where A_{jp} is the j-th antecedent node associated with the p-th input node that is connected to the rule node R_k; Next C_{mq} is the m-th consequent node for deriving the q-th output node.

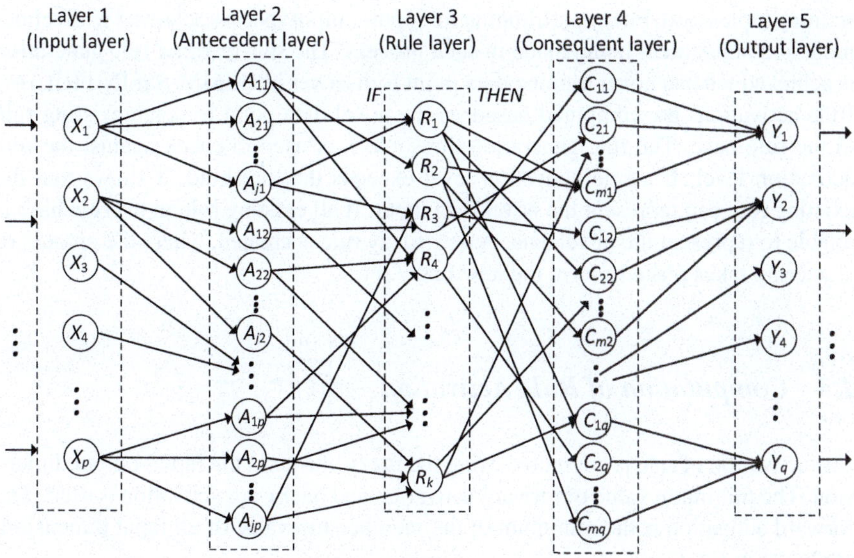

Fig. 1 Example networks of GEMM-SaFIN(FRIE)++ architecture

2.2 *Self-Learning Rule Generation*

The reasoning process and learning method of GEMM-SaFIN(FRIE)++ are similar to those of SaFIN [18]. The learning mechanism of GEMM-SaFIN(FRIE)++ is through automated rule generations according to operation scenarios. Created fuzzy rules are learned from numerical data without involvements of experts and users on predetermined initial rule base. Employing the self-automated rule generation technique addresses the inconsistency and exponential expansion problems encountered in the rule base formulation. Whenever there is a novel data tuple in operating environments, the GEMM-SaFIN(FRIE)++ creates a new corresponding rule and add into the rule base. The inconsistency check is then performed to avoid duplicated or conflicted rules, thus maintaining a compact rule base. Besides, a forgetting factor is incorporated to keep updated with new knowledge while retaining old knowledge. It is to simulate human memory to refreshed by most recent knowledge and gradually phase out old knowledge which are not used for long time. It is to avoid the indefinite growing rule base by just adding all new rules into the rule base. A fuzzy rule is created to capture the knowledge from each of the incoming training tuples. Following that, each of the fuzzy rules in the rule base is assigned a weightage represent its significance in the modelling of the application environment. Conflicting rules with low affect are deemed as outliers and subsequently be removed from the GEMM-SaFIN(FRIE)++ system.

The first fuzzy cluster in each input and output dimension is initialise in the GEMM-SaFIN(FRIE)++. New fuzzy cluster and rule are established from incoming training tuples next. Fuzzy partitioning for input–output space is achieved with refinement and adjustment being made in such process. The self-learning rule generation is achieved when a novel training data point is discover in GEMM-SaFIN(FRIE)++. Rule activations are computed based on the novel data point for each existing rule in the rule base. The rule generation threshold λ is used to check against the rule activation level. If a rule activation level exceeds the threshold, it shows that the existing rule can represent the novel data input. If all existing rule in the rule base is unable to represent the novel data. A new fuzzy rule is created, where the process of clustering takes place to form the new fuzzy set.

2.3 Computation of Rule Activation

The activation of rule is computed using the fuzzy cluster in the input–output dimension. The minimum value of forward activation and backward activation is used. The forward activation is the minimum of the membership value of all input dimension, as shown in Eq. (2).

$$FA_k = \min_{i=1...I}(SV(R_{ki}, X_i)) \tag{2}$$

The backward activation is the minimum of membership value of all output dimension, as shown in Eq. (3).

$$BA_k = \min_{i=1...I}(SV(R_{ki}, Y_i)) \tag{3}$$

where, the notation FA_k represents the forward activation of the i-th fuzzy rule; BA_k represents the backward activation of the i-th fuzzy rule; SV represents the membership functions; R_{ki} is for fuzzy set of rules k in the i-th input dimension; X_i is the input value in the i-th dimension; and Y_i is the output value in the i-th dimension.

Rule activation is the minimum value of the forward and backward activation. If all the rule activation maximum values are larger than the rule generation threshold λ, existing rule can represent new input data point. The weight of the rules will be increased by one unit. Otherwise, it will be updated with the decaying factor.

A new rule is generated when the existing rules are unable to represent the new data. There are three possible cases of creating a new rule.

(a) Creating a new rule based on current fuzzy sets:
It happens when the computation of similar values between the new data and the current fuzzy set in the dimension which is higher than the rule generation threshold λ. That, the current fuzzy set is used to represent the new data.

(b) Modifying current fuzzy to cover new data point:
This happens when a data value and fuzzy set in the dimension falls below the rule generation threshold λ. A condition is called to modify the fuzzy set by

increasing the cluster spread $\sigma(t)$ in Eq. (4). It is to check if the modified fuzzy set can cover the new data point.

$$\sigma(t+1) = \sigma(t) + \eta(1 - SV(R_{ki}, X_i))\sigma(t) \tag{4}$$

where, η is the modification rate.

(c) Interpolating / Extrapolating from the nearest fuzzy rules to create new rule: It happens when the membership value between the data point and the best matched cluster is less than the rule generation threshold λ, while greater than the distance between the new data point and the nearest fuzzy set.

2.4 *Rules Obsoletion*

Rule obsoletions are required for the GEMM-SaFIN(FRIE)++, in order to prevent explosive number of rules stored in the rule base. During training, some rules are created in the rule base that are no longer relevant. There is a need to remove such rules from the rule base when the rules are inactive. It can be achieved by introducing a deletion threshold γ. When a rule is activated, there will be an increase in the rule's weight. When a rule is not activated, its weight is reduced by a forgetting factor. In the experiments of this chapter, the value of the forgetting factor is set as 0.99. It means the weight of an inactivated rule keeps reducing by the value of (0.99 $\times$ *current rule weight*). Hence, It allows the GEMM-SaFIN(FRIE)++ to notice which rule is inactive. Once the rule falls to the deletion threshold γ, it will be removed from the rule base.

2.5 *GEMM Mechanism*

The general episodic memory mechanism of GEMM-SaFIN(FRIE)++ is to store important rules through event detection and episodic recall. The GEMM-SaFIN(FRIE)++ architecture is able to detect concept drifts and shifts [15]. When detecting such concept drifts, it triggers interpolation or extrapolation (IE) to formulate new rules accordingly. Therefore, new rule formed is a result of IE caused by an event being detected. Similar to [19], every event detection by the GEMM-SaFIN(FRIE)++ is considered as an episode, each of which has interpolated or extrapolated rules.

If an event is detected, episode search is conducted to check if same or similar episodes have been stored in the long-term episode memory. If no existing episodes are found, the new rules associated with the episode will be stored in long-term episode memory. These new IE rules are also copied into the existing rule base. However, if same or similar episodes are found in the long-term episode memory, it means previous similar events have been detected and captured by the system.

There is no need to store the episodes and rules. Instead, episodic recall occurs to retrieve stored relevant episodes, that are updated by computing weighted averages with the existing rules and new rules [19]. The updated rules are stored back into the long-term episode memory, as well as into the existing rule base.

If the episodic rules in the rule base have been inactivated for long period of time, they will be deleted from the rule base due to the rule obsoletions. However, all existing episodic rules and information are still retained in the long-term episode memory.

3 Explainable AI Visualization System for GEMM-SaFIN(FRIE)++

One of the main barriers in AI and machine learning models is the explanability. The neuro-fuzzy systems, such as the GEMM-SaFIN(FRIE)++ have complex structures which are difficult to comprehend by human experts or users. Therefore, an effective visualization system is needed to explain and visualize how networks are trained, how nodes are connected, and how fuzzy rules are formed, etc. The explainable AI visualization system is proposed and designed for the GEMM-SaFIN(FRIE)++. Its design process, graphics user interface (GUI), features, and limitations will be described in this section.

3.1 Development Process

The purpose of developing the explainable AI visualization system in the learning process is to give users the idea of how the GEMM-SaFIN(FRIE)++ makes sense of the incoming data, forms its membership function and generates the rules base on its neural fuzzy architecture.

The explainable AI visualization system of the GEMM-SaFIN(FRIE)++ is developed using MATLAB App Designer, that allows applications being built in MATLAB environment. The MATLAB App Designer allows drag and drop of visual components to the design canvas and generates corresponding object-oriented codes.

The features are integrated in the development of the explainable AI visualization system of the GEMM-SaFIN(FRIE)++ are listed as follows.

- Tunable parameters.
- Colour Scheme for each cluster.
- Membership functions graph for input, output, deleted rules and interpolation or extrapolation.
- Display of deleted clusters.
- Indication event of interpolation or extrapolation.
- Merging of clusters.

- Display of generated rules and its weights.
- Display of deleted rules and its weights.

3.2 GUI of Explainable AI Visualization System

The explainable AI visualization system is able to get an insight of how the data moves in the GEMM-SaFIN(FRIE)++ architecture. Getting the right attribute such as the membership centroid and spread is the key of visualizing the fuzzy clusters. Example membership functions to be visualised are shown in Fig. 2. As one of the limitations of MATLAB App Designer, the graph axes currently do not support annotation. If annotation is supported it will be a good indicator of identifying the clusters by users, even if some clusters having the same colour. An ideal representation of the graph with annotation is shown in Fig. 3.

The GUI of the explainable AI visualization system for the GEMM-SaFIN(FRIE)++ is shown in Fig. 4. The fields on the left-hand side are for the inputs to be fed into the system, that include training dataset file, testing dataset file, parameters such as number of input and output dimensions, epochs, etc. Some of tuneable parameters and their meanings are shown as follow:

- Forgetting Factor: value to reduce the rules' weight when they are not fired.
- Lamda (λ): rule generation threshold value.
- Rate (η): modification rate for sigma $\sigma(t)$ (i.e., membership function spread).

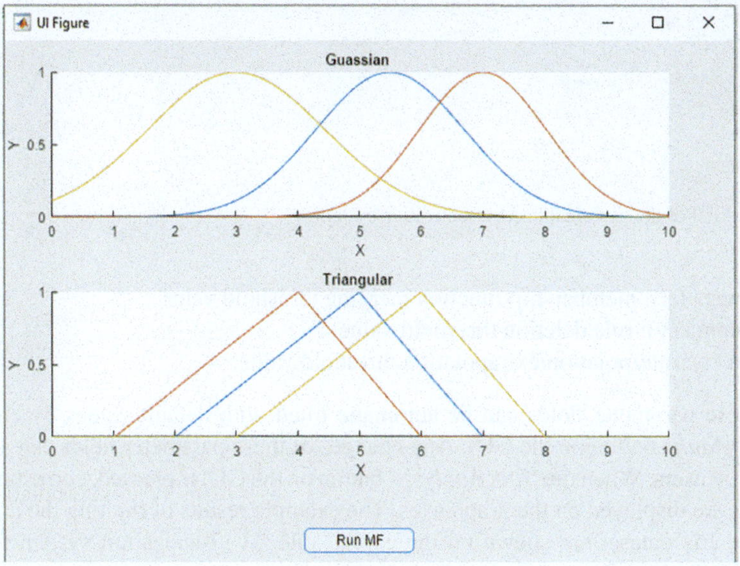

Fig. 2 Example Gaussian and triangular membership functions

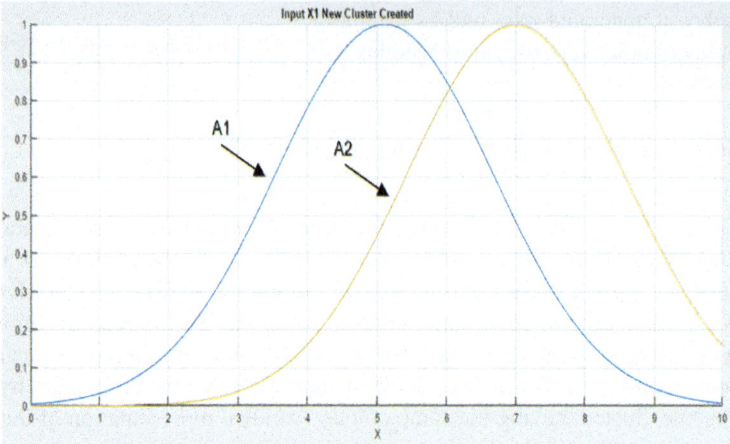

Fig. 3 Membership functions cluster with annotations

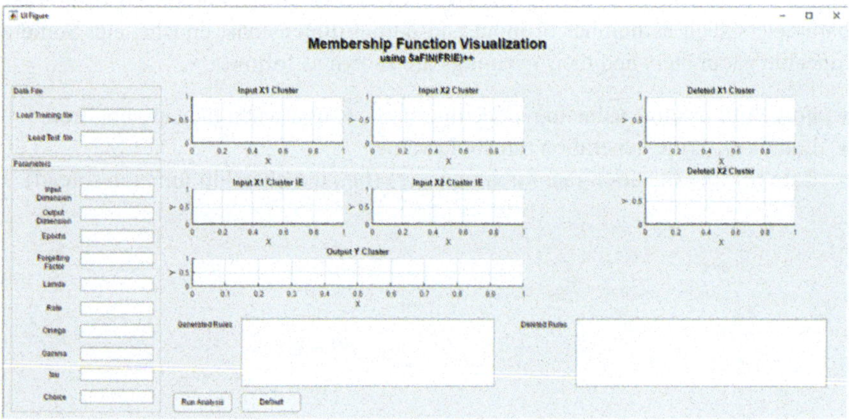

Fig. 4 GUI of the explainable AI visualization system

- Omega (ω): membership function merging threshold value.
- Gamma (γ): rule deletion threshold value.
- Tau (τ): interpolation/extrapolation threshold value.

These parameter fields can be automatic filled with default values by clicking the '*Default*' button on the GUI. Any changes of these parameter fields can also be made by users. When the '*Run Analysis*' button on the GUI is pressed, corresponding results are displayed on the graph axes. The example results of running the analysis for the Iris dataset are shown on the explainable AI visualization system for the GEMM-SaFIN(FRIE)++, in Fig. 5.

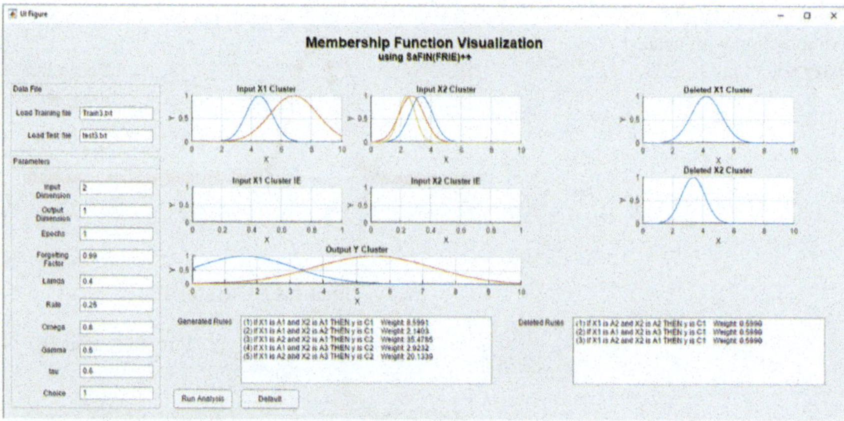

Fig. 5 Analyze Iris dataset in the explainable AI visualization system

The tuneable parameters play huge roles of controlling the clustering, interpolation/extrapolation and deletion of fuzzy *IF–THEN* rules during the learning process for the GEMM-SaFIN(FRIE)++. Changing any of these tuneable parameters will affect its behaviour during learning process. With different combination of threshold values in the tuneable parameters, there will be different results presented to users even if a same dataset is used in an experiment. Therefore, the explainable AI visualization system gives users the idea of how the GEMM-SaFIN(FRIE)++ reacts based on the tuneable parameters and how results are displayed while going through its learning process.

3.3 *Features of Interpolation/Extrapolation*

When the IE happens, the explainable AI visualization system displays a contrasting curve that marks on the input clusters for interpolation, shown in Figs. 6 and 7. The 'red' contrasting Gaussian curve shown in the input cluster window of Fig. 6 is identified as the nearest fuzzy set used for the IE. When the computation of IE is the done, a 'green' contrasting Gaussian curve is shown in the input cluster IE window to indicate the interpolated fuzzy set. Similarly, the IE happens when the data point has a left and right neighbour of nearest fuzzy set marked as "red" in Fig. 7. The interpolated results are marked as "green" in the input cluster IE window.

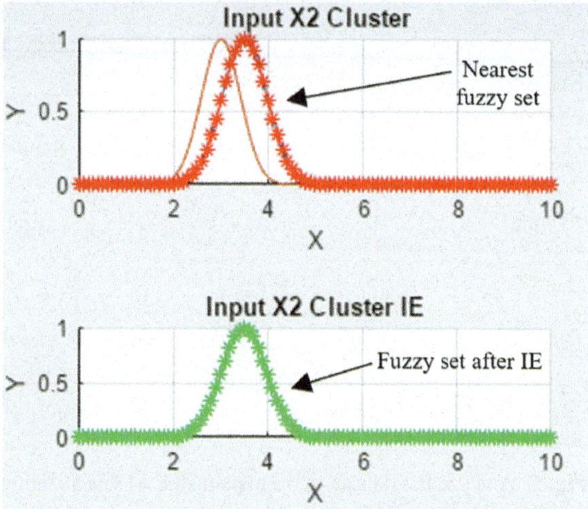

Fig. 6 Interpolation/extrapolation with nearest fuzzy set

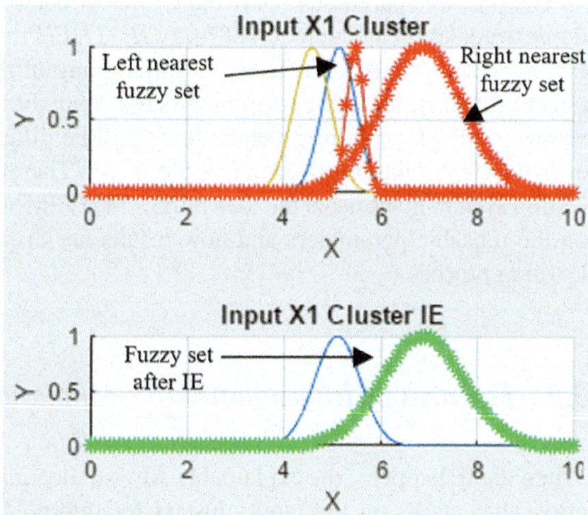

Fig. 7 Interpolation/extrapolation with left and right nearest fuzzy set

3.4 Merging of Membership Functions

The merging of membership functions occurs when the neighbouring clusters are above the membership function merging threshold value ω (i.e., value of Omega). The process can be illustrated by the explainable AI visualization system. Example membership functions of two clusters before merging and after merging into one fuzzy set are shown in Figs. 8 and 9.

Similarly, the output clusters are handled by the same principal of the merging of membership functions. The output membership functions are merge when one of the fuzzy sets overlays one another, as shown in Fig. 10.

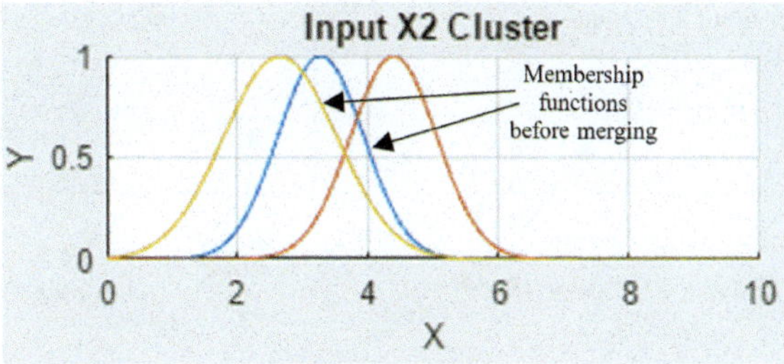

Fig. 8 Example membership functions before merging

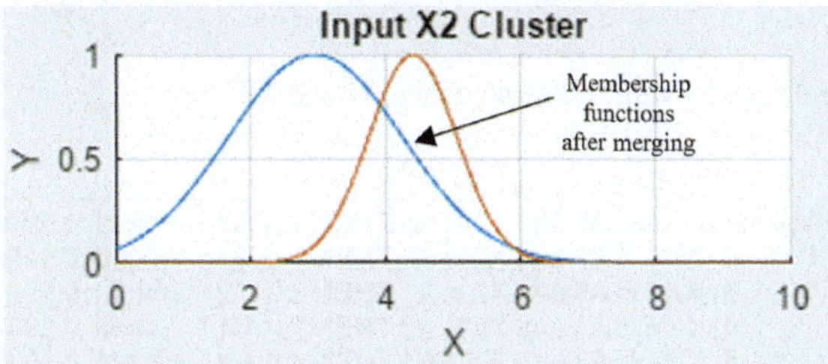

Fig. 9 Example membership functions after merging

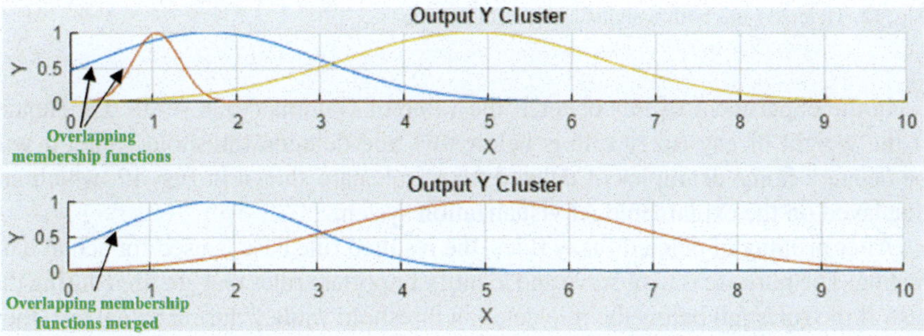

Fig. 10 Overlapping membership functions before and after merging in the output cluster

3.5 Deletion of Rules

An *IF–THEN* fuzzy rule is deleted from the rule base, when its weight is lower than the rule deletion threshold value γ (i.e., the value of Gamma). When the fuzzy rule is deleted, the input cluster graphs show the membership functions with respect to the deleted cluster, giving a sense to users of which antecedents belonging to the rule being deleted. Examples of deleted clusters are shown on the explainable AI visualization system, as in Fig. 11.

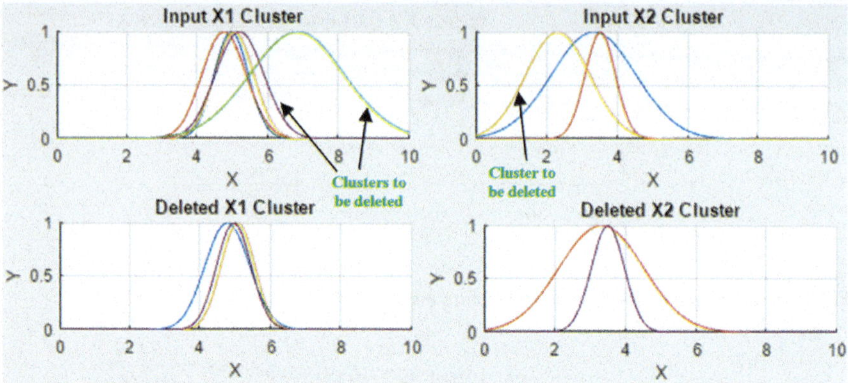

Fig. 11 Deleted rule cluster with respect to inputs X1 and X2

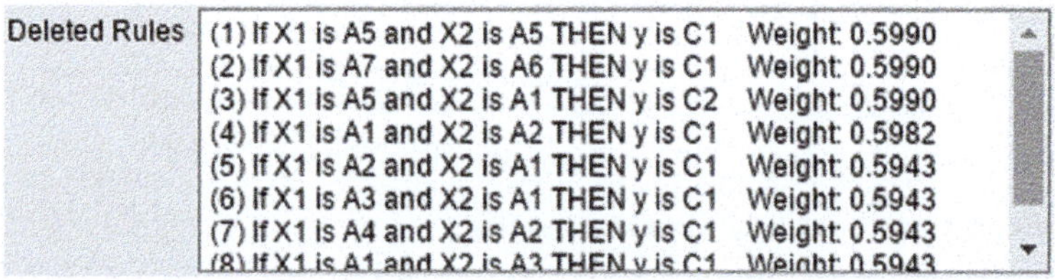

Fig. 12 Deleted rules with weights lower than γ value

In the experiment of this chapter, the value of Gamma (γ) is set to 0.6. Hence, if the weight of any fuzzy rule is below this rule deletion threshold value, it will be deleted. Some examples of deleted fuzzy rules are shown in Fig. 12, which are displayed on the explainable AI visualization system.

After pruning of deleted fuzzy rules, the resulted rule base is used for actual data testing. The purpose is to observe and identify important rules that are fired during the test. If the rules fall below the rule deletion threshold value γ during actual test, they will also be deleted. After running through the test dataset, some examples of rules being kept are shown in Fig. 13 with their weights. In this case, the third rule shown in Fig. 13 is identified as the most important rule, as its weight is significantly high compared to other rules. A possible reason is that this rule has been fired numerous times in the operations. Each time when it gets fired, its weight increases and gets stronger.

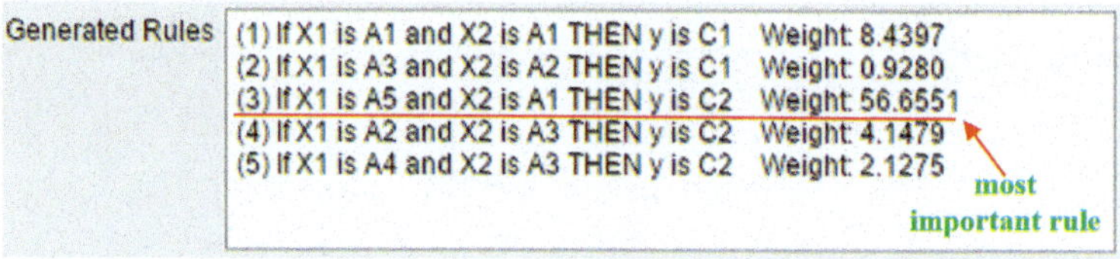

Fig. 13 Rules with their weight after training and testing

3.6 Neuro-fuzzy Network in Explainable AI Visualization System

Multiple layer neuro-fuzzy networks of the GEMM-SaFIN(FRIE)++ will be constructed and connected in the learning process. With the need of explainable AI features to users, the visualization system also needs to animate such connecting process dynamically. But attempts of developing realistic animation processes in MATLAB is not easy and not straightforward. Web application programming provides powerful libraries and animation design, which is an ideal method for creating the visualization. Therefore, web application programming using JavaScript is selected for this part of the explainable AI visualization system.

The features of the explainable AI visualization system illustrating the neuro-fuzzy networks of the GEMM-SaFIN(FRIE)++ are shown as follows.

- Constructing the five layers of neuro-fuzzy networks.
- Creating nodes based on the number of fuzzy clusters from the GEMM-SaFIN(FRIE)++.
- Linking of nodes of the layers based on the fuzzy rules generated from the GEMM-SaFIN(FRIE)++.
- Updating dynamically the thickness of links according to varying weights associated to the rules.
- Animating the rules being activated (i.e., fired) in the neuro-fuzzy networks.

The explainable AI visualization system of the GEMM-SaFIN(FRIE)++ presents and simulates the neuro-fuzzy networks to users how the rules are formed during learning process by linking nodes in these five layers. After the learning process, GEMM-SaFIN(FRIE)++ constructs corresponding nodes at each layer. The number of nodes in each layer reflects the number of membership fuzzy set that GEMM-SaFIN(FRIE)++ creates.

The neuro-fuzzy networks connect the nodes from Layer 1 (the input layer) to Layer 5 (the output layer), according to each rule in its rule base. To better show the linking procedure of the nodes for each rule, the dynamic network connections are illustrated using the Iris dataset. With this dataset, the fuzzy rules are generated and listed in Fig. 14. It is observed that the first rule generated is "K_1 : *If* X_1 is A_1 and X_2 is A_4 *Then* y is C_1". Its linking procedure is demonstrated and visualized in Fig. 14. The node X_1 at the input layer is connected to the node A_1 at the antecedent layer;

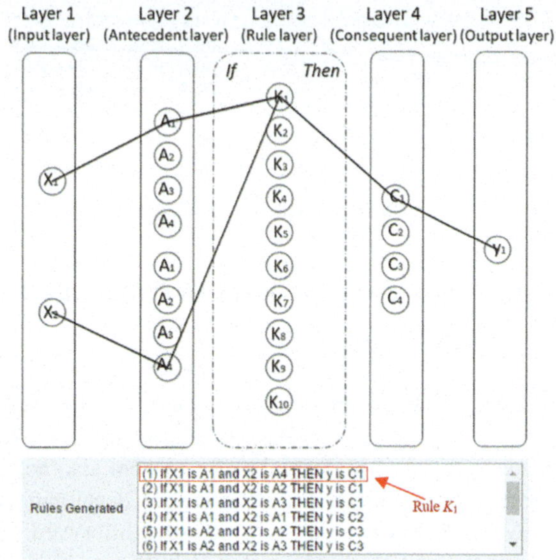

Fig. 14 Linking the first rule from the rule base

while the node X_2 at the input layer is connected to the node A_4 at the antecedent layer. They are further connected to the node K_1 at the rule layer, that is next connected to the node C_1 at the consequent layer, and the node Y_1 at the output layer.

Following the same procedure in the explainable AI visualization system, the linking of the second rule generated, i.e., "K_2 : *If* X_1 is A_1 and X_2 is A_2 *Then* y is C_1" is visualized in Fig. 15. The nodes X_1 and X_2 at the input layer are connected to the nodes A_1 and A_2 at Layer 2, respectively. These two links are connected in sequence to the node K_2 at Layer 3, the node C_1 at Layer 4, and the node Y_1 at Layer 5.

Similarly, the linking procedure of all rules in the rule base are conducted that produce a fully connected neuro-fuzzy network by the explainable AI visualization system. A fully connected neuro-fuzzy network of the GEMM-SaFIN(FRIE)++ for the Iris dataset is shown in Fig. 16.

For rules with different weights in the GEMM-SaFIN(FRIE)++ neuro-fuzzy networks, the proposed explainable AI visualization system displays the links associated with corresponding thickness. In the learning progress, the weight of a rule will be increased dynamically if a constant firing of that rule occurs. Hence, the explainable AI visualization system changes the link thickens of rules, giving a visual meaning to users as that the weights of rules are changed. The thicker of the link width, the stronger weight of the rule is, as shown in Fig. 17. The thinner of the link width, the weaker weight of the rule is.

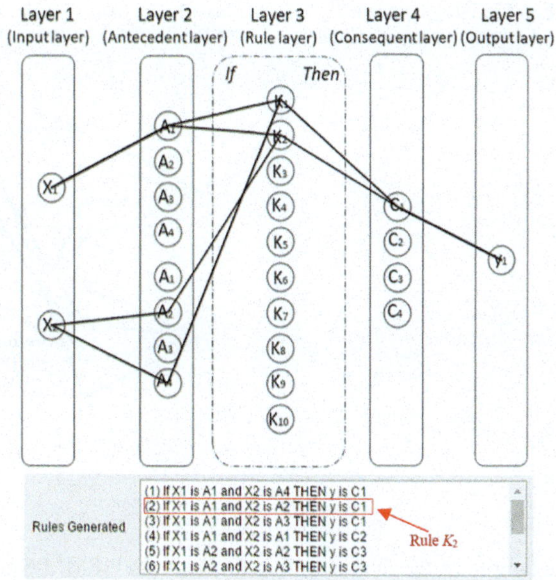

Fig. 15 Linking the second rule from the rule base

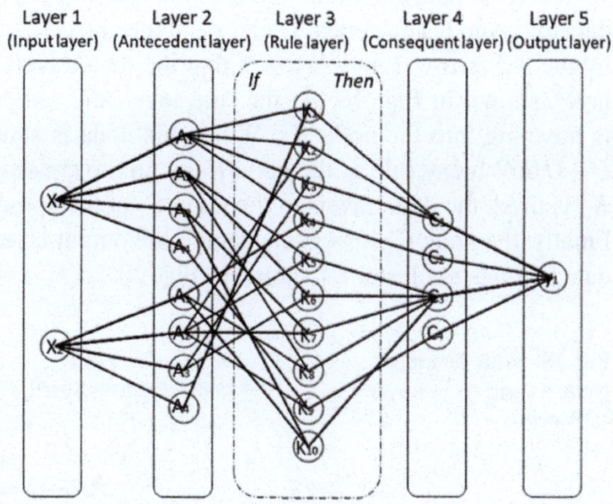

Fig. 16 Networks with fully connected rules for the Iris dataset

3.7 *Animating Activation of Rules in Explainable AI Visualization System*

The animating feature of the proposed visualization system illustrates the activation proces of rules that is being fired. Showing a traversing of data point from the input layer to the output layer through three hidden layers. If the weight of a rule is strong enough, it will trigger the node and fires that rule. The procedure is illustrated for an

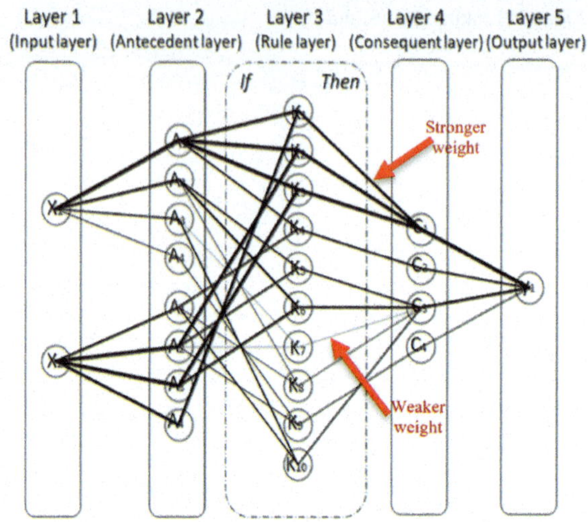

Fig. 17 Fully connected rules with dynamic weights

example of the data traveling through the GEMM-SaFIN(FRIE)++ and activating a rule in Figs. 18, 19 and 20.

When the data is fed into the GEMM-SaFIN(FRIE)++, the input layer injects the data to Layer 2. The nodes A_1 and A_2 at the antecedent layer are activated, indicating by the red Arrow 1 and Arrow 2 that the data travels through the link into the rule layer, shown in Fig. 18. At the rule layer, the red Arrow 3 shows the data point is traveling into the node K_2. When this node is activated, it will be complete the *IF–THEN* fuzzy rule as the node K_2 is link to consequent node C_1. When the node K_2 is fired, the data travels to the node C_1 at the consequent layer shown in Fig. 19. Finally, the node C_1 injects the data to the output layer, giving the conclusion of the data belongs to cluster C_1 shown in Fig. 20.

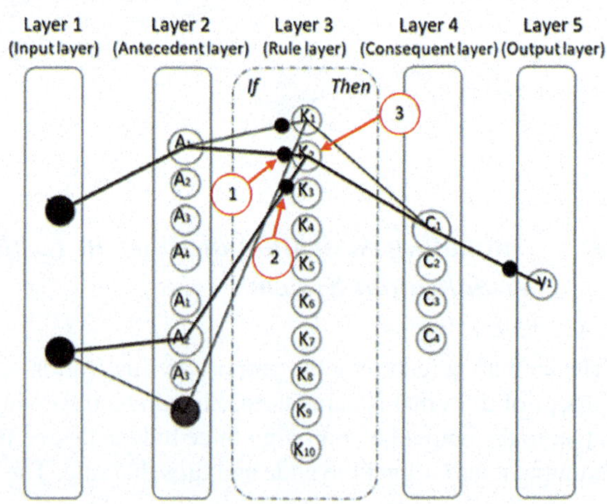

Fig. 18 With data input, Node A_1 and A_2 in Layer 2 is activated

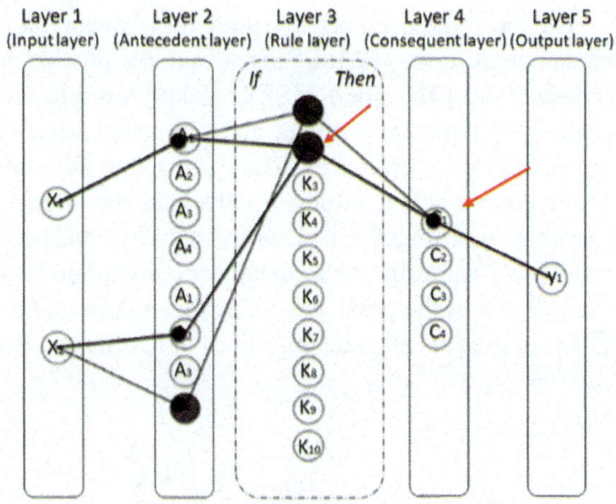

Fig. 19 Node K_2 in Rule layer is activated next

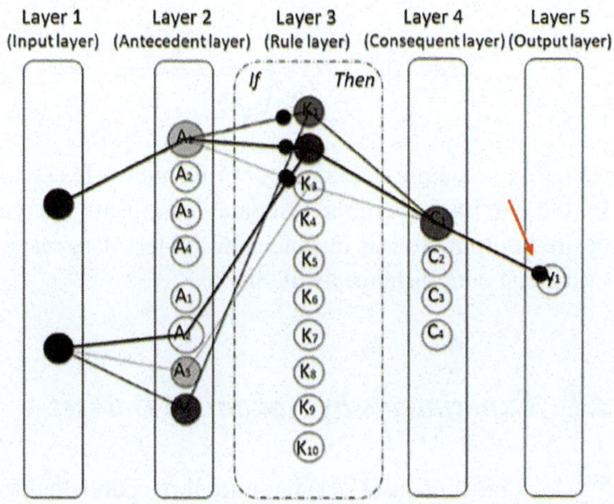

Fig. 20 Node C_1 in Consequent layer activated, before to output layer

In this research, the animations of the entire procedure have been incorporated into the explainable AI visualization system. It brings great convenience to users to visualise and better understand how the GEMM-SaFIN(FRIE)++ works in the training and testing with input data in real operations.

4 Experimental Analysis and Benchmarking

To evaluate the performance of the GEMM-SaFIN(FRIE)++ with the support of the explainable AI visualization system, several experiments and benchmark tests are conducted. Experiments include Nakinishi dataset [21] and event detection for stock markets crisis, such as events during COVID-19, subprime, and 9–11.

The performance metrics used to evaluate the neuro-fuzzy models are root mean-square error (RMSE) and Pearson's product-moment correlation coefficient (Pearson's r) [22]. The RMSE is usually used to measure the differences between values predicted by a model and the actual values. The lower the RMSE value, the smaller the differences. The Pearson's r measures the linear correlation of the predicted and actual values. It calculates the strength of relationship between two variables, and the effect to one variable when the other changes. The higher the Pearson's r value, the better the linear correlation. A perfect positive correlation indicates that both the predicted and actual values move in the same direction together. The formulas to calculate the values of RMSE and Pearson's r are shown in Eqs. (5) and (6).

$$RMSE = \left(\frac{1}{P} \sum_{p=1}^{P} \left(o_p - y_p\right)^2 \right)^{\frac{1}{2}} \tag{5}$$

$$Pearson's\ r = \left(\frac{\sum_{p=1}^{P} \left(o_p - \overline{o}\right)\left(y_p - \overline{y}\right)}{\sqrt{\sum_{p=1}^{P} \left(o_p - \overline{o}\right)^2} \sqrt{\sum_{p=1}^{P} \left(y_p - \overline{y}\right)^2}} \right)^2 \tag{6}$$

where o_p is the predicted output of the neuro-fuzzy system; y_p is the desired output for the p-th instance of the test data; $\overline{o}$ and $\overline{y}$ are the mean values of the predicted and desired outputs for all instances of the test data respectively; P is the total number of the test data instance available.

4.1 Experiments by Nakanishi Dataset

The Nakanishi dataset [21] includes three published sub datasets on: (1) a non-linear system; (2) the human operation of a chemical plant; and (3) the daily price of stock in a stock market. The datasets are small containing only 25, 35 and 50 training data tuples respectively, where predicting could be relatively difficult. The three sub-datasets are used in three experiments, respectively. The GEMM-SaFIN(FRIE)++ is benchmark against with another three neuro-fuzzy systems: PIE-RSPOP [23], eMFIS(FRI/E), and eMFIS(FRI/E)++. PIE-RSPOP is a Rough Set Pseudo Outer Product (POP) fuzzy neural network which consist of rough set rule and attribute reduction [23]. eMFIS(FRI/E) is an Evolving Mamdani Fuzzy Inference System with Fuzzy Rule Interpolation and Extrapolation [19, 24], that can detect concept drifts or shift. While eMFIS(FRI/E)++ is an improved version of eMFIS(FRI/E), where the notation '++' means that the model can perform online learning.

In these experiments, the parameters for the GEMM-SaFIN(FRIE)++ used are the same as shown in Fig. 5 in *Sect.* 3.2: Forgetting Factor = 0.99, Lamda = 0.4, Rate = 0.25, Omega = 0.8, Gamma = 0.6, Tau = 0.6. In the experiments for the

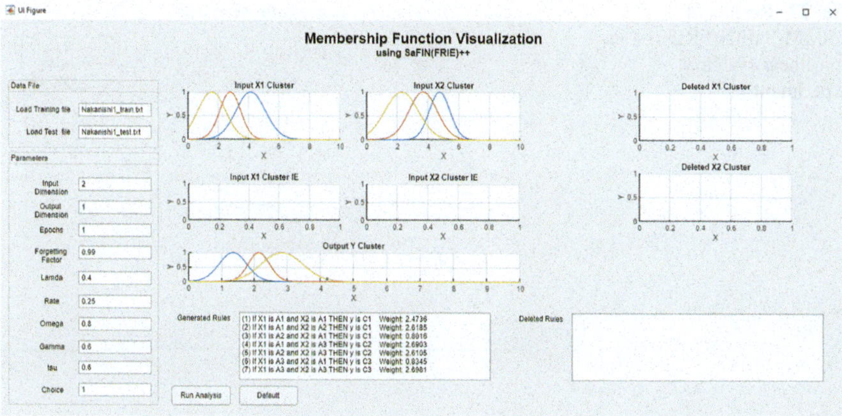

Fig. 21 The explainable AI visualization system for non-linear system experiment

explainable AI visualization system, the training and test data are normalized into the range [0, 10].

4.1.1 Experiment 1: A Non-linear System

This experiment is to identify and model the basic principles of a non-linear system. The original dataset [21] consists of four input variables (x_1, x_2, x_3, x_4), one output (y) variable, and 25 training data tuples. The variables x_1 and x_2 are used for inputs of the GEMM-SaFIN(FRIE)++ in the experiment. The semantic labels of antecedent and consequent clusters are set as {*A*1—*Low*, *A*2—*Medium*, *A*3-*High*} and {*C*1—*Low*, *C*2—*Medium*, *C*3-*High*}, respectively. The dimensions of both input spaces are set to be three clusters. The dimension of the output space is set as three clusters. The clusters generated for input and output spaces and the induced rules are shown in Fig. 21 during the training and test of the GEMM-SaFIN(FRIE)++. The output results of the GEMM-SaFIN(FRIE)++ are shown in Fig. 22. It is observed that the interpolations/extrapolations are triggered to formulate new rules at the green plots by the GEMM-SaFIN(FRIE)++.

The result comparisons of GEMM-SaFIN(FRIE)++ with other three neuro-fuzzy systems are shown in Table 1. It is observed that GEMM-SaFIN(FRIE)++ achieves the best RMSE value compared to the rest. For the value of Pearson's r, eMFIS(FRI/E) has the best result, with just slightly higher than that of GEMM-SaFIN(FRIE)++.

4.1.2 Experiment 2: Human Operation of a Chemical Plant

GEMM-SaFIN(FRIE)++ is employed to model the human operation of a chemical plant in this experiment. The Nakanishi dataset [21] consists of five input variables

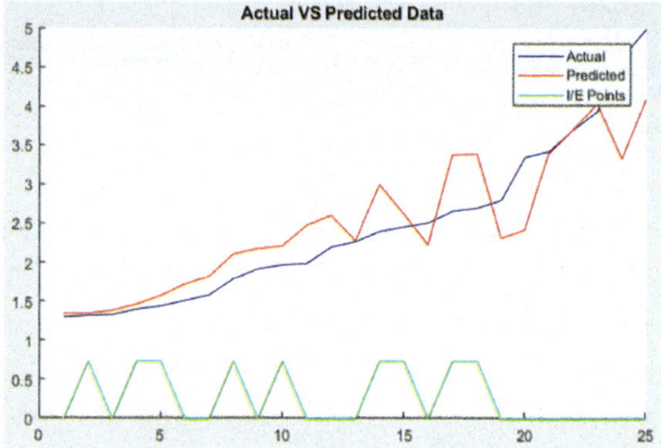

Fig. 22 Results of GEMM-SaFIN(FRIE)++ in non-linear system experiment

Table 1 Result comparisons in Nakanishi non-linear system experiment

Model	RMSE	Pearson's r	No. Rules
GEMM-SaFIN(FRIE)++	0.483	0.883	20
PIE-RSPOP [23]	0.740	0.766	32
eMFIS(FRI/E) [24]	0.555	0.896	3
eMFIS(FRI/E)++	0.777	0.861	2

$(x_1, x_2, x_3, x_4, x_5)$, one output (y) variable, and 35 training data tuples. The semantic labels of antecedent and consequent clusters are set as $\{A1$—*Low*, *A2*—*Medium*, *A3*—*High*, *A4*—*Very High*$\}$ and $\{C1$—*Low*, *C2*—*Medium*, *C3*—*High*, *C4*—Very *High*$\}$, respectively. The dimension of one input space is set to be three clusters, with the other input space being set to four clusters. The dimension of the output space is set as four clusters. The clusters are generated for input and output spaces, and the rules are induced as shown in Fig. 23 during the training and test of the GEMM-SaFIN(FRIE)++. The experiment results of GEMM-SaFIN(FRIE)++ are shown in Fig. 24. With the increment of training data, predicted results can better follow the actual data, through the rules interpolation/extrapolation.

The experiment result comparisons are shown in Table 2. GEMM-SaFIN(FRIE)++ achieves the second best RMSE value, and also produces a second highest result in the Pearson's r value. However, it loses out to other models in high number of rules generated. eMFIS(FRI/E) achieves best RMSE value, and fewest number of rules generated. PIE-RSPOP achieves the highest Pearson's r value.

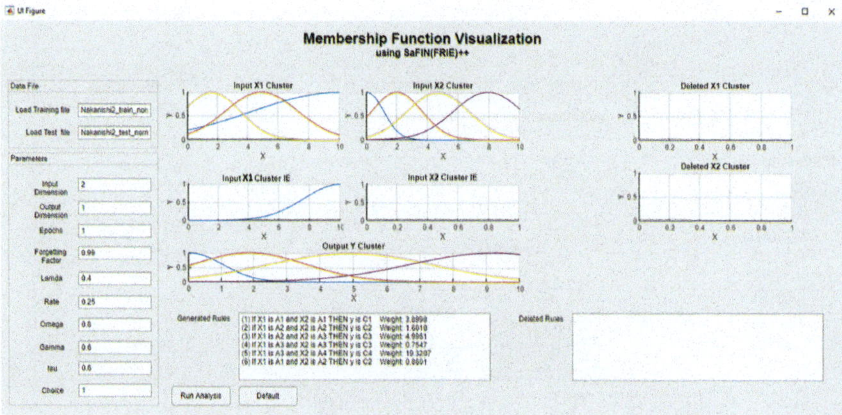

Fig. 23 The explainable AI visualization system for the chemical plant experiment

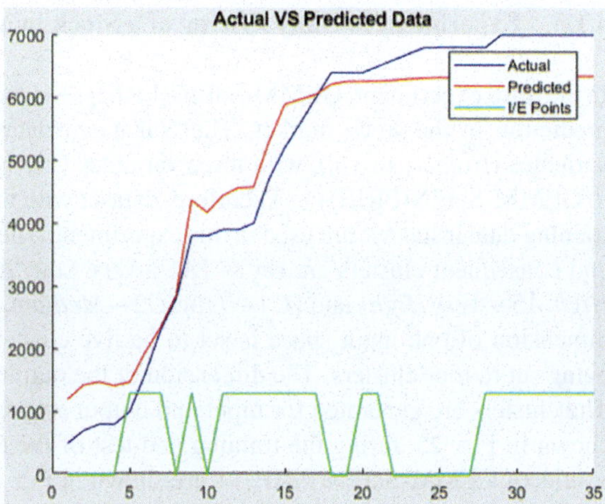

Fig. 24 Results of GEMM-SaFIN(FRIE)++ in human operation of chemical plants experiment

Table 2 Result comparisons in Nakanishi human operation of chemical plants experiment

Model	RMSE	Pearson's r	No. Rules
GEMM-SaFIN(FRIE)++	512.42	0.988	29
PIE-RSPOP [23]	526.84	0.990	14
eMFIS(FRI/E) [24]	488.34	0.985	3
eMFIS(FRI/E)++	1887.00	0.593	3

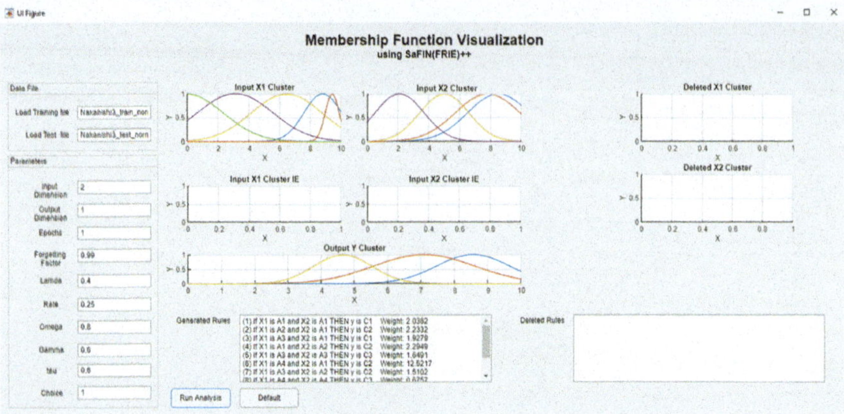

Fig. 25 The explainable AI visualization system for the stock price prediction experiment

4.1.3 Experiment 3: Daily Pricing of a Stock in a Stock Market

In the third experiment, GEMM-SaFIN(FRIE)++ is employed to perform stock price prediction in the stock market. The Nakanishi dataset [21] consists of ten input variables (from x_1 to x_{10}), with three variables (x_4, x_5, x_8) being selected as inputs to GEMM-SaFIN(FRIE)++. Therefore, dataset with three inputs, one output, and 50 training data instances are used in this experiment. The semantic labels of antecedent and consequent clusters are set as $\{A1$—*Very Low*, $A2$—*Low*, $A3$—*Medium*, $A4$—*High*, $A5$—*Very High*$\}$ and $\{C1$—*Low*, $C2$—*Medium*, $C3$—*High*$\}$, respectively. The dimension of one input space is set to be five clusters, with the other input space being set to four clusters. The dimension of the output space is set as three clusters. The clusters are generated for input and output spaces, and the rules are induced as shown in Fig. 25 during the training and test of the GEMM-SaFIN(FRIE)++. The results of GEMM-SaFIN(FRIE)++ are shown in Fig. 26.

Observed from Table 3, noticed that eMFIS(FRI/E) has the lowest RMSE value and highest Pearson's r value. However, it generated too high number of rules. On the other hand, GEMM-SaFIN(FRIE)++ produce a reasonably well result. Even though it has just slightly lower RMSE value and Pearson's r value than those of eMFIS(FRI/E), it achieves good accuracy with only 24 number of rules generated.

It is observed in the results of these three experiments shown in Tables 1, 2 and 3, overall GEMM-SaFIN(FRIE)++ achieves good performance in high Pearson's r value, low in RMSE, with reasonable number of rules generated.

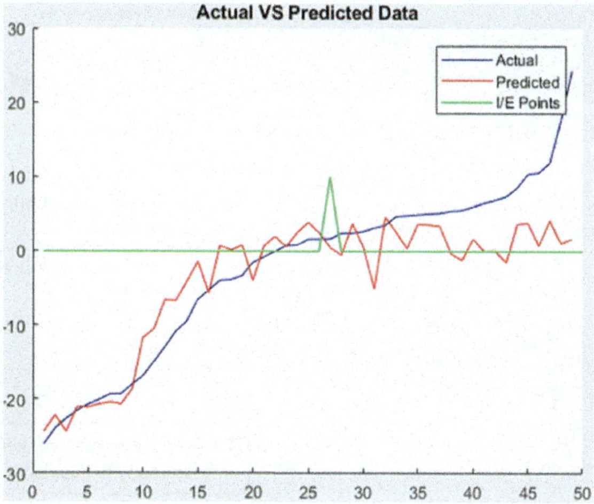

Fig. 26 Results of GEMM-SaFIN(FRIE)++ in daily stock price experiment

Table 3 Result comparisons in Nakanishi daily stock price experiment

Model	RMSE	Pearson's r	No. rules
GEMM-SaFIN(FRIE)++	5.803	0.886	24
PIE-RSPOP [23]	7.649	0.868	56
eMFIS(FRI/E) [24]	4.767	0.922	64
eMFIS(FRI/E)++	20.732	0.408	3

4.2 *Event Detection of Stock Market Crisis*

In this section, GEMM-SaFIN(FRIE)++ is used to explore stock market index for event detection. S&P 500 and DJI index are used in this experiment. S&P 500 is a stock market index that tracks the stocks of 500 large companies listed on stock exchanges in the United States. Dow Jones Industrial Average (DJIA) is a widely watched benchmark index tracking 30 larges listed companies trading in the New York Stock Exchange (NYSE). The stock index dataset contains historical data where it has records through the period during COVID-19 (Coronavirus Disease of 2019), subprime and September 11 attack. These events have caused stock market crashing. It is good to test the ability of GEMM-SaFIN(FRIE)++ to detect market trend reversal for such events.

4.2.1 Event of COVID-19

On March 12, 2020, The World Health Organization (WHO) declared COVID-19 as global pandemic which affected the stock market turning into a bear market. The S&P 500 stock index dataset are processed by the GEMM-SaFIN(FRIE)++. The results of the event detection for S&P 500 data in the period of COVID-19 is shown in Fig. 27.

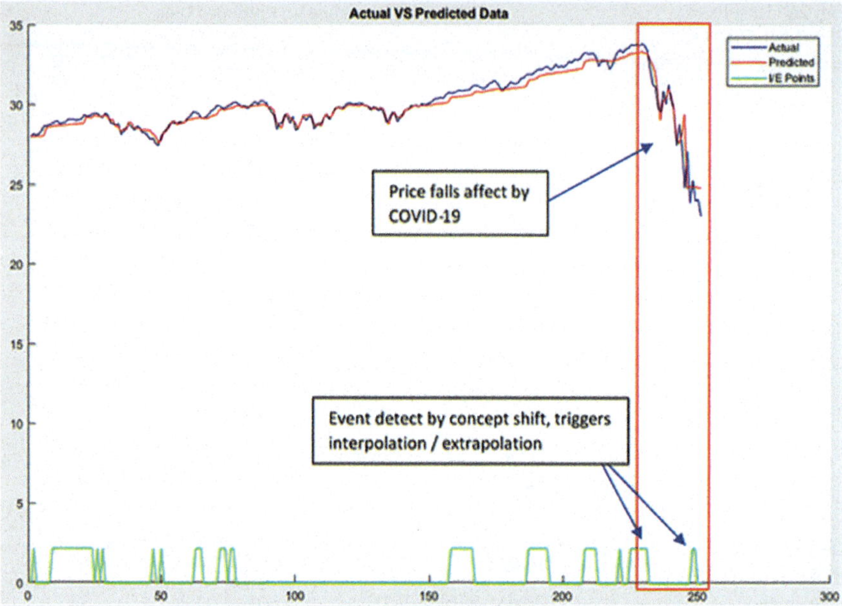

Fig. 27 Detection of S&P500 period from March 2019–March 2020 by GEMM-SaFIN(FRIE)++

The reversal of the index price is detected by the GEMM-SaFIN(FRIE)++, that triggers the interpolation/extrapolation to formulate new rules accordingly, shown in the green plots and red rectangle of Fig. 27. These generated rules are stored in the episodic memory of the neuro-fuzzy system.

4.2.2 Event of Subprime

The financial crisis of 2007–2008, known as the global financial crisis was a severe worldwide economic crisis. It is considered as the most serious financial crisis considered by many economists. The crisis began in 2007 with a depreciation in the subprime mortgage market in the United States. The collapse of the investment bank Lehman Brothers developed into an international banking crisis. The results of the event detection using the GEMM-SaFIN(FRIE)++ for S&P 500 data in the period of subprime is shown in Fig. 28.

S&P 500 contains stock data from Jan 2007 to March 2012. Observed in the red rectangle in Fig. 28, there is a noticeable steep fall of stock price which is caused by subprime. Such fall of the prices is detected by GEMM-SaFIN(FRIE)++ as the concept shifts. Hence, the detection of this event triggers the interpolation/ extrapolation to formulate rules for this crisis, represented by the green plots. All the interpolated/extrapolated rules are store in the episodic memory, that can be recalled if similar event occurs in the future.

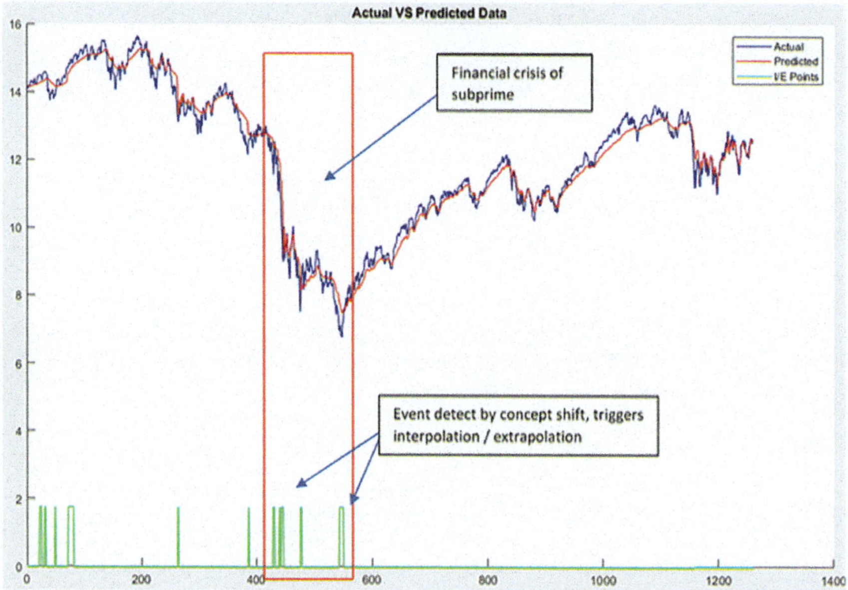

Fig. 28 Detection of S&P500 period from Jan 2007–Jan 2012 by GEMM-SaFIN(FRIE)++

4.2.3 Event of September 11 Crisis

The 9–11 attack caused global stock markets to drop sharply. The NYSE did not open for trading to prevent a stock market meltdown during September 11 - 17, 2001. The Dow Jones fell down sharply in five days. The event detected from DJIA data by the GEMM-SaFIN(FRIE)++ during the period of September 11 crisis is shown in Fig. 29. Observed that GEMM-SaFIN(FRIE)++ is able to detect the concept shifts from the sharp stock price falls during this period, that triggers interpolation/extrapolation to formulate rules of such event shown by the green plots in this figure.

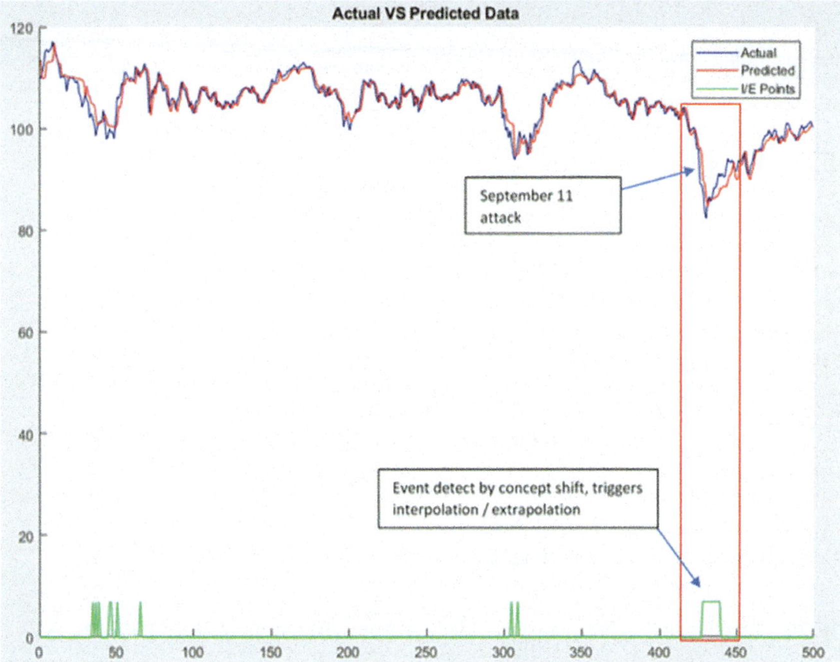

Fig. 29 Detection of DJIA period from Jan 2000–Jan 2002 by GEMM- SaFIN(FRIE)++

5 Conclusions and Future Work

In this chapter, the explainable AI visualization system has been proposed for the GEMM-SaFIN(FRIE)++. The proposed explainable AI visualization system provides user options to tune parameters and choose training/testing data for the learning inference of the GEMM-SaFIN(FRIE)++. It is able to produce clear procedural illustrations during its learning and categorical phase. It gives transparency to users on how clustering and interpolation/extrapolation are performed, how rules are formed, and which rules are generated. The proposed explainable AI visualization system also provides the visualising the connections in the neuro-fuzzy structure of GEMM-SaFIN(FRIE)++. It illustrates how results are produced from the GEMM-SaFIN(FRIE)++ reflected as nodes in five layers, including the input, Antecedent, Rule, Consequent and output layers. Linking of the nodes is done when a rule is formed. The proposed visualization system is able to present how data flow in the neuro-fuzzy network and how the rules are being fired based on the strength (i.e., the weight) of the links.

The GEMM-SaFIN(FRIE)++ is benchmarked against with other three existing neural system. It demonstrates good experiment results, in terms of RMSE value, Pearson's r value, and number of rules generated. In addition, the GEMM-SaFIN(FRIE)++ has interpolation/extrapolation features coupled with episodic memory for event detections. Experiments have been conducted for the GEMM-SaFIN(FRIE)++ to detect events for crisis in stock index prices. Experiment results for event detections are encouraging and able to generate rules to capture a shift in stock prices. These rules are stored separately from the rule base of the system to prevent deletion if the rule is inactive for a long period of time. Rules in the episodic memory are useful when similar event happens in the future that can be recalled for corresponding predictions.

The proposed explainable AI visualization system provides insightful information about the GEMM-SaFIN(FRIE)++. Future research could be conducted to further improve the user interactivity of the visualization system. It could also be improved with a better episodic memory mechanism. The current method of episodic memory is storing rules from detection of the concept shift. An improved scheme of episodic memory can be designed to not only store the generated rules, but also categorize these rules base on the severity of the events.

References

1. P. V. Souza, "Fuzzy Neural Networks and Neuro-Fuzzy Networks: A Review the Main Techniques and Applications used in the Literature," Applied Soft Computing, vol. 92, 2020.
2. Z. Pezeshki, and S. M. Mazinani, "Comparison of Artificial Neural Networks, Fuzzy Logic and Neuro Fuzzy for Predicting Optimization of Building Thermal Consumption: A Survey," Artificial Intelligence Review, vol. 52, 2019.
3. W. L. Tung, and H.C. Quek, "eFSM - A Novel Online Neural-Fuzzy Semantic Memory Model," IEEE Transactions on Neural Networks, vol. 21, no. 1, pp. 136-157, 2010.
4. G. Tiruneh, A. Fayek, and V. Sumati, "Neuro-fuzzy Systems in Construction Engineering and Management Research," Automation in Construction, vol. 119, 2020.
5. L. L. X. Yeo, Q. Cao, and C. Quek, "Dynamic Portfolio Rebalancing with Lag-Optimised Trading Indicators using SeroFAM and Genetic Algorithms," Expert Systems with Applications, vol. 216, 2023.
6. T. Miller, "Explanation in Artificial Intelligence: Insights from the Social Sciences," Artificial Intelligence, vol. 267, pp. 1–38, 2019.
7. S. Sachan, J. B. Yang, D. L. Xu, D. E. Benavides, and Y. Li, "An Explainable AI Decision-Support-System to Automate Loan Underwriting," Expert Systems with Applications, vol. 144, 2020.
8. Q. Zhang, Y. Nian Wu, S. C. Zhu, "Interpretable Convolutional Neural Networks," IEEE Conference on Computer Vision and Pattern Recognition, 2018.
9. P. Dassanayake, A. Anjum, A. K. Bashir, *et al.*, "A Deep Learning Based Explainable Control System for Reconfigurable Networks of Edge Devices," IEEE Transactions on Network Science and Engineering, 2021.
10. J. M. Mendel, and P. P. Bonissone, "Critical Thinking About Explainable AI (XAI) for Rule-Based Fuzzy Systems," IEEE Transactions on Fuzzy Systems, vol. 29, no. 12, pp. 3579 – 3593, 2021.
11. S. Löbner, W. B. Tesfay, T. Nakamura, and S. Pape, "Explainable Machine Learning for Default Privacy Setting Prediction," IEEE Access, vol. 9, 2021.
12. H. F. Cheng, R. Wang, Z. Zhang, *et al.*, "Explaining Decision-Making Algorithms through UI: Strategies to Help Non-Expert Stakeholders," CHI Conference on Human Factors in Computing Systems, 2019.
13. J. Gama, I. Žliobaitė, A. Bifet, M. Pechenizkiy, and A. Bouchachia, "A Survey on Concept Drift Adaptation," ACM Computing Surveys, vol. 46, no. 4, pp. 1–37, 2014.

14. B. Halstead, Y. S. Koh, P. Riddle, R. Pears, M. Pechenizkiy, A. Bifet, G. Olivares, and G. Coulson, "Analyzing and Repairing Concept Drift Adaptation in Data Stream Classification," Machine Learning, 2021.
15. K. Nishida, and K. Yamauchi, "Learning, Detecting, Understanding, and Predicting Concept Changes," International Joint Conference on Neural Networks, 2009, pp. 2280–2287.
16. M. Alzubi, Z. Johanyák, and K. Szilveszter, "Fuzzy Rule Interpolation Methods and FRI Toolbox," Journal of Theoretical and Applied Information Technology, vol. 96, no. 2, 2018.
17. C. Chen, N. Parthaláin, Y. Li, C. Price, H. C. Quek, and Q. Shen, "Rough-fuzzy Rule Interpolation," Information Sciences, vol. 351, pp. 1-17, 2016.
18. S. W. Tung, H. C. Quek, and C. Guan, "SaFIN: A Self-Adaptive Fuzzy Inference Network," IEEE Transactions on Neural Networks, vol. 22, no. 12, pp. 1928-1940, 2011.
19. S. W. Pang, H. C. Quek and D. K. Prasad, "GEMM-eMFIS (FRI/E): A Novel General Episodic Memory Mechanism for Fuzzy Neural Networks," International Joint Conference on Neural Networks, 2020, pp. 1–8.
20. F. Liu, H. C. Quek, and G. S. Ng, "A Novel Generic Hebbian Ordering-Based Fuzzy Rule Base Reduction Approach to Mamdani Neuro-Fuzzy System," Neural Computation, vol. 19, no. 6, pp. 1656-1680, 2007.
21. H. Nakanishi, I. B. Turksen, and M. Sugeno, "A Review and Comparison of Six Reasoning Methods," Fuzzy Sets and Systems, vol. 57, pp. 257-294, 1993.
22. R. Goldman, and J. S. Weinberg, "Statistics: An Introduction," Prentice-Hall: Lebanon, Indiana, USA, 1985.
23. A. R. Iyer, D. K. Prasad, and H. C. Quek, "PIE-RSPOP: A Brain-Inspired Pseudo-Incremental Ensemble Rough Set Pseudo-Outer Product Fuzzy Neural Network," Expert Systems with Applications, vol. 95, pp. 172-189, 2018.
24. Susanti, "The Evolving Mamdani Fuzzy Inference System with Fuzzy Rule Interpolation and Extrapolation (eMFIS (FRI/E))," FYP Report, Nanyang Technological University, Singapore, 2014.

CI in Vehicles, Smart Cities/Energy, and Networking

Traffic Sign Recognition Robustness in Autonomous Vehicles Under Physical Adversarial Attacks

Kyriakos D. Apostolidis, Emmanouil V. Gkouvrikos, Eleni Vrochidou, and George A. Papakostas

Abstract Nowadays we are all witnesses of the technological development of the so-called 4th industrial revolution (Industry 4.0). In this context, a daily living environment of smart cities is formed in which artificial intelligence applications play a dominant role. Autonomous (pilotless) vehicles are a shining example of the application of artificial intelligence, based on which vehicles are allowed to move autonomously in both residential and rural areas. The proposed article examines the robustness, in adversarial attacks in the physical layer, of the deep learning models used in autonomous vehicles for the recognition of road traffic signals. As a case study the roads of Greece, having traffic signs highly contaminated not on purpose, is considered. Towards investigating this direction, a novel dataset with clear and attacked images of traffic signs is proposed and used in the evaluation of popular deep learning models. This study investigates the level of readiness of autonomous vehicles to perform in noisy environments that affect their ability to recognize road signs. This work highlights the need for more robust deep learning models in order to make the use of autonomous vehicles a reality with maximum safety for citizens.

Keywords Autonomous vehicles · Self-driving cars · Traffic sign recognition · Artificial intelligence · Computer vision · Deep learning

---

K. D. Apostolidis · E. V. Gkouvrikos · E. Vrochidou · G. A. Papakostas (✉)
MLV Research Group, Department of Computer Science, International Hellenic University, Kavala, Greece
e-mail: gpapak@cs.ihu.gr

K. D. Apostolidis
e-mail: kyriapos1@cs.ihu.gr

E. V. Gkouvrikos
e-mail: emgkouv@teiemt.gr

E. Vrochidou
e-mail: evrochid@cs.ihu.gr

K. Daimi et al. (eds.), *Cutting Edge Applications of Computational Intelligence Tools and Techniques*, Studies in Computational Intelligence 1118,
https://doi.org/10.1007/978-3-031-44127-1_13

1 Introduction

In recent years, great achievements have been made in the field of artificial intelligence thanks to Deep Learning (DL). DL has actually enabled Computer Vision (CV) in real-world problems. Some of the most representative disciplines that use DL for CV tasks are robotics [1], biometrics [2], object detection [3], autonomous cars [4], etc. Autonomous cars are one of the most challenging tasks of Artificial Intelligence (AI) with great development in recent years. The evolution of hardware has allowed researchers to implement several DL models on cars in order to deal with numerous demanding problems.

One of the most important tasks of a pilotless car is to detect and recognize traffic signs as it is crucial for transportation safety. CV for traffic signs detection, tracking and classification methodologies have been studied for several reasons like Advanced Driver Assistance Systems (ADAS) and Auto Driving Systems (ADS) [5]. In general Traffic Sign Detection (TSD) and Traffic Sign Recognition (TSR) consists of detecting the location of traffic signs in the environment and recognizing their categories. DL has provided us with many models for this work with promising results [6–10].

However, recent works by Szegedy et al. [11] demonstrated that Deep Neural Networks (DNN) are vulnerable to imperceptible perturbations called adversarial attacks. Several adversarial attacks, which carefully modify the images in order to lead models to misclassification, have been proposed. According to Goodfellow et al. [12], these attacks are effective due to the nonlinearity of deep learning models. Adversarial attacks on traffic signs raised questions about the safety of autonomous cars as they could lead to serious accidents. Moreover, the majority of proposed attacks, involve the digital layer of the deep learning models such as medical image analysis [13]; however, attacking road signs in the physical layer is of great interest because it is more realistic and at the same time leads towards building robust DL models for road traffic classification that is crucial because they can be easily modified by human, affecting transportation safety.

In this chapter, some of the state-of-the-art DL models are evaluated under unintentional physical attacks. A typical example is the Greek roads where many signs have various stickers or spray paintings. For this purpose, a new dataset with traffic signs from Greece is designed, which is divided into five classes, as it is shown in Fig. 1, and it consists of clean and attacked images. Our work raises an important question: can the models, no matter how good they are, be applied to the existing conditions?

Therefore, the contribution of this work is summarized in the following items:

(1) For the first time the case of physical adversarial attacks on DL models for traffic sign recognition is investigated. In this context, the performance of the DL models is quantified.
(2) A new dataset with physical adversarial attacked traffic images is proposed.
(3) The maturity of the automatic traffic sign recognition technology towards implementing self-driving cars is determined.

Fig. 1 The five classes of the dataset: no entry, no parking, no right turn, no left turn and stop

The rest of this chapter is organized as follows. Section 2 presents an introduction to traffic sign recognition in autonomous vehicles. In Sect. 3 an overview of adversarial attacks on computer vision is provided. Section 4 introduces adversarial attacks on traffic sign recognition and information about the proposed dataset. Section 5 presents the experimental study, while Sect. 6 discusses the current status and challenges. Finally, Sect. 7 concludes this study.

2 Traffic Signs Recognition in Autonomous Vehicles

In recent years, the great development of artificial intelligence has made the concept of autonomous cars almost a reality. Many auxiliary driving systems have already been implemented in conventional cars. Traffic Sign Recognition (TSR) is one of the basic systems and concerns the detection and classification of road signs. Nowadays, the main function of TSR is to warn and help drivers making transportation safer because sometimes, drivers may not notice the road signs. An interesting study by Costa et al. [14] showed that different types of signs attract different attention from drivers and this can be eliminated through TSR systems in vehicles. Moreover, it is obvious that TSR is a vital function for fully autonomous cars and therefore needs to be optimized in order to operate under any circumstances. However, there are numerous difficulties in TSR because road sign visibility is usually not ideal. Bad weather conditions, the angle of the camera, illumination and distorted signs are some of the difficulties that need to be addressed.

A typical TSR system consists of three stages [15] detection, tracking and classification. The detection stage deals with the possible location of a road sign and a Region of Interest (ROI) box encloses it. The tracking stage enhances the robustness and accuracy of the information that was acquired from the detection stage, and it is significant in real-time applications. The last, but not least stage, processes the detected ROI from the first stage and then predicts the category of the road sign.

Traffic signs usually have specific color and shape so the first algorithms exploited this visual information [16]. Nevertheless, these methods failed to cope real time conditions because of camera angle, driving speed, illumination, etc. Classical machine learning algorithms such as Support Vector Machines (SVM) and Neural Networks (NN) depend on specific features extracted from several extractors like SIFT, HAAR Cascade, HOG, etc., which perform poorly in real time conditions. The great evolution of hardware enabled the implementation of DL in embedded systems for autonomous cars performing high accuracy under various conditions. CNNs are the most popular approaches of DL in computer vision and as a consequence in TSR. Some of the most used models in the TSR are You Only Look Once (YOLO) [17, 18] and Single Shot MultiBox Detector (SSD) [19, 20] which detect and classify simultaneously.

A significant problem in literature for TSR is the fact that most countries have different traffic signs therefore it is difficult for TSR methods to be compared. Table 1 presents information for some of the most known public datasets for traffic sign recognition [13], including the number of classes and the total number of images, as well as the country of origin of the datasets. It should be noted most of the existing traffic sign datasets include a great number of total images and many classes, referring to multiple traffic signs. This is due to the reason for their creation, which is for traffic sign detection/recognition tasks. To this end, the contribution of the proposed dataset, namely GReek Adversarial Traffic Signs (GRATS), lies to the introduction of a novel dataset suitable for adversarial training whose images are not artificially made but physically attacked. This dataset is the first with traffic signs *in the wild* as it presents real traffic sign conditions of Greek roads. The proposed dataset is designated to allow other researchers to investigate the robustness of their pattern recognition models due to the substantial data distortion, referring to real-world physically attacked traffic sign images. As it can be notices from Table 1, the proposed dataset is limited in terms of classes and instances compared with other traffic sign datasets; however, this limitation is to be lifted in future research by introducing more images of more classes.

Table 1 Datasets for traffic sign detection and recognition

Reference	Dataset	Task	Classes	Total images	Country
[21]	GTSDB	Detection	43	900	Germany
[22]	GTSRB	Recognition	43	50.000+	Germany
[23]	BTSD	Detection	62	25.634	Belgium
[23]	BTSC	Recognition	62	7.125	Belgium
[24]	TT100K	Detection/Recognition	45	100.000	China
[25]	LISA	Detection/Recognition	49	7.855	United States
[26]	STS	Detection/Recognition	7	20.000	Sweden
[27]	RUG	Detection/Recognition	3	48	Netherlands
[28]	Stereopolis	Detection/Recognition	10	847	France
[29]	FTSD	Detection/Recognition	–	4.239	Sweden
[30]	MASTIF	Detection/Recognition	–	4.875	Croatia
[31]	ETSD	Recognition	164	82.476	Europe
Proposed	GRATS	Adversarial training/ Detection/ Recognition	5	850	Greece

3 Adversarial Attacks in Computer Vision

Deep learning is the most widely used tool for computer vision as it provides high performance in several tasks and fields such as classification, segmentation, reconstruction medical image analysis, autonomous vehicles, etc. However, Szegedy et al. [11] discovered adversarial attacks in which imperceptible carefully crafted noise can manipulate model's decision. These attacks have shown that can lead models to wrong predictions with high confidence (Fig. 2) which makes them a serious threat. In general, the problem can be formulated as follows [32]:

$$M(I+\rho) \rightarrow \tilde{l} \quad \text{s.t} \quad \tilde{l} \neq l, \ \|\rho\|_p < \eta \tag{1}$$

where M(.) is the DL model, I is the clean image, ρ is the perturbation which is restricted by $\|.\|_p$ that denotes the l_p-norm and η is a predefined scalar.

Maliamanis and Papakostas [33] proposed a taxonomy for adversarial attacks in computer vision, which is based on three axes:

Knowledge Axis

- *White-box attacks* where adversary knows everything about the model such as parameters, weights, architecture, and dataset.
- *Grey-box attacks* in which adversary has partial knowledge of the target model like dataset and architecture.
- *Black-box attacks* where adversary knows nothing about the target model and dataset.

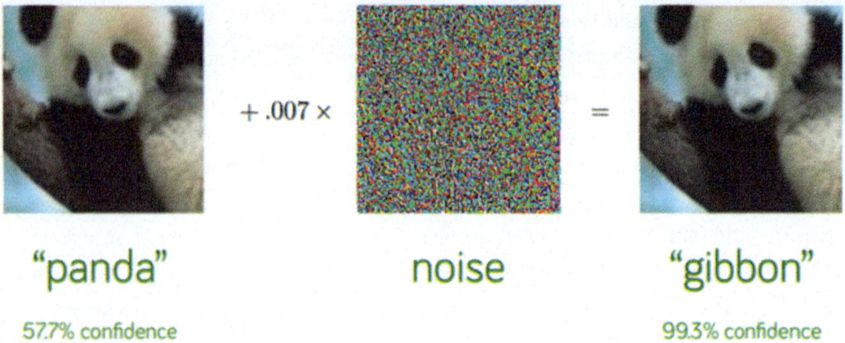

Fig. 2 Prediction before and after attack

Specificity Axis

- *Targeted attacks* aim to misclassify the input sample in a specific class.
- *Untargeted attacks* just aim for the sample data to be misclassified.

Applicability Axis

- *Implementable attacks* that can be materialized in physical form in order to be applied in the real world.
- *Virtual attacks* which can be found and applied only in digital data forms.

Numerous virtual adversarial attacks have been proposed [14] but in this paper are presented some of the most known.

Fast Gradient Sign Method (FGSM) [12] is one of the first attacks which compute efficiently adversarial perturbation according to Eq. 2:

$$x\prime = x + \epsilon * sign(\nabla x J(\theta, x, y)) \tag{2}$$

where x is the initial image, y is the label of the image and θ represents the weights of the model. Also, ϵ is a scalar value of perturbation magnitude, $J(\theta, x, y)$ is the gradient loss, sign (.) is the sign function and ∇x is the gradient w.r.t x.

Projected Gradient Descent (PGD) [34] is one of the most powerful first order attacks and attempts to find the perturbation, according to L_2 and L_∞ norm, that maximizes the loss function of the model. PGD can be considered as a variant of iterative FGSM.

Jacobian-based Saliency Map Attack (JSMA) [35] perturbate a small region of an image and in contrary to conventional attacks which compute backward gradients of a model, JSMA computes forward gradients to create perturbation.

Carlini & Wagner (C&W) [36] is one of the most efficient attacks and consists of three methods, $C\&W L_\infty$, $C\&W L_0$, $C\&W L_2$ each minimizing the L_∞, L_0 and L_2

Fig. 3 Adversarial patch attack examples that made the road sign classifier recognize them as "Speed Limit 80" [41]

norm respectively. C&W is also the first attack that broke defensive distillation [37] which was the most prominent defensive method.

DeepFool [38] is an iterative attack that finds the minimum amount of perturbation required to cause misclassification by measuring the distance from the original input to the decision boundary.

Adversarial Patch Attack [39] is a method that adds a patch of specially designed pixels to the input image to cause misclassification (Fig. 3).

Universal Adversarial Perturbation (UAP) [40] is a type of attack that generates a single perturbation that can cause misclassification across a wide range of input data. UAP can be used to attack models in real-world settings where the attacker does not have access to the specific input data used to train the model.

Adversarial learning is the most widely used method for creating robust models, in which adversarial samples are provided in the training set of models in order to learn these features. However, this method is robust only under attacks that are used for adversarial learning. According to Akhtar and Mian [42], the defense methods are divided into three categories. The first arises from robustification of the target model, the second from pre-processing of data such as removing perturbation and the third from adding modules to the models like detectors.

This study researches a case of an *implementable* attack as we test some state-of-the-art DL models in road traffic signs under perturbations having a physical form. Kurakin et al. [43] was the first study that proved the existence of adversarial attacks in real world as it is shown in Fig. 4. Physical world attacks are more challenging than digital attacks as they are subjected to some restrictions [44] such as:

- ***Environmental Conditions.*** Sensor viewing angle, light and distance are some significant factors that affect the efficacy of the attack. The real world is in a

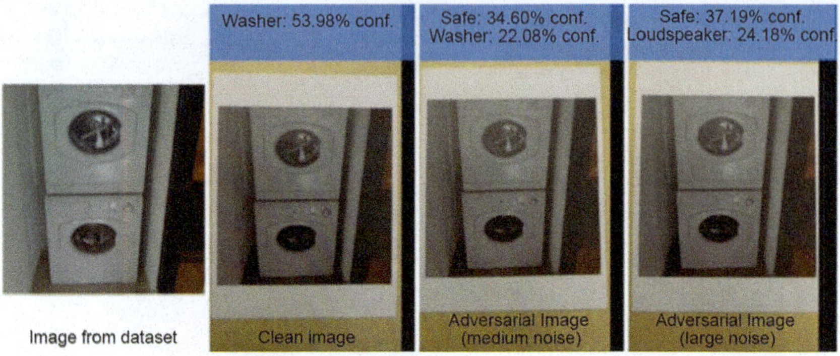

Fig. 4 Adversarial attacks on printed out images [23]

three-dimensional space and the aforementioned factors change continuously. A successful attack should create efficient perturbations in all perspectives.

- ***Spatial Constraints.*** In digital images, perturbations can be added to any part of the image. On the contrary in the physical world, we cannot use the background of a road sign.
- ***Fabrication Error.*** It is a significant factor in physical world attacks because it is common for modern printers to not print accurately adversarial examples due to the limitation of the color range. That is why Sharif et al. [45] proposed the non-printable score (NPS) in order to measure this error.
- ***Perturbation Smoothness.*** Another challenge in physical world attacks is smoothness as in natural images the colors change smoothly. For this reason, perturbation must not be intense so that citizens do not suspect and that is why Sharif et al. [45] proposed a smooth limiting function.
- ***Physical Limits on Imperceptibility.*** In digital attacks, perturbations can be imperceptible, however, in the physical world attacks the perturbation should be captured from the camera to affect the model's accuracy. Thus, a balance must be found between perceptibility and smoothness.

4 Towards Attacking Traffic Signs Recognition Systems

In the past years, several works have attempted to attack traffic sign recognition tasks in the digital and physical layers. Kumar et al. [46] proposed a query-based attack called Modified Simple Black Box Attack (M-SimBA) for traffic sign recognition using GTSRB dataset. Results have shown that this attack significantly decreases the efficacy of the models. Lengyal et al. [47] introduced an attack that is based on stickers that can be printed and then deployed on traffic signs. Zhong et al. [48] investigated the robustness of models in TSR under a natural phenomenon that is *shadows*. They generated optical adversarial examples by simulating shadows on

Fig. 5 Left image is a real graffiti in a stop sign while right image is under RP2 attack

traffic signs, achieving 98.23% and 90.47% success rates on LISA and GTSRB datasets, respectively. Eykholt et al. [49] proposed a general attack algorithm which is called Robust Physical Perturbations (RP2). One case study of this algorithm was TSR (Fig. 5) where it achieves 100% targeted attack success rates in the laboratory and approximately 85% in driving field tests.

In this chapter, a new dataset for traffic sign recognition is introduced, which consists of five different classes, *Stop, No entry, No parking, No right turn* and *No left turn.* In what follow, details about the creation and characteristics of the introduced dataset are presented.

A. ***Dataset Creation***

The proposed dataset called GReek Adversarial Traffic Signs (GRATS)[1] consists of traffic signs from Greek roads collected from 5 different cities. In order for images to be captured, three individuals used their smartphone camera with 12MP resolution. All images had 3000 × 4000 pixels resolution and used jpeg compression. In camera settings, HDR was enabled while all others A.I. retouching features were disabled, without any further editing. Moreover, almost all images were captured during morning hours in order to take advantage of the sunlight and to avoid the appearance of digital noise.

B. Dataset Characteristics

The dataset splits into two parts and consists of five traffic signs, which are crucial for the safe navigation of the autonomous vehicles: *Stop, No entry, No parking, No right turn* and *No left turn* (Fig. 1). Both parts of the dataset are balanced; each class consists of 85 images, therefore 425 images in total. The first part consists of clean images without stickers or spray paintings. Furthermore, all images were cropped

[1] The dataset is provided via the GitHub account of the MLV Research Group. (https://github.com/MachineLearningVisionRG/GRATS).

in such a way that only the road sign is depicted and then they resized to 224,224 pixels.

The second part of the dataset consists of the same classes but with "dirty" images with stickers, graffiti, etc. (Fig. 6). Again, each class has 85 images, resulting in 425 images in total. This dataset has been divided to training and testing datasets, consisting of 800 and 50 images, respectively. To ensure representative samples, the datasets for both training and testing were formed by keeping a balance; 80 images from each class for both datasets were used for training (80 × 5 × 2), while five images from each class for both dataset were used for testing (5 × 5 × 2).

This dataset is the first with traffic signs *in the wild* as it presents real conditions of Greek roads. In this way, the robustness of the models under different circumstances can be tested. Training models with "dirty" images can learn robust features for each category and overcome difficult conditions.

5 Experimental Study

In order to evaluate the introduced dataset, four DL models, VGG-16 [50], ResNet-50 [51], GoogLeNet [52] and DenseNet-161 [53] were trained on it. The four models were selected due to their popularity and their state-of-the-art reported performances in image classification applications. All DL models were trained in Google Collab with Pytorch library. The training was carried out with categorical cross-entropy loss function, gradient descent optimizer and 0.001 learning rate for 25 epochs. Selected experimental setup was determined after trial and error. The experiments with the aforementioned models were done in two ways. The first was with Transfer Learning (TL) by unfreezing the last layers of networks and the second without transfer learning by training the entire networks. In the case of transfer learning, the models were pre-trained on ImageNet and then they were finetuned in the proposed dataset.

In addition to TL, our experiments were divided into three additional categories in terms of the training setup. In the first setup, the models were trained with clear images and then tested on both clean and attacked (dirty) (Table 2) ones. In the second setup, the models were trained with attacked images and tested on clean and attacked (Table 3) images, while in the third setup the training was done with mixed images and tested on clear and attacked (Table 4) ones.

The following images (Figs. 7, 8, 9 and 10) illustrate some typical examples for which the trained models failed to predict the correct class.

Fig. 6 Clean and dirty images of each class **a** stop, **b** no right turn, **c** no parking, **d** no left turn and **e** no entry

Table 2 Training with clean images

Models	Unfreeze	Clear signs accuracy (%)	Dirty signs accuracy (%)
VGG-16	No	76	40
ResNet-50	No	72	60
GoogLeNet	No	76	64
DenseNet-161	No	80	68
VGG-16	Yes	**100**	60
ResNet-50	Yes	84	76
GoogLeNet	Yes	84	64
DenseNet-161	Yes	96	**84**

Table 3 Training with dirty images

Models	Unfreeze	Clear signs accuracy (%)	Dirty signs accuracy (%)
VGG-16	No	76	52
ResNet-50	No	76	**72**
GoogLeNet	No	72	56
DenseNet-161	No	**84**	60
VGG-16	Yes	84	64
ResNet-50	Yes	80	**72**
GoogLeNet	Yes	80	68
DenseNet-161	Yes	**84**	52

Table 4 Training with mixed images

Models	Unfreeze	Clear signs accuracy (%)	Dirty signs accuracy (%)
VGG-16	No	80	60
ResNet-50	No	80	52
GoogLeNet	No	80	64
DenseNet-161	No	84	56
VGG-16	Yes	96	88
ResNet-50	Yes	**100**	**92**
GoogLeNet	Yes	80	68
DenseNet-161	Yes	**100**	80

6 Discussion

Autonomous vehicles are one of the most important challenges of AI and the 4th industrial revolution. A significant part of them is the traffic sign recognition task, which allows cars to transport safely. Numerous models have been developed in

Fig. 7 Examples of misclassification from VGG-16

Fig. 8 Examples of misclassification from ResNet-50

Fig. 9 Examples of misclassification from GoogLeNet

Fig. 10 Examples of misclassification from DenseNet-161

Fig. 11 Example of a "dirty" stop sign

order to recognize road signs with high accuracy based on datasets from several countries. However, the previous datasets consist of flawless road signs that are not representative. The proposed dataset provides both clean and attacked (dirty) images for developing models that are efficient and safe at the same time in real conditions. Robustness is one of the most critical challenges in artificial intelligence, and this work contributes in this way, as it aims to help the scientific community evaluate computer vision models under challenging conditions.

In order for the DL models to be deployed into the autonomous cars should be able to overcome numerous challenges and one of them is the "dirty" traffic signs. On the other hand, there are traffic signs that are totally hidden from stickers and even humans cannot recognize them. In Fig. 11 is presented a Stop sign, which is almost completely erased. This raises a significant question about the readiness of autonomous vehicles to perform well in noisy environments. Maybe, the creation of new infrastructures that can support the implementation of autonomous cars should be our priority.

Moreover, the proposed dataset could be used as a benchmark for the model's robustness. Developing robust models which learn only strong features from each sign could recognize the majority of dirty images. According to Table 2, presenting models that have been trained in clean images and tested in dirty, the DenseNet-161 seems to learn the most robust features because presents relatively high accuracy in both clean and dirty images. On the other hand, VGG-16 presents 100% accuracy in clean images but failed to keep its high accuracy in dirty images. Also, transfer learning is beneficial for small datasets but this study proves that it is more vulnerable to attacks. From Table 3 we understand that training with only dirty images does not help models learn robust features. This may be because these physical attacks do not have a specific format for decoding by the models. Finally, Table 4 shows that adversarial learning helps models to learn strong features as the accuracy is maintained relatively high even under attacks. In mixed training, ResNet-50 was the most efficient in both clear and dirty images.

Furthermore, it is noticed that all models are prone to confusing the *no right turn* with *no left turn* sign and vice versa. This happens very often even in clear images. These two signs are quite similar; however, this is worrying as it can cause significant problems in real navigation conditions.

7 Conclusion

Deep learning has dramatically improved computer vision tasks including the application of autonomous cars. Traffic sign recognition is an important part of pilotless cars and deep learning algorithms provide promising results. Nevertheless, adversarial attacks undermine the proper functionality of these algorithms. Numerous studies have proved that imperceptible noise can lead to the misclassification of road signs. This raises significant questions about the safety of AI and industry 4.0 applications. How can we trust DL models when they can be easily fooled? In this paper, the task of traffic sign recognition under real conditions on Greek roads was investigated. The results showed that state of the art algorithms failed to recognize traffic signs with stickers and graffiti to a great extent. At the same time, a new dataset for road sign recognition from Greek roads was herein proposed, which consists of clean and "dirty" images in order to investigate the robustness of the DL models. Future work includes the development of more stable defense strategies by investigating alternative DL architectures able to conclude into more robust features, as well as more data to augment the proposed dataset in term of classes and instances.

References

1. A. Geiger, P. Lenz, C. Stiller, and R. Urtasun, "Vision meets robotics: The KITTI dataset," *The International Journal of Robotics Research*, vol. 32, no. 11, pp. 1231–1237, Sep. 2013, doi: https://doi.org/10.1177/0278364913491297.
2. K. Apostolidis, P. Amanatidis, and G. Papakostas, "Performance Evaluation of Convolutional Neural Networks for Gait Recognition," in *24th Pan-Hellenic Conference on Informatics*, Athens Greece, Nov. 2020, pp. 61–63. doi: https://doi.org/10.1145/3437120.3437276.
3. Z. Zou, Z. Shi, Y. Guo, and J. Ye, "Object detection in 20 years: A survey," *arXiv preprint* arXiv:1905.05055, 2019.
4. J. Janai, F. Güney, A. Behl, and A. Geiger, "Computer vision for autonomous vehicles: Problems, datasets and state of the art," *Foundations and Trends® in Computer Graphics and Vision*, vol. 12, no. 1–3, pp. 1–308, 2020.
5. C. Liu, S. Li, F. Chang, and Y. Wang, "Machine Vision Based Traffic Sign Detection Methods: Review, Analyses and Perspectives," *Machine Vision*, vol. 7, p. 19, 2019.
6. D. Tabernik and D. Skočaj, "Deep Learning for Large-Scale Traffic-Sign Detection and Recognition," arXiv:1904.00649 *[cs]*, Apr. 2019, Accessed: Apr. 28, 2022. [Online]. Available: http://arxiv.org/abs/1904.00649.
7. K. Bayoudh, F. Hamdaoui, and A. Mtibaa, "Transfer learning based hybrid 2D-3D CNN for traffic sign recognition and semantic road detection applied in advanced driver assistance

systems," *Appl Intell*, vol. 51, no. 1, pp. 124–142, Jan. 2021, doi: https://doi.org/10.1007/s10489-020-01801-5.
8. Z. Liu, J. Du, F. Tian, and J. Wen, "MR-CNN: A Multi-Scale Region-Based Convolutional Neural Network for Small Traffic Sign Recognition," *IEEE Access*, vol. 7, pp. 57120–57128, 2019, doi: https://doi.org/10.1109/ACCESS.2019.2913882.
9. Y. Yuan *et al.*, "VSSA-NET: Vertical Spatial Sequence Attention Network for Traffic Sign Detection," *IEEE Trans. on Image Process.*, vol. 28, no. 7, pp. 3423–3434, Jul. 2019, doi: https://doi.org/10.1109/TIP.2019.2896952.
10. A. Vennelakanti, S. Shreya, R. Rajendran, D. Sarkar, D. Muddegowda, and P. Hanagal, "Traffic Sign Detection and Recognition using a CNN Ensemble," in *2019 IEEE International Conference on Consumer Electronics (ICCE)*, Jan. 2019, pp. 1–4. doi: https://doi.org/10.1109/ICCE.2019.8662019.
11. C. Szegedy *et al.*, "Intriguing properties of neural networks," arXiv:1312.6199 *[cs]*, Feb. 2014, Accessed: Jun. 04, 2021. [Online]. Available: http://arxiv.org/abs/1312.6199.
12. I. J. Goodfellow, J. Shlens, and C. Szegedy, "Explaining and Harnessing Adversarial Examples," arXiv:1412.6572 *[cs, stat]*, Mar. 2015, Accessed: Jun. 04, 2021. [Online]. Available: http://arxiv.org/abs/1412.6572.
13. K. D. Apostolidis and G. A. Papakostas, "A Survey on Adversarial Deep Learning Robustness in Medical Image Analysis," *Electronics*, vol. 10, no. 17, p. 2132, Sep. 2021, doi: https://doi.org/10.3390/electronics10172132.
14. M. Costa, A. Simone, V. Vignali, C. Lantieri, and N. Palena, "Fixation distance and fixation duration to vertical road signs," *Applied Ergonomics*, vol. 69, pp. 48–57, May 2018, doi: https://doi.org/10.1016/j.apergo.2017.12.017.
15. S. B. Wali *et al.*, "Vision-Based Traffic Sign Detection and Recognition Systems: Current Trends and Challenges," *Sensors*, vol. 19, no. 9, 2019, doi: https://doi.org/10.3390/s19092093.
16. Y. Zhu, "Traffic sign recognition based on deep learning," *Multimedia Tools and Applications*, p. 13.
17. J. Redmon, S. Divvala, R. Girshick, and A. Farhadi, "You Only Look Once: Unified, Real-Time Object Detection," arXiv:1506.02640 *[cs]*, May 2016, Accessed: Apr. 28, 2022. [Online]. Available: http://arxiv.org/abs/1506.02640.
18. C. Dewi, R.-C. Chen, and H. Yu, "Weight analysis for various prohibitory sign detection and recognition using deep learning," *Multimedia Tools and Applications*, vol. 79, no. 43–44, pp. 32897–32915, 2020.
19. W. Liu *et al.*, "SSD: Single Shot MultiBox Detector," arXiv:1512.02325 *[cs]*, vol. 9905, pp. 21–37, 2016, doi: https://doi.org/10.1007/978-3-319-46448-0_2.
20. S. You, Q. Bi, Y. Ji, S. Liu, Y. Feng, and F. Wu, "Traffic sign detection method based on improved SSD," *Information*, vol. 11, no. 10, p. 475, 2020.
21. S. Houben, J. Stallkamp, J. Salmen, M. Schlipsing, and C. Igel, "Detection of traffic signs in real-world images: The German traffic sign detection benchmark," in *The 2013 International Joint Conference on Neural Networks (IJCNN)*, Dallas, TX, USA, Aug. 2013, pp. 1–8. doi: https://doi.org/10.1109/IJCNN.2013.6706807.
22. J. Stallkamp, M. Schlipsing, J. Salmen, and C. Igel, "Man vs. computer: Benchmarking machine learning algorithms for traffic sign recognition," *Neural Networks*, vol. 32, pp. 323–332, Aug. 2012, doi: https://doi.org/10.1016/j.neunet.2012.02.016.
23. M. Mathias, R. Timofte, R. Benenson, and L. Van Gool, "Traffic sign recognition — How far are we from the solution?," in *The 2013 International Joint Conference on Neural Networks (IJCNN)*, Dallas, TX, USA, Aug. 2013, pp. 1–8. doi: https://doi.org/10.1109/IJCNN.2013.6707049.
24. Z. Zhu, D. Liang, S. Zhang, X. Huang, B. Li, and S. Hu, "Traffic-Sign Detection and Classification in the Wild," in *2016 IEEE Conference on Computer Vision and Pattern Recognition (CVPR)*, Las Vegas, NV, USA, Jun. 2016, pp. 2110–2118. doi: https://doi.org/10.1109/CVPR.2016.232.
25. A. Mogelmose, D. Liu, and M. M. Trivedi, "Detection of U.S. Traffic Signs," *IEEE Trans. Intell. Transport. Syst.*, vol. 16, no. 6, pp. 3116–3125, Dec. 2015, doi: https://doi.org/10.1109/TITS.2015.2433019.

26. F. Larsson and M. Felsberg, "Using Fourier Descriptors and Spatial Models for Traffic Sign Recognition," in *Image Analysis*, vol. 6688, A. Heyden and F. Kahl, Eds. Berlin, Heidelberg: Springer Berlin Heidelberg, 2011, pp. 238–249. doi: https://doi.org/10.1007/978-3-642-21227-7_23.
27. C. Grigorescu and N. Petkov, "Distance sets for shape filters and shape recognition," *IEEE Transactions on Image Processing*, vol. 12, no. 10, pp. 1274–1286, Oct. 2003, doi: https://doi.org/10.1109/TIP.2003.816010.
28. R. Belaroussi, P. Foucher, J.-P. Tarel, B. Soheilian, P. Charbonnier, and N. Paparoditis, "Road Sign Detection in Images: A Case Study," in *2010 20th International Conference on Pattern Recognition*, Istanbul, Turkey, Aug. 2010, pp. 484–488. doi: https://doi.org/10.1109/ICPR.2010.1125.
29. H. Fleyeh, "Traffic and Road Sign Recognition," p. 255.
30. S. Segvic *et al.*, "A computer vision assisted geoinformation inventory for traffic infrastructure," in *13th International IEEE Conference on Intelligent Transportation Systems*, Funchal, Madeira Island, Portugal, Sep. 2010, pp. 66–73. doi: https://doi.org/10.1109/ITSC.2010.5624979.
31. C. Gamez Serna and Y. Ruichek, "Classification of Traffic Signs: The European Dataset," *IEEE Access*, vol. 6, pp. 78136–78148, 2018, doi: https://doi.org/10.1109/ACCESS.2018.2884826.
32. N. Akhtar, A. Mian, N. Kardan, and M. Shah, "Advances in Adversarial Attacks and Defenses in Computer Vision: A Survey," vol. 9, p. 36, 2021.
33. T. Maliamanis and G. Papakostas, "Adversarial computer vision: a current snapshot," in *Twelfth International Conference on Machine Vision (ICMV 2019)*, Amsterdam, Netherlands, Jan. 2020, p. 121. doi: https://doi.org/10.1117/12.2559582.
34. A. Madry, A. Makelov, L. Schmidt, D. Tsipras, and A. Vladu, "Towards Deep Learning Models Resistant to Adversarial Attacks," arXiv:1706.06083 *[cs, stat]*, Sep. 2019, Accessed: Jun. 04, 2021. [Online]. Available: http://arxiv.org/abs/1706.06083.
35. N. Papernot, P. McDaniel, S. Jha, M. Fredrikson, Z. B. Celik, and A. Swami, "The Limitations of Deep Learning in Adversarial Settings," arXiv:1511.07528 *[cs, stat]*, Nov. 2015, Accessed: Jun. 04, 2021. [Online]. Available: http://arxiv.org/abs/1511.07528.
36. N. Carlini and D. Wagner, "Towards Evaluating the Robustness of Neural Networks," arXiv: 1608.04644 *[cs]*, Mar. 2017, Accessed: Jun. 04, 2021. [Online]. Available: http://arxiv.org/abs/1608.04644.
37. N. Papernot, P. McDaniel, X. Wu, S. Jha, and A. Swami, "Distillation as a Defense to Adversarial Perturbations against Deep Neural Networks," arXiv:1511.04508 *[cs, stat]*, Mar. 2016, Accessed: Feb. 27, 2022. [Online]. Available: http://arxiv.org/abs/1511.04508.
38. S.-M. Moosavi-Dezfooli, A. Fawzi, and P. Frossard, "DeepFool: A Simple and Accurate Method to Fool Deep Neural Networks," in *2016 IEEE Conference on Computer Vision and Pattern Recognition (CVPR)*, Las Vegas, NV, USA, Jun. 2016, pp. 2574–2582. doi: https://doi.org/10.1109/CVPR.2016.282.
39. T. B. Brown, D. Mané, A. Roy, M. Abadi, and J. Gilmer, "Adversarial Patch." arXiv, May 16, 2018. Accessed: Mar. 19, 2023. [Online]. Available: http://arxiv.org/abs/1712.09665.
40. S.-M. Moosavi-Dezfooli, A. Fawzi, O. Fawzi, and P. Frossard, "Universal Adversarial Perturbations," in *2017 IEEE Conference on Computer Vision and Pattern Recognition (CVPR)*, Honolulu, HI, Jul. 2017, pp. 86–94. doi: https://doi.org/10.1109/CVPR.2017.17.
41. H. Yakura, Y. Akimoto, and J. Sakuma, "Generate (non-software) Bugs to Fool Classifiers." arXiv, Nov. 19, 2019. Accessed: Mar. 19, 2023. [Online]. Available: http://arxiv.org/abs/1911.08644.
42. N. Akhtar and A. Mian, "Threat of Adversarial Attacks on Deep Learning in Computer Vision: A Survey," *IEEE Access*, vol. 6, pp. 14410–14430, 2018, doi: https://doi.org/10.1109/ACCESS.2018.2807385.
43. A. Kurakin, I. Goodfellow, and S. Bengio, "Adversarial examples in the physical world," arXiv: 1607.02533 *[cs, stat]*, Feb. 2017, Accessed: Jun. 04, 2021. [Online]. Available: http://arxiv.org/abs/1607.02533.
44. H. Ren, T. Huang, and H. Yan, "Adversarial examples: attacks and defenses in the physical world," *Int. J. Mach. Learn. & Cyber.*, vol. 12, no. 11, pp. 3325–3336, Nov. 2021, doi: https://doi.org/10.1007/s13042-020-01242-z.

45. M. Sharif, S. Bhagavatula, and L. Bauer, "Accessorize to a Crime: Real and Stealthy Attacks on State-of-the-Art Face Recognition," p. 13.
46. K. N. Kumar, C. Vishnu, R. Mitra, and C. K. Mohan, "Black-box Adversarial Attacks in Autonomous Vehicle Technology," arXiv:2101.06092 *[cs]*, Jan. 2021, Accessed: Apr. 28, 2022. [Online]. Available: http://arxiv.org/abs/2101.06092.
47. H. Lengyel, V. Remeli, and Z. Szalay, "EASILY DEPLOYED STICKERS COULD DISRUPT TRAFFIC SIGN RECOGNITION," p. 9.
48. Y. Zhong, X. Liu, D. Zhai, J. Jiang, and X. Ji, "Shadows can be Dangerous: Stealthy and Effective Physical-world Adversarial Attack by Natural Phenomenon," arXiv:2203.03818 *[cs]*, Mar. 2022, Accessed: Apr. 28, 2022. [Online]. Available: http://arxiv.org/abs/2203.03818.
49. K. Eykholt *et al.*, "Robust Physical-World Attacks on Deep Learning Visual Classification," in *2018 IEEE/CVF Conference on Computer Vision and Pattern Recognition*, Salt Lake City, UT, USA, Jun. 2018, pp. 1625–1634. doi: https://doi.org/10.1109/CVPR.2018.00175.
50. K. Simonyan and A. Zisserman, "Very Deep Convolutional Networks for Large-Scale Image Recognition," arXiv:1409.1556 *[cs]*, Apr. 2015, Accessed: Jun. 04, 2021. [Online]. Available: http://arxiv.org/abs/1409.1556.
51. K. He, X. Zhang, S. Ren, and J. Sun, "Deep Residual Learning for Image Recognition," in *2016 IEEE Conference on Computer Vision and Pattern Recognition (CVPR)*, Las Vegas, NV, USA, Jun. 2016, pp. 770–778. doi: https://doi.org/10.1109/CVPR.2016.90.
52. C. Szegedy *et al.*, "Going Deeper with Convolutions," arXiv:1409.4842 *[cs]*, Sep. 2014, Accessed: Apr. 28, 2022. [Online]. Available: http://arxiv.org/abs/1409.4842.
53. G. Huang, Z. Liu, L. van der Maaten, and K. Q. Weinberger, "Densely Connected Convolutional Networks," arXiv:1608.06993 *[cs]*, Jan. 2018, Accessed: Sep. 14, 2021. [Online]. Available: http://arxiv.org/abs/1608.06993.

Computational Intelligence in Smart Cities and Smart Energy Systems

Yi Wang and Jian Fu

Abstract As cities become more interconnected and energy systems become smarter, the role of computational intelligence (CI) becomes increasingly significant. In this chapter, the focus is on investigating the involvement of computational intelligence (CI) within the realm of smart connected systems, with a particular emphasis on smart citites and smart energy systems. The chapter focuses on the Margin Setting Algorithm (MSA), a cutting-edge machine learning-based CI technique, and its potential applications in two critical areas: human activity recognition in smart homes and False Data Injection Attacks (FDIA) detection in smart grids. The MSA technique has demonstrated its efficacy and it has shown promise in accurately recognizing human activities through sensor data and has also proven to be effective in detecting anomalies in energy data to identify False Data Injection Attacks (FDIA). By highlighting these two applications, the chapter demonstrates the vast potential of CI in addressing complex challenges in modern urban and energy systems. The chapter suggests that CI holds the promise of opening up avenues for developing more effective, secure, and sustainable solutions in the dynamic realm of smart cities and smart energy systems.

Keywords Margin setting algorithm · Human activity recognition · False data injection attack · Smart homes · Smart grids

---

Y. Wang (✉)
Manhattan College, Riverdale, NY, USA
e-mail: yi.wang@manhattan.edu

J. Fu
Alabama A & M University, Alabama, AL, USA
e-mail: jian.fu@aamu.edu

K. Daimi et al. (eds.), *Cutting Edge Applications of Computational Intelligence Tools and Techniques*, Studies in Computational Intelligence 1118,
https://doi.org/10.1007/978-3-031-44127-1_14

1 Introduction

Smart Connected Systems (SCS), which is considered as the next generation of cyber physical systems (CPS) and Internet of Things (IoT). SCS interacts with human and the surroundings by integration of embedded sensing, communication devices, networked data processing, and physical infrastructure, which also employs computational intelligence (CI) using data sources from both from physical objects and virtual components to optimize efficiency, comfort, safety and security. By 2050, 66% of the world population will be living in urban areas while the number of "mega-cities" with 10 million inhabitants or more is expanding in the same pace [13]. As more people move to urban areas, there is an increasing demand for efficient and sustainable infrastructure, services, and solutions to manage the complexities of urban living. Smart cities and smart energy systems, two crucial components of SCS are emerging as key solutions to address the challenges by leveraging computational intelligence tools and Internet of Things (IoT). IoT has created opportunities for the development the smart cities and smart energy systems through the utilization of the connectivity and data sharing capabilities of physical devices, sensors, vehicles, and software. Computational Intelligence (CI) tools include artificial intelligence (AI), machine learning, and big data analytics is able to monitor, analyze and control the system to better meet the needs of the people in today's modern world. In addition, both smart cities and smart energy systems, as crucial components of smart connected communities (S&CC), that combines computational intelligent technologies with its natural and built environments, including infrastructure, in a way that improves the social, economic, and environmental well-being of its residents, workers and travelers [27].

Smart cities are urban areas that connect physical entities and devices, such as smart homes, smart hospitals, smart traffic lights, smart transportation, energy grids, creating an interconnected complex adaptive system for the citizens [15]. This system allows citizens to exchange data with both human beings and physical devices, enabling them to understand how these entities interact and change in both spatial and temporal domains. One the other hand, smart energy system specifically focusing on integrating smart electivity, thermal, and gas grids, along with storage technologies, which aims to attain an ideal outcome for each sector separately, while optimizing the performance of entire energy system as a whole [21, 24].

In this chapter, we explore the significant uses of computational intelligence (CI) in the dynamic and evolving field of smart connected systems, with a particular focus on securing and advancing their capabilities of SCS. In particular, we highlight the potential of a state-of-the-art new machine learning-based CI technique, the Margin Setting Algorithm (MSA), in two crucial areas: advancing SCS capabilities through human activity recognition in smart homes and securing SCS through the detection of False Data Injection Attacks (FDIA) in smart grids [14, 31]. By exploring these two applications, we aim to shed light on the vast potential of CI to address complex challenges in modern urban and energy systems, paving the way for more efficient, secure, and sustainable solutions.

2 Margin Setting Algorithm

Margin setting algorithm (MSA) is a new sphere-based classification algorithm. It employs an artificial immune system approach to construct a number of hyperspheres that cover each class of a given set of data. Margin setting algorithm creates hyperspheres (sphere in three-dimensional space, circle in two-dimensional space) as decision boundaries. The decision boundary of MSA are called prototypes as (centroid, radius, class) triplet:

$$G_i = \left(\omega_i, R_i, C_p\right), 1 \leq i \leq N, p = 1, \ldots, P. \tag{1}$$

where δ_k is the centroid of G, R_k is the radius of G, C_p is the class label. i and k are natural numbers. N is the number of prototypes belonging to class C_p. P is the total number of classes ($P > 1$).

MSA has an algorithmic parameter called margin, which affects the classification performance. It is a negative reinforcement procedure that reduces the area of coverage of hyperspheres. However, change margin provides a room for test data variations, which tends to improve generalization ability. Specifically, margin controls the magnitude of decreasing the radius of hyperspheres.

Definition 1 Margin: it is defined as the percentage χ that the radius of hypersphere shrinks, so radius of χ margin is of the quantity:

$$R_\chi = (1 - \chi)R_0. \tag{2}$$

where $0 \leq \chi < 1$. R_χ is the radius that shrinks χ from R_0. R_0 is called zero margin when $\chi = 0$.

A. **Training**

The process of generating final decision boundaries is the training procedure of MSA. There are two processes involved, namely evolution process and partition process as shown in Fig. 1. They execute iteratively until the stopping conditions are satisfied.

(1) **Evolution Phase**

Evolution process employed an artificial immune system approach to generate hyperspheres classifiers. In two-dimension space, decision boundaries are circles. In three-dimension space, decision boundaries are spheres. The immune system continually develops new antibodies to fight off new pathogens. Pathogens are antigens that stimulate the immune system to generate antibodies. First, create prototypes. Start with N random points as antibodies. The training samples points are antigens. For each antibody, find the closest member of antigen. Suppose this antigen belongs to class C, find the distance from the antibody to the nearest member of a different class C'. This distance is the radius of a prototype. The antibody is the centroid of a prototype. The prototype is constructed as a hypersphere classifier that covers class

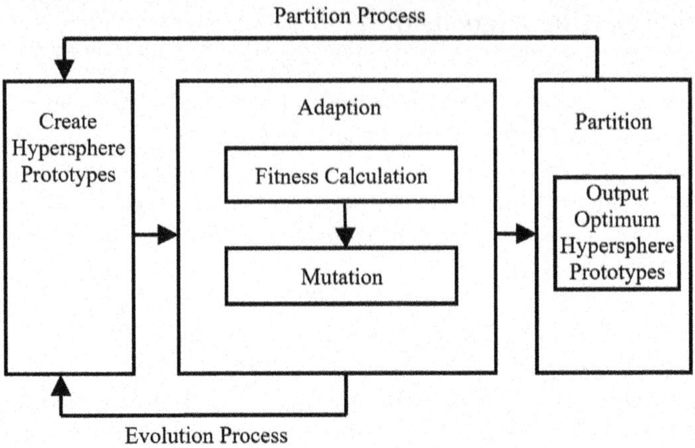

Fig. 1 MSA training process

C antigens. Second, Adaption. Calculate the fitness of the classifier called figure of merit. It equals the number of antigens falling inside of the hypersphere. Then, generate N new antibodies by mutating the old ones, i.e., the centroid of the prototypes. For each class, doing this by stochastically selecting one prototype's centroid to mutate. Selection is based on a probability distribution function proportional to the prototype's figure of merit value. Repeat this step until a maximum figure of merit is found.

The constructed initial prototypes contain redundancy. Let us consider Fig. 2 consisting of two classes of sample sets. The twelve red and twelve blue points are the points we want to classify. The constructed prototypes in two dimensional spaces are seven red circles and three blue circles shown in Fig. 2a. The disjunction of the areas that the red circles covers are initial classifiers for red points. The disjunction of the areas that the blue circles covers are initial classifiers for blue points. It is obvious that some red circles are redundant, since some red points can be covered by one big red circle instead of two smaller circles.

To eliminate redundancy, MSA utilizes a fitness calculation and mutation step to identify the optimal prototypes during each partition phase. These optimal prototypes are characterized by having the highest number of sample points within them, known as the Figure of Merit (FOM). To achieve this, each mutation generates its own FOM and strives to produce optimal prototypes with higher FOM until a stopping condition is met during the evolution stage. In the mutation process, each class is separately mutated by randomly selecting one prototype's centroid to mutate, based on a probability distribution function proportional to the prototype's FOM value. The algorithm then iteratively generates prototypes in the neighborhood areas to identify the ones with the highest FOM for each class. The prototypes with the highest FOM for each class are presented in Fig. 2b, which cover four red points and four blue points, both having an FOM of 4.

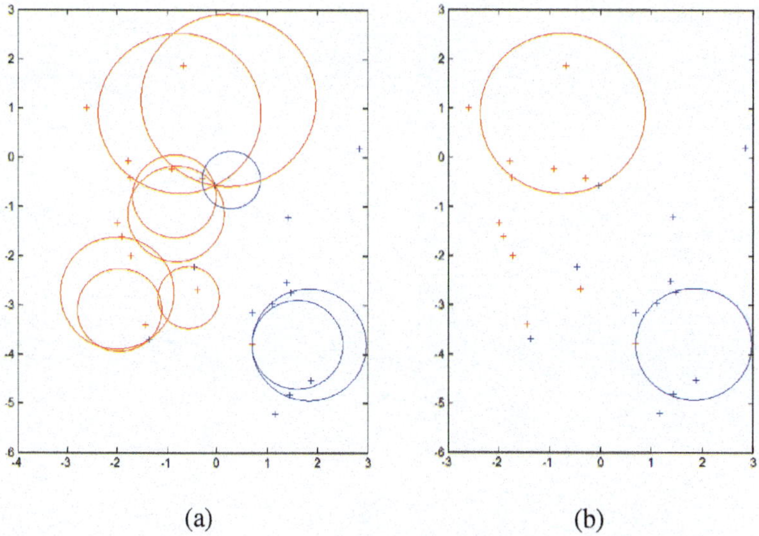

Fig. 2 MSA Prototypes Construction. **a** Initial Prototypes with redundancy; **b** Prototypes of the highest FOM

(2) **Partition Phase**

During the partitioning phase, the training datasets are split into smaller subsets and the hypersphere decision boundaries are trained using an algorithmic parameter known as Margin.

Definition 1 Margin: it is defined as the percentage χ that the radius of hypersphere shrinks, so radius of χ margin is of the quantity:

$$R_\chi = (1 - \chi)R_0. \tag{2}$$

where $0 \leq \chi < 1$. R_χ is the radius that shrinks χ from R_0. R_0 is called zero margin when $\chi = 0$.

It can be seen in Fig. 3, the twelve red and twelve blue points are partially classified by prototypes as red circles and blue circles. The big red circle's margin is 0. After we shrink the circle's radius by 20%, i.e., the smaller red circle has a margin = 0.2. Margin parameter can be adjusted to improve the performance.

Partition process splits the training set T into several subsets T_1, T_2, T_3.... The subsets are non-empty and non-overlapping. The training set can be viewed as a union of its subsets as below:

$$T = U_{i>1}T_i.$$

where

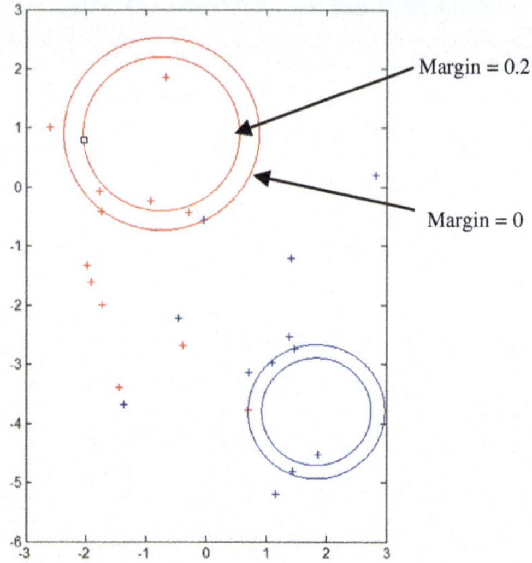

Fig. 3 Margin in MSA

$$T_i \neq \varnothing \; and \; T_i \cap T_j = \varnothing (i \neq j). \tag{3}$$

Once the optimal prototypes are obtained, the sample points within them are eliminated from the sample dataset. Then, we randomly select points from the remaining sample points and begin constructing initial prototypes again. This process is repeated until no sample points are left. During the testing phase, it is possible to encounter misclassified or unclassified data points. To handle this, we can set the number of generations as a parameter to keep track of the number of partition phases executed. Once it reaches the specified number of generations, the algorithm terminates.

B. **Testing**

After the MSA training process is complete, multiple prototypes are obtained as classification decision boundaries through a few iterations, referred to as generations. In each generation, MSA divides the training set T into two subsets: the training points covered by the prototypes, and the remaining unclassified class points. MSA typically retains only one prototype with the largest figure of merit for each class. If all training points of a class are classified after several generations, no more prototypes will be generated for that class. However, MSA continues to generate prototypes to cover the remaining unclassified training points. To assign class labels to test data, the Euclidean distance between the centroid of the hypersphere and the test data is calculated. The class label is assigned based on the hypersphere it falls inside.

The goal is for all data samples to be correctly classified, but sometimes this is not achievable due to misclassifications or unclassified points. To end the algorithm, we can set two stopping conditions: (1) the training set size reaches a user-defined limit; or (2) the number of generations G reaches a user-defined limit.

C. **Algorithm: MSA**

Notation:

MU: Maximum Mutation, $MU = 20$;

MG: Maximum Generation, $MG = 20$;

N: Number of random points, $N = 20$;

χ: margin, $\chi = 0$;

M: number of iterations during mutation, $M = 0$;

G: number of partitions/generations, $G = 0$;

Input:

(1) Training set $T = \{(x_1, \ldots, x_m)\}$, consists of m training samples as antigens. Each training sample $x_k (1 \leq k \leq m)$ is n-dimension vector $(n \geq 2)$ with class label $C_p (p = 1, \ldots, P)$. $(P > 1)$ classification problem.

(2) Testing set $S = \{(y_1, \ldots, y_n)\}$., consists of n unknown label testing samples.

For $(T' \neq \emptyset \& Q < MG)$ **do**

1. Normalize T into [0, 1] space. Generate N random points as antibodies that follows uniform distribution in [0, 1] space, denoted as $R = \{(\omega_1, \ldots, \omega_N)\} \in \mathrm{Unif}[0, 1]$.

For $(M \leq MU \ \& \ \mathrm{LF}^M \leq \mathrm{LF}^{M+1})$ **do**

2. Build initial prototypes $G_i = \left(\omega_i, R_i, C_p\right), 1 \leq i \leq N, p = 1, 2, \ldots, P\}$. Each random point $\omega_i \in \mathrm{R}(1 \leq i \leq N)$ is n-dimensional $(n \geq 2)$ vector. ω_i will serve as the centroid of G. For each classC_i, find the ω_i that is the minimum Euclidean distance from ω_i to x_k:

$$min\|\omega_i - x_k\|. \tag{4}$$

The radius of prototypes R_k equals to the following minimum Euclidean distance from ω_i to x_j:

$$R_k = min\|\omega_i - x_j\|, j \neq k. \tag{5}$$

3. Compute the fitness, measured by figure of merit of the prototype. The figure of merit of prototype G_i is denoted as F_{G_i}, i.e., the number of class C_p data samples inside of G_i geometrically. Suppose class label C_p contains total h prototypes for during the current iteration. The largest figure of merit among all h prototypes is LF:

$$LF = \max\left\{\mathrm{F_{G_1}}, \mathrm{F_{G_2}}, \ldots, \mathrm{F_{G_h}}\right\}. \tag{6}$$

4. Stochastically select a center $\omega_i{}'$ of prototype G_i to mutate. $\omega_i{}'$ is selected based on a probability distribution function proportional to the prototype's figure of merit value. Calculate the proportional of the porotypes of figure of merit f_p:

$$f_p = \frac{F_{G_i}}{\sum_1^h F_{G_i}}. \tag{7}$$

If i in $\omega_i{}'$ satisfy the following probability distribution function and $\zeta \in$ Unif[0, 1]:

$$\sum_{\xi=1}^{i-1} f_\xi < \zeta \le \sum_{\xi=1}^{i} f_\xi. \tag{8}$$

Select $\omega_i{}'$ to mutate to another N points. The mutated N points are $\omega_i \in R'(1 \le i \le N)$:

$$\omega_i' + \varepsilon\alpha U. \tag{9}$$

where ε is random sign symbol $\{-1,1\}$. $\alpha \in$ Unif [0, 1]. U is the maximum perturbation:

$$U = \begin{cases} \omega_i{}' & if\ \omega_k \le \frac{\min\{x_k\}+\max\{x_k\}}{2} \\ \max\{x_k\} - \omega_i{}' & Otherwise \end{cases} (1 \le k \le m). \tag{10}$$

5. M = m + 1; R = R′;

End for

6. Partition the training set and yield the optimum prototypes $G_i{}^o$. The number of generations Q increment by 1 to$Q+1$. Store the prototype $G_i{}^o$ with largest figure of merit LF^M, and radius $R_{i,Q}$. Remove all data samples falling inside of $G_i{}^o$ of the current generation Q to get reduced training set T′. Set the margin χ, the radius of $G_i{}^Q$ is $R_{i,Q}$:

$$R_{i,Q} = (1-\chi)R_{i,Q} \tag{11}$$

7. G = g + 1; $T = T'$;

End for

8. The classifiers are the disconjunction of prototypes for all p classes generated in Total Q generations:

$$G' = \bigcup_{i=1}^{Q}\bigcup_{j=1}^{P} G_i. \tag{12}$$

For each $y_i \in S$ **do**

For each $G_i \in G'$ **do**

9. If $||y_i - \omega_i|| \leq R_i$ then

$$c_i = C_p$$

End for

End for

Output c_i, the class label of y_i.

3 Smart Cities Application: Human Activity Recognition

A. Background

The study of Human Activity Recognition (HAR) has gained significant attention due to its vital role in IoT-based smart environments, such as smart cities, smart homes, smart commercial buildings, factories, vehicles, and energy management [4, 6, 9]. These smart environments form a network of sensors, actuators, and computing devices that work together to enhance human comfort and convenience. The progress in sensor technology, data management, and efficient algorithms has enabled these systems to move beyond data collection and communication to include data processing and activity recognition. This technology has transformed different fields, such as healthcare, security, and elderly care, by improving patient care, personalizing assistance, and detecting abnormal behaviors and threats [8, 12]. Activity recognition requires appropriate sensor selection, data collection and analysis, computational model development, and algorithmic inference.

HAR in smart homes to continuously monitor and anticipate the future activities of elderly or physically impaired individuals. This approach enhances the quality of life for older individuals and relieves family caregivers [18, 23]. Recent research that focuses on HAR using sensors have been tackled using mostly the generative and discriminative modelling approach. The generative modeling approach tries to use a probabilistic method to build a perfect description of an input model. One of the limitations of these methods is its use of large dataset to learn every probabilistic outlook that may be required for optimum results. Artificial Neural Networks (ANN) techniques have been successfully applied to various human activities recognition. Researchers use ANN to make indoor relative humidity and temperature prediction which is essential in smart home energy efficiency and indoor air quality [20, 26].

Much research has been conducted into human activity recognition using Support Vector Machines (SVM) as well [4]. Recently, researchers introduced two novel incremental SVM techniques that enhance the performance of SVM classification for human activity recognition tasks [22]. However, MSA, a supervised algorithm that is similar to SVM and ANN, has not yet been fully explored for human activity recognition. Therefore, the application of MSA to this task can provide a more comprehensive and thorough examination of its potential in improving the performance of human activity recognition systems.

B. Methodology

The proposed methodology for recognizing human activity is shown in Fig. 4. The raw data comprises sensor readings with corresponding annotated activities at different timestamps, indicating the beginning and end of activity occurrences. A preprocessing step is then carried out to transform the data into the required format of an activity label vector for MSA. This vector is then split into training and test data using a four-fold cross validation. More details regarding the data preprocessing can be found in the next section. The activity label vector is fed into the MSA training stage, which generates hypersphere prototypes and optimizes them through two phases: evolution and partition. The resulting optimized prototypes constitute the MSA model, which is used to classify test data. The predicted activity label is then compared with the true label to output the performance accuracy.

C. Experiments and Results

Smart homes play a crucial role as a foundational component of Smart Cities, which has been a vision for many decades [15]. To evaluate the effectiveness of the proposed MSA algorithm for human activity recognition, we selected two real-world data sets in the smart home environment.

(1) Datasets

To evaluate the effectiveness of the proposed MSA algorithm for human activity recognition, two real-world data sets were chosen in the context of a smart home environment.

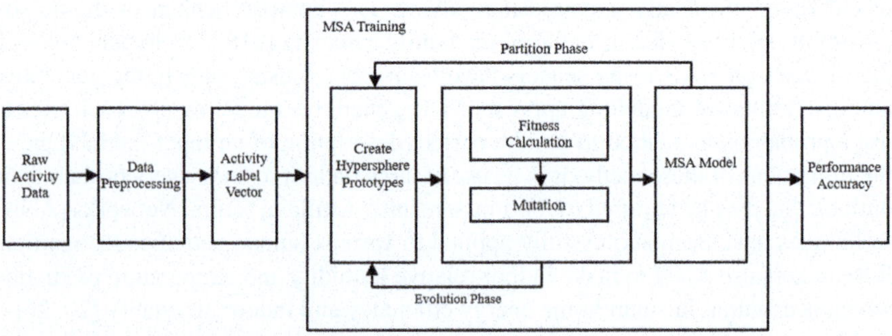

Fig. 4 Human activity recognition using MSA

(1) ARAS dataset [1]. This dataset contains two houses A and B, with four residents, two in each of the houses, of which 27 different human activities were recorded which was performed for 30 days on 20 different sensors. The sensor ID, type and location are demonstrated as the ambient sensor descriptions. Sensor ID is a unique identifier for the sensor. Among these sensors, they are classified as seven types. They are photocell to detect open drawers and wardrobes activities, which are located in the drawers, the wardrobes and the refrigerator. IR (infrared) sensor is located near the TV to detect watch TV activity. Force sensor or Pressure Mat are sensors under the bed and the couches to detect sleeping, sitting, and napping actions. Contact sensors are put in the door frames, shower cabins and cupboard to detect actions of opening and closing of the doors and cupboards. Sonar distance sensor is detecting presence, which are located on the walls and door frames. Temperature sensors recognize cooking activity which are near the oven in the kitchen.

(2) The CASAS dataset was obtained from the Tulum test bed of the Washington State University CASAS smart home project [7]. This smart home consists of three bedrooms, one bathroom, one kitchen, and a living/dining room, and it recorded the normal daily activities of two married residents from April to July 2009. The dataset includes 18 motion sensors and two temperature sensors that collected data on 11 human activities, such as cooking breakfast, cooking lunch, entering the home, having group meetings, leaving the home, eating breakfast, having snacks, washing dishes, watching TV, and other activities.

When it comes to selecting locations for sensors in human activity recognition, matching the sensors to the specific activity is taken into consideration. Human activities may require one or more sensors to be identified. It is common for several actions to take place concurrently or successively during the recognized activity period. For instance, when preparing breakfast (an activity listed in Table 2), actions may involve opening the fridge or kitchen drawer to retrieve utensils and food. In such cases, sensor readings should be active in at least one of the four sensors: Ph3, Ph4, So2, or Te1. Table 1 displays that Ph3 is the photocell sensor located in the fridge, Ph4 is the photocell sensor in the kitchen drawer, So2 is the sonar distance sensor in the kitchen, and Te1 is the temperature sensor in the kitchen.

(2) **Results**

In our experiments, we assessed the effectiveness of the MSA on two benchmark datasets, in comparison with two other popular algorithms: SVM and ANN. While SVM uses a hyperplane and ANN uses a perceptron to classify points on a plane, MSA applies a hypersphere to correctly classify class samples. The study was conducted over a ten-day period and default parameters were selected for all three algorithms. Specifically, the number of random points at initialization (NA) was set to 20, the maximum number of mutations for every generation (MU) was set to 20, and the maximum number of generations (MG) was set to 20. The margin parameter (χ) was set to zero as the default value for MSA. For SVM, we used the Radius Basis Function (RBF) kernel with a regularization parameter C of 1, and set the kernel

Table 1 Ambient sensor descriptions of ARAS dataset

Column	House A			House B		
	Sensor ID	Sensor type	Location	Sensor ID	Sensor type	Location
1	Ph1	Photocell	Wardrobe	Co1	Contact sensor	Kitchen cupboard
2	Ph2	Photocell	Convertible	Co2	Contact sensor	Kitchen cupboard
3	Ir1	IR	TV receiver	Co3	Contact sensor	House door
4	Fo1	Force sensor	Couch	Co4	Contact sensor	Wardrobe door
5	Fo2	Force sensor	Couch	Co5	Contact sensor	Wardrobe door
6	Di3	Distance	Chair	Co6	Contact sensor	Shower cabinet
7	Di4	Distance	Chair	Di2	Distance	Tap
8	Ph3	Photocell	Fridge	Fo1	Force sensor	Chair
9	Ph4	Photocell	Kitchen drawer	Fo2	Force sensor	Chair
10	Ph5	Photocell	Wardrobe	Fo3	Force sensor	Chair
11	Ph6	Photocell	Bathroom cabinet	Ph1	Photocell	Fridge
12	Co1	Contact sensor	House door	Ph2	Force sensor	Kitchen drawer
13	Co2	Contact sensor	Bathroom door	Pr1	Pressure mat	Couch
14	Co3	Contact sensor	Shower cabinet	Pr2	Pressure mat	Couch
15	So1	Sonar distance	Hall	Pr3	Pressure mat	Bed
16	So2	Sonar distance	Kitchen	Pr4	Pressure mat	Bed
17	Di1	Distance	Tap	Pr5	Pressure mat	Arm chair
18	Di2	Distance	Water closet	So1	Sonar distance	Bathroom door
19	Te1	Temperature	Kitchen	So2	Sonar distance	Kitchen
20	Fo3	Force sensor	Bed	So3	Sonar distance	Closet

Table 2 Activity descriptions in the datasets

ARAS dataset		CASA dataset	
Activity	Number of events	Activity	Number of events
Other	238	Cook breakfast	80
Going out	683	Cook lunch	71
Preparing breakfast	10	Enter home	73
Having breakfast	21	Group meeting	11
Sleeping	559	Leave home	75
Watching TV	207	R1 eat breakfast	66
Studying	52	R1 snack	491
Toileting	41	R2 eat breakfast	47
Using internet	145	Wash dishes	71
Laundry	22	Watch TV	528
Brushing teeth	10		
Talking on the phone	62		
Changing clothes	21		

parameter γ as the reciprocal of the number of features in the datasets. For ANN, we implemented a two-layer network with 10 neurons in its hidden layer and used default values for all other parameters. The results of the analysis are presented in Table 3.

The effectiveness of MSA in activity recognition is demonstrated in the performance analysis, which compares it to SVM and ANN. The results presented in

Table 3 Activity recognition results in accuracy (%) between MSA, SVM and ANN

Days	ARAS dataset						CASAS dataset		
	House A			House B			Tulum		
	MSA	SVM	ANN	MSA	SVM	ANN	MSA	SVM	ANN
1	60.32	58.51	59.97	88.56	85.64	86.25	72.24	70.13	70.96
2	76.29	75.14	75.68	91.25	90.21	90.89	70.54	69.52	70.21
3	65.55	63.89	64.44	98.71	97.86	97.22	57.74	56.23	57.35
4	66.85	63.25	63.85	96.84	95.23	96.14	54.15	53.19	53.88
5	69.75	66.56	67.23	97.36	95.15	96.98	75.90	74.21	75.05
6	63.24	61.59	62.84	99.55	98.46	99.21	68.12	66.93	67.55
7	69.84	68.98	68.86	99.28	98.68	98.91	71.32	69.38	70.11
8	76.57	73.24	72.15	95.84	93.87	94.68	67.45	65.25	66.45
9	61.55	59.87	60.22	98.21	97.55	98.08	67.22	66.24	66.95
10	78.59	77.98	78.01	96.83	95.46	95.89	75.36	74.95	75.21
Average	68.85	66.90	67.32	96.24	94.81	95.42	68.00	66.60	67.37

Table 3 and Fig. 5a show that MSA outperforms SVM and ANN in detecting human activities over a period of 10 days in House A. MSA has the lowest number of mis-detected activities, with a significant increase of 3.6% and 3.33% performance accuracy compared to SVM on days 4 and 8, respectively. The recognized activities for residents 1 and 2 on day 4 and day 8 are listed in detail. Additionally, MSA shows superiority in performance accuracy of 76.57% on day 8 compared to ANN, which is 72.15%. MSA also yields a 2.52% performance gain on day 5 compared to ANN, and its average performance accuracy is 1.95% and 1.53% better than SVM and ANN, respectively.

Fig. 5 Activity recognition performance accuracy of the proposed MSA, SVM and ANN on ARAS dataset: **a** house A, **b** house B

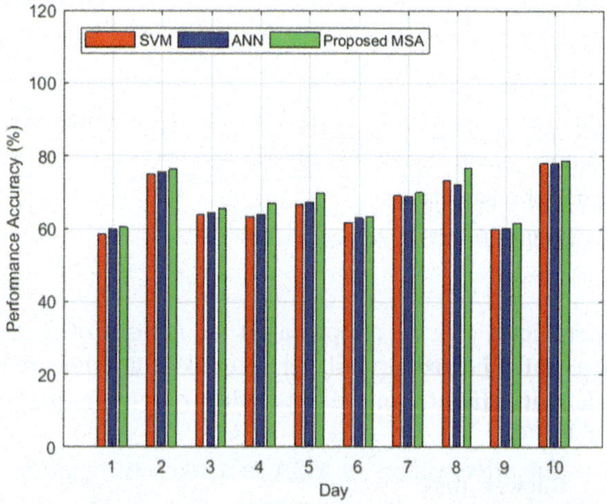

(a)

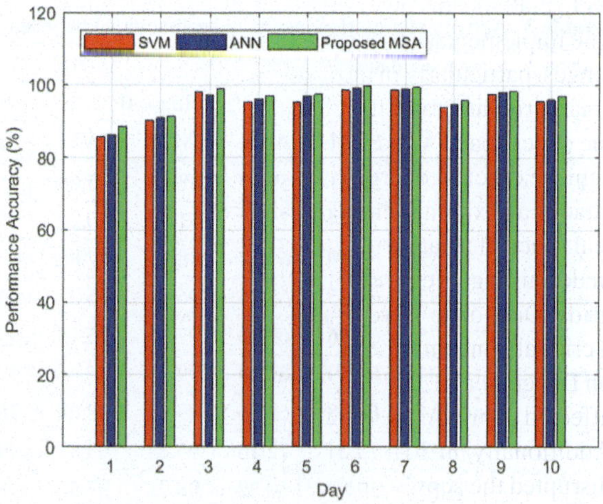

(b)

In contrast, the detection accuracy for the married couple in House B is generally higher than that for the two male residents in House A. This is because the couple has fewer but more regular daily activities. Table 1 and Fig. 5b show the performance analysis of House B, which has much better accuracy than House A. On day 1, MSA's performance accuracy is 88.56%, compared to SVM and ANN with 85.64% and 86.52%, respectively, with an increase of 2.92%. On day 5, MSA has a recognition accuracy of 97.36%, while SVM has a performance loss of 2.21%. The recognized activities for residents 3 and 4 on day 5 are also listed. MSA outperforms both SVM and ANN in all experiments, and its average performance is 1.43% and 0.82% better than SVM and ANN, respectively.

Finally, Table 3 also reports the activity recognition performance on the CASAS dataset, where MSA again performs better than SVM and ANN in all ten days' experiments, with an average performance of 68%, 66.60%, and 67.37% for MSA, SVM, and ANN, respectively. MSA significantly improves the recognition performance on some days, such as days 1, 7, and 8, with a 2.11%, 1.94%, and 2.2% increase compared to SVM. MSA also performs better than ANN on day 1, with an accuracy of 72.24% compared to 70.96%. Overall, MSA's classification performance is close to or better than SVM and ANN in all experiments.

4 Smart Energy Systems Application: False Data Injection Detection

A. Background

Smart energy systems comprise smart grids as a key component of the electricity sector, alongside other constituents, namely heating, cooling, and transportation. The implementation of modern smart grid systems has introduced new security challenges, particularly in the form of cyber-physical attacks or threats. This is because the smart grid integrates the physical infrastructure of the traditional power network with the cyber space, which encompasses information sensing, processing, and control to achieve efficient energy consumption and transmission. As a result, a cyber-physical attack can exploit vulnerabilities in cyber space, causing significant harm to the physical space of smart grids. The consequences of such attacks can be severe, potentially undermining or entirely disrupting the control systems that underpin electric power grids. Due to the increasing occurrence of cyber-physical attacks, they have become a critical concern for both industrial control system users and vendors. For instance, on December 23, 2015, Ukraine's power grid experienced several power outages that affected approximately 225,000 customers due to a Trojan called "BlackEnergy" [5]. Additionally, in 2019, a ransomware cyber-attack on the power grid in South Africa disrupted the supply of electricity to millions of people. There have also been reports of attempted cyber-attacks on power grids in other countries, including the United States, Israel, and Saudi Arabia [3].

The False Data Injection Attack (FDIA) is considered one of the most perilous cyber-attacks, as it involves injecting false data to execute a strategic attack that can result in substantial harm. As a result, the research community has shown significant interest in this type of attack in recent years [10, 19, 25, 31]. State-of-the-art CI techniques, including SVM and Adaboost, and other machine learning algorithm have been utilized to detect and prevent FDIA. Hink et al. [11] detected the power system faults and cyber-attacks using several batch processing-based machine learning and data mining algorithms, including Random Forests, Naïve Bayes, SVM, Adaboost, etc. [10]. Landford et al. proposed a machine learning approach to detect FDIAs using a two-class SVM. This method analyzed the change of correlation between two PMU parameters using Pearson correlation coefficient [17].

Synchronized phasor measurement units (PMUs) are a key sensing component that can be a source of False Data Injection Attacks (FDIAs) in smart grids [2, 16]. PMUs provide a solution for time-synchronization of phase and sequence measurements from dispersed nodes, enabling monitoring, control, evaluation, and protection of the smart grid system. However, the traditional defense approaches for FDIAs are ill-prepared for the data challenges posed by the large-scale deployment of PMUs in future smart grid cyber-physical systems (CPS) [30]. The high volume of data generated by PMUs presents real-time computational and storage challenges. Nonetheless, this challenge presents an opportunity for machine learning-based data analytical techniques to detect and prevent FDIAs.

Detecting and preventing FDIA is a critical challenge in securing smart grid systems. CI techniques, such as Machine learning algorithms have already found applications in various cybersecurity domains such as sensor networks, vehicular networks, and smart grids due to the increasing complexity of cybersecurity threats that cannot be addressed by traditional manual and signature-based approaches [28, 29, 32]. Machine learning can process large amounts of data, learn from patterns beyond human comprehension, and capture the non-linear and complex relationships between measurements to detect false PMU data injection, making it an attractive approach for PMU data analytics under FDIAs. While existing computation intelligence techniques such as SVM and Adaboost have been used for FDIA detection, they have limitations in terms of accuracy and efficiency. This has motivated researchers to explore novel machine learning-based CI techniques for more effective FDIA detection. Margin Setting algorithm (MSA) is one such technique that has shown promise in detecting FDIA with higher accuracy and efficiency than traditional CI methods. In this context, MSA has emerged as a key tool in the fight against cyber-attacks on smart grid systems, and its potential for detecting and preventing FDIA makes it an essential technique for securing the future of smart energy systems.

B. Methodology

The proposed methodology of using MSA for FIDA detection is shown as follows in Fig. 6. Online PMU data is gathered from PDC as the input of the MSA algorithm. MSA build initial classification boundaries called prototypes. Then the prototypes are trained by MSA to generate the optimal prototypes as the output. The output can classify abnormal data and normal data. Abnormal data are results from FDIAs.

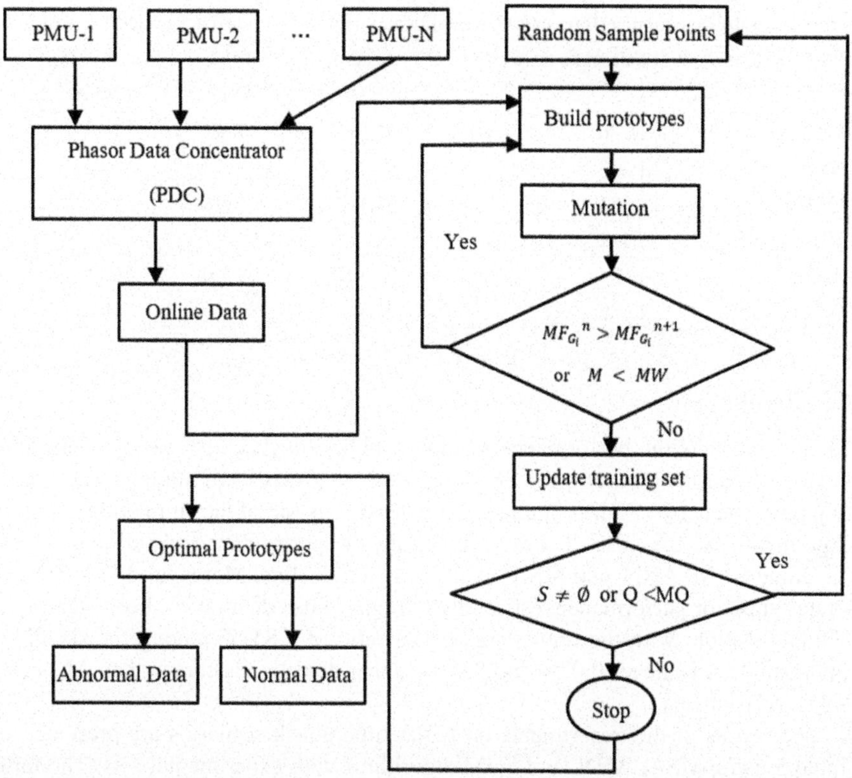

Fig. 6 FDIA detection using MSA

C. Experiments and Results

In this section, the performance of the proposed MSA is demonstrated by comparing it with another two state of the art machine learning data analytical methods—SVM and ANN. Extensive experiments are conducted on both the real-world PMU data sets.

(1) Data Sets

For our experiment to detect FDIAs, we have utilized PMU data from the Texas Synchrophasor Network, obtained from a real-PMU dataset [2]. Due to the sampling rate of 30 Hz, only low frequency oscillations below 15 Hz could be analyzed. Our analysis was conducted using an hourly PMU data comprising 108,000 data points, with each point including three signals: voltage magnitude, angle, and frequency. All measurements were taken at the customer-level (120-V). The network consisted of six PMU stations with the labels McDonald, Harris, UT Pan, UT 3, Austin, and WACO. Our experiments were performed under playback attack and time attack.

Table 4 FDIA detection performance of experimental data sets for playback attack

PMU stations	False data ratio (%)	Accuracy (in %)		
		SVM	ANN	MSA
McDonald	1.736	97.665	97.680	97.693
Harris	1.253	98.246	98.261	98.271
UT Pan	1.851	97.472	97.489	97.511
UT 3	1.142	98.372	98.385	98.394
Austin	1.034	98.492	98.501	98.507
WACO	1.039	98.501	98.508	98.513

(2) **Results**

Table 4 presents the performance results of FDIA playback attack, revealing that the proposed MSA outperforms SVM and ANN in all six PMU stations. The detection performance of FDIA exhibits a similar trend to that of the simulation data sets experiment. As the false data ratio increases, the performances of all algorithms, including SVM, ANN, and MSA, exhibit a linear decline. Moreover, MSA demonstrates superior performance, especially at higher false data ratios. For instance, at UT Pan station, with the highest false data ratio of 1.851% among the six PMU stations, MSA achieves 0.02% and 0.06% higher detection accuracy than ANN and SVM, respectively.

The results of the experiments on FDIA time attack scenarios are presented in Table 5 using MSA, ANN, and SVM algorithms with experimental sets. The false data ratios for time attack scenarios are detailed in Table VII. Four different scenarios were simulated by changing the resampling rate factor of 7/6, 3/2, 2, and 4, slower than real-time in the final 30 min of the time-series datasets. The MSA algorithm outperforms ANN and SVM in all six PMU stations for all four scenarios. As the false data ratio decreases, the detection performance tends to increase. For example, in the factor of 7/6 scenario, the McDonald PMU yields 96.543% accuracy when the false data ratio is 1.784%, whereas the performance goes up to 97.522% for a false data ratio of 0.663% in the WACO PMU. Although the false data ratio varies in the six PMU stations for the four different cases, it is noteworthy that the slower the resampling rate, the higher the detection accuracy.

Table 5 FDIA detection performance of experimental data sets for time attack

Accuracy (in %)	Factor of 7/6 slower			Factor of 3/2 slower			Factor of 2 slower			Factor of 4 slower		
	SVM	ANN	MSA	SVM	ANN	MSA	SVM	ANN	MSA	SVM	ANN	MSA
McDonald	96.54	96.55	96.57	97.47	97.48	97.50	98.04	98.05	98.07	98.76	98.78	98.79
Harris	96.72	96.74	96.75	97.66	97.67	97.68	98.22	98.24	98.25	98.95	98.96	98.97
UT pan	97.03	97.04	97.05	97.96	97.97	97.98	98.53	98.54	98.55	99.25	99.26	99.27
UT 3	97.15	97.16	97.17	98.08	98.10	98.10	98.65	98.66	98.67	99.37	99.38	99.39
Austin	97.28	97.30	97.30	98.21	98.23	98.23	98.78	98.79	98.80	99.51	99.51	99.52
WACO	97.52	97.53	97.53	98.45	98.46	98.46	99.02	99.03	99.03	99.74	99.75	99.75

5 Conclusion

In this chapter, we explored the critical applications of CI, specifically the Margin Setting Algorithm, to advance and secure SCS in two crucial areas: human activity recognition in smart homes and the detection of False Data Injection Attacks in smart grids. These applications demonstrate the vast potential of CI to address complex challenges in modern urban and energy systems. The first application demonstrates that MSA can accurately recognize and classify the activities of occupants of a smart home. Another critical application of the MSA is in the detection of False Data Injection Attacks (FDIA) in smart grids. It is important for smart grid systems as it helps ensure the security and reliability of the system, preventing potential cyber-attacks that could have catastrophic consequences. Both two applications demonstrate the importance of leveraging advanced technologies such as machine learning and big data analytics to create more efficient, secure, and sustainable solutions for the smart cities and smart energy systems of the future.

References

1. Alemdar, H., Ertan, H., Incel, O. D., & Ersoy, C. (2013). *ARAS human activity datasets in multiple homes with multiple residents.* Paper presented at the Proceedings of the 7th International Conference on Pervasive Computing Technologies for Healthcare.
2. Allen, A., Singh, M., Muljadi, E., & Santoso, S. (2014). *PMU data event detection: A user guide for power engineers.* Retrieved from.
3. Analytica, O. (2019). South Africa power cyberattack underlines acute risks. *Emerald Expert Briefings*(oxan-es).
4. Arshad, M. H., Bilal, M., & Gani, A. (2022). Human Activity Recognition: Review, Taxonomy and Open Challenges. *Sensors, 22*(17), 6463.
5. Case, D. U. (2016). Analysis of the cyber attack on the Ukrainian power grid. *Electricity Information Sharing and Analysis Center (E-ISAC), 388*, 1–29.
6. Chen, K., Zhang, D., Yao, L., Guo, B., Yu, Z., & Liu, Y. (2021). Deep learning for sensor-based human activity recognition: Overview, challenges, and opportunities. *ACM Computing Surveys (CSUR), 54*(4), 1–40.
7. Cook, D. J., Krishnan, N. C., & Rashidi, P. (2013). Activity discovery and activity recognition: A new partnership. *IEEE transactions on cybernetics, 43*(3), 820–828.

8. Dahmen, J., Thomas, B. L., Cook, D. J, Wang, X. (2017). Activity learning as a foundation for security monitoring in smart homes. *Sensors, 17*(4), 737. https://doi.org/10.3390/s17040737
9. Dang, L. M., Min, K., Wang, H., Piran, M. J., Lee, C. H., & Moon, H. (2020). Sensor-based and vision-based human activity recognition: A comprehensive survey. *Pattern Recognition, 108*, 107561.
10. Habib, A. A., Hasan, M. K., Alkhayyat, A., Islam, S., Sharma, R., & Alkwai, L. M. (2023). False data injection attack in smart grid cyber physical system: Issues, challenges, and future direction. *Computers and Electrical Engineering, 107*, 108638.
11. Hink, R. C. B., Beaver, J. M., Buckner, M. A., Morris, T., Adhikari, U., & Pan, S. (2014). *Machine learning for power system disturbance and cyber-attack discrimination.* Paper presented at the 2014 7th International symposium on resilient control systems (ISRCS).
12. Holzinger, A., Röcker, C., & Ziefle, M. (2015). From smart health to smart hospitals. *Smart Health: Open Problems and Future Challenges*, 1–20.
13. Hui, T. K., Sherratt, R. S., & Sánchez, D. D. (2017). Major requirements for building Smart Homes in Smart Cities based on Internet of Things technologies. *Future Generation Computer Systems, 76*, 358–369.
14. Igwe, O. M., Wang, Y., Giakos, G. C., & Fu, J. (2020). Human activity recognition in smart environments employing margin setting algorithm. *Journal of Ambient Intelligence and Humanized Computing*, 1–13.
15. Khan, A., Aslam, S., Aurangzeb, K., Alhussein, M., & Javaid, N. (2022). Multiscale modeling in smart cities: A survey on applications, current trends, and challenges. *Sustainable cities and society, 78*, 103517.
16. Khare, G., Mohapatra, A., & Singh, S. (2021). A real-time approach for detection and correction of false data in PMU measurements. *Electric Power Systems Research, 191*, 106866.
17. Landford, J., Meier, R., Barella, R., Wallace, S., Zhao, X., Cotilla-Sanchez, E., & Bass, R. B. (2016). *Fast sequence component analysis for attack detection in smart grid.* Paper presented at the 2016 5th International Conference on Smart Cities and Green ICT Systems (SMARTGREENS).
18. Lentzas, A., & Vrakas, D. (2020). Non-intrusive human activity recognition and abnormal behavior detection on elderly people: A review. *Artificial Intelligence Review, 53*(3), 1975–2021.
19. Li, Y., Wei, X., Li, Y., Dong, Z., & Shahidehpour, M. (2022). Detection of false data injection attacks in smart grid: A secure federated deep learning approach. *IEEE Transactions on Smart Grid, 13*(6), 4862–4872.
20. Lu, T., Viljanen, M. J. N. C., & Applications. (2009). Prediction of indoor temperature and relative humidity using neural network models: model comparison. *18*(4), 345.
21. Lund, H., Østergaard, P. A., Connolly, D., & Mathiesen, B. V. (2017). Smart energy and smart energy systems. *Energy, 137*, 556–565.
22. Nawal, Y., Oussalah, M., Fergani, B., & Fleury, A. (2022). New incremental SVM algorithms for human activity recognition in smart homes. *Journal of Ambient Intelligence and Humanized Computing*, 1–18.
23. Ramanujam, E., Perumal, T., & Padmavathi, S. (2021). Human activity recognition with smartphone and wearable sensors using deep learning techniques: A review. *IEEE Sensors Journal, 21*(12), 13029–13040.
24. Razmjoo, A., Mirjalili, S., Aliehyaei, M., Østergaard, P. A., Ahmadi, A., & Nezhad, M. M. (2022). Development of smart energy systems for communities: Technologies, policies and applications. *Energy, 248*, 123540.
25. Reda, H. T., Anwar, A., & Mahmood, A. (2022). Comprehensive survey and taxonomies of false data injection attacks in smart grids: Attack models, targets, and impacts. *Renewable and Sustainable Energy Reviews, 163*, 112423.
26. Shi, X., Lu, W., Zhao, Y., & Qin, P. J. I. A. (2018). Prediction of indoor temperature and relative humidity based on cloud database by using an improved BP neural network in Chongqing. *6*, 30559–30566.

27. Sun, Y., Song, H., Jara, A. J., & Bie, R. (2016). Internet of things and big data analytics for smart and connected communities. *IEEE access, 4*, 766–773.
28. Tan, K., Bremner, D., Le Kernec, J., Zhang, L., & Imran, M. (2022). Machine learning in vehicular networking: An overview. *Digital Communications and Networks, 8*(1), 18–24.
29. Tsimenidis, S., Lagkas, T., & Rantos, K. (2022). Deep learning in IoT intrusion detection. *Journal of network and systems management, 30*, 1–40.
30. Wallace, S., Zhao, X., Nguyen, D., Lu, K.-T., Buyya, R., Calheiros, R., & Dastjerdi, A. (2016). Big data analytics on smart grid: Mining pmu data for event and anomaly detection. *Big data: principles and paradigms, 17*, 417–429.
31. Wang, Y., Amin, M. M., Fu, J., & Moussa, H. B. (2017). A novel data analytical approach for false data injection cyber-physical attack mitigation in smart grids. *IEEE access, 5*, 26022–26033.
32. Wei, J., & Mendis, G. J. (2016). *A deep learning-based cyber-physical strategy to mitigate false data injection attack in smart grids*. Paper presented at the 2016 Joint Workshop on Cyber-Physical Security and Resilience in Smart Grids (CPSR-SG).

Ontology-Based Similarity Estimates for Fuzzy Data: Semantic Wiki Approach

Julia Rogushina and Anatoly Gladun

Abstract We analyze main types of dirty data processed by computing intelligence, criteria of their classification and means of their detection. Results of this analysis are represented by ontological model that contains taxonomy of classical and non-classical data and knowledge-oriented methods of their transformation. Special attention is paid to semantically incorrect data that corresponds to vague knowledge. Such data cannot be reconciled with rules and restriction imported from external knowledge sources such as domain ontologies. Incorrectness of such data can be caused by irrelevant attribute, incorrect type of value, inadmissible relation between individuals of classes, etc. We propose to use various estimations of semantic similarity between domain concepts to detect these types of data incorrectness and transform it into classical data. Sources and types of fuzzy data are considered on example of representation possibilities of the Wiki technological environment and its semantic extension Semantic MediaWiki. Wiki resources are interesting from the point of view of the fuzzy data processing by the heterogeneous information objects, knowledge schemas and subjects of data entering. We analyze types of dirty data are detected automatically or semi-automatically in this technology and define what external means of knowledge management can be used for processing of these data.

Keywords Computational intelligence · Fuzzy data · Semantic similarity · Ontology · Wiki technology

---

J. Rogushina
Institute of Software Systems of the National Academy of Sciences of Ukraine, 40, Ave Glushkov, Kyiv 03181, Ukraine

A. Gladun (✉)
International Research and Training Center for Information Technologies and Systems of NAS and MES of Ukraine, GSP, 40, Ave Glushkov, Kyiv 03680, Ukraine
e-mail: glanat@yahoo.com

K. Daimi et al. (eds.), *Cutting Edge Applications of Computational Intelligence Tools and Techniques*, Studies in Computational Intelligence 1118,
https://doi.org/10.1007/978-3-031-44127-1_15

1 Introduction

Fuzzy computing is oriented on the analysis of data that contain various types of uncertainty, incompleteness, and errors. Pre-processing of data allows transforming "raw" representations into "smart" ones that are more suitable for automated acquisition of useful information. Depending on uncertainty type, different preprocessing procedures can be applied. Therefore, we need in classification of existing types of data uncertainty that can be represented by corresponding taxonomic model that can be used for selection of processing means.

We have to take into account that the choice of processing methods depends on subject domain of "raw" data itself and the aims of data use. Some types of uncertainty can be remove by automatic transformation methods, and some others require direct human involvement. But we can detect a lot of situations where data pre-processing can automatically use external sources of knowledge to reduce the involvement of vague knowledge of human experts to the selection of such pertinent sources or the formulation of conditions for their search.

Knowledge about solved tasks and about structure of the information objects that are typical for this task allows to ensure the data transformation at the semantic level, for example, by semantics of relations and evaluating the semantic proximity between different domain concepts.

Functioning of the Web-oriented information systems is based on the open information environment paradigm where all information objects can be constantly changed and replenished, and therefore we propose to focus on open standards of knowledge representation, for example, languages developed in the Semantic Web project—RDF [1] and OWL [2]. The use of ontological models of knowledge representation provides an unambiguous interpretation of information from external sources, but requires the creation of rather complex methods for their use.

Computational intelligence (CI) is a heterogeneous field of research that combines, harmonizes and coordinates several technologies of intelligent data processing, such as probabilistic reasoning, artificial neural networks, fuzzy systems, evolutionary algorithms, etc.

CI paradigm is aimed to develop systems with intelligent behavior in complex open environment [3].

The integration of areas such as *machine learning* (ML), *artificial intelligence* (AI), *decision support systems* (DSS), *multi-agent systems* (MAS) and database management systems (DBMS) increases the power and impact of CI to solve many applied engineering tasks.

The term "computational intelligence" refers to the ability of a computer to solve a specific problem with use of data or experimental observations. CI is usually considered a synonym for soft computing, although there is no general definition of computational intelligence [4]. Soft computing can be considered as a set of methodologies for processing of uncertainty and partial trust [5] but it some specific features in MAS and DSS.

CI is usually considered as a set of computational methodologies and approaches for solving complex real-world problems where other types of traditional modeling are not effective for a number of reasons: (1) modeled processes too complex for mathematical reasoning and contain some uncertainties during the process, or (2) the process is stochastic in nature. Many real-life problems cannot be translated into binary data, but CI is aimed to provide solutions to them. Therefore, we can consider CI as an instrument for processing of non-classical data.

CI methods are close to the human way of reasoning, i.e. use imprecise and incomplete knowledge, and allow for adaptive decision-making. CI helps to deal with inaccuracy, uncertainty and incompleteness of data based on various combinations (defined by task specifics) of the following AI elements:

- fuzzy logic;
- artificial neural networks,
- evolutionary calculations,
- methods of machine learning and
- probabilistic methods.

Information in the most general form can be considered a statement conveyed by means of signs, most often symbols, about objects from some domain. The transition from information to knowledge occurs when we trust the information and intend to act on it.

If the response to the observation and evaluation of symbols is some action or intention to act, then these symbols can be regarded as knowledge. The criteria for evaluating certain information are its coherence defined by the state of logical sequential connections that provides a logical structure for the intelligent integration of various elements. Another criterion for information evaluating is its ambiguity that causes frequently errors of incorrect identification of objects and relations. That is why disambiguation is an important task in semantic systems. Context can be used to resolve ambiguities: the same information can be used in different ways or have different importance depending on the circumstances.

Classic data (CD) can be defined of as precise, defined, consistent, clear one, with no missing or missing values, etc. Classical data prevailed in many sciences for a long time before L. Zadeh began to form the mathematical foundations of fuzzy logic (in the 1960s) [6].

Models of CD do not allow the representation or manipulation of data (or knowledge) that is imprecise, uncertain, vague, etc. However, such data and knowledge are increasingly used in modern information systems, databases, data warehouses, knowledge bases, since they are oriented on domains that generate data with various forms of vagueness. Therefore, we need to identify the forms of such data and knowledge, recognize them, use them as input raw datasets in different representation form, transform and even delete them if necessary.

In this article, the whole set of heterogeneous forms of data, which are imprecise, vague, uncertain, inconsistent, incomplete, etc., and cannot be consider as CD we name *non-classical data* (NCD).

Fuzzy computing is an instrument that allows analyzing such non-classical data and performing operations of logical inference on them. They include a collection of facts, linguistic variables and corresponding functions of membership, fuzzy "if-else" statements, and fuzzy production rules. This apparatus is a valuable resource for describing fuzzy concepts, intelligent data analysis, and decision-making in various fields of science, business, and manufacturing.

The nature and origin of such information varies, and we need in different technologies to handle each form of non-classical data.

In most cases, important information for an information system comes from two sources:

(1) from human experts who describe their domain knowledge by means that include natural language (NL) that causes subjectivity and ambiguity of information;
(2) from external technical devices (sensors, counters, mathematical calculations, etc.) that can be imprecise or contain transmission error that in general cause uncertainty and inconsistency of data.

Therefore, the store of expert assessments and opinions, unreliable data requires means for representation of non-classical data and the ability to work with them. Such processing includes data mining and interpreting from fuzzy databases and knowledge bases that often contain non-classical data. In general, most modern systems, databases and knowledge management systems, controllers, devices and software applications require a mechanism to support and manage non-classical data and fuzzy knowledge, as well as the ability to extract and analyze them.

Today, the range of tasks that are solved in CI with the help of the apparatus of fuzzy sets and fuzzy logic has significantly expanded and covers such areas as data analysis and intelligent data analysis, pattern recognition, operations research, modeling of complex systems, decision support, etc. Those properties of data that differ from classic data are the basis for choosing methods of their analysis and processing. Therefore, it is advisable to analyze the main types of NCDs and the possibility of their transformation into CDs.

2 Classification of Non-classical Data Types

The life cycle of data includes its collection, storage, updating, transmission, access, archiving, recovery, deletion and cleaning, etc. Data is considered dirty if a user or a properly functioning application is unable to obtain the result of its processing or obtains an incorrect result due to some problem with the data. Analysis of dirty data includes two different aspects—why the data became dirty and what can be done to make it suitable for analysis.

For example, the sources of dirty data can be an error in entering or updating data (by a person or a computer system), data transmission errors, or an incorrectly selected form of data submission.

Incomplete, inconsistent, undefined, ambiguous, vague, imprecise, null data represent different types of non-classical data forms. Below we represent the characteristics, examples and sub-forms of these forms. It should be noted that each non-classical form of data is itself a manifestation of something in real world (e.g., incompleteness, contradiction, imprecision, ambiguity, vagueness, etc.).

Incomplete data. Data incompleteness usually means absence of value or inaccurate information where the set of possible values covers the entire range of possible values. Incompleteness of data can be caused by lost updates, incorrect reading, and lack of access to information, etc. [7]. For example, the phone number can be entered with insufficient number of characters or not be entered at all. For incomplete data, an important aspect of the analysis is understanding whether a value exists at all even if it is currently unknown (for example, email of some person exists but we does not know it) or whether it cannot be obtained at all at the current time (for example, a date of death for alive person). Different logic systems use different notations to identify the type of data incompleteness. In some cases, incomplete data do not allow to identify uniquely the value, but allow to narrow the range of possible values (for example, if last name of person is entered incorrectly in the database, then various correction variants allow to associate the corresponding record with one of 10 people and not with all others).

Inconsistent data. The concept of inconsistency is more related to data storing with different models than the data itself. Inconsistency is a semantic conflict where the same aspect or the same meaning of data elements has such representations that their interpretations can not be true simultaneously. For example, one data source defines the year of birth of person X as 1985, and another one as 1988. Inconsistency of information is usually caused by the process of integration (combination) of information from different input sources.

One of the inconsistency reasons may be the use of different units of measurement (for example, the distance between A and B is given in kilometers or in miles) or different order of information parts (for example, the date format "11.05" and "05.11"). In such cases, data transformation and reconciliation can be automated after analyzing the semantics of the source. Another inconsistency reason is the entry of the data value with mistake: for example, the date of birth "33.41.77" cannot be interpreted in any date formats). To choose the right way of data correction we have to understand the source of inconsistency and distinguish data with errors from data with wrong interpretation model.

In addition, we have to take into account discrepancies in data values caused by the time of their entry. For example, in different sources, the number of publications for person X is equal to 55 and 78, but in the first case, the information is entered for 2015 year, and in the second—for 2020. In such cases, data integration can be based on the selection of the most recent data. But in this case, we also need to take into account the semantics of the data—for example, the values of some data can decrease but never increase.

It is much more difficult to process data that uses the same (or similar) parameter names but has different meanings. For example, two sources show the number of publications for person X, but the first source counts all publications, and the second—only publications in English.

Uncertain data. Data uncertainty arises if an estimate of the truth of a fact is indicated. For example, an expert gives a subjective assessment of some statement to estimate the probability that such information is true or false on some infinite interval of values (usually the intervals [0, 1] and [0, 100] are used, where the first and last values identify 100% true and 100% false information, respectively) [8]. The probability of truth depends on the number of inconsistent records in the database, on the rating of experts, on statistical forecasts, on the individual accuracy of measurement tools, on the amount of processed data, etc. In addition, uncertain data can be caused by the processing of other uncertain data.

Ambiguous data. We consider some data as ambiguous, if they can be interpreted in different ways. In general, ambiguity means that some data due to certain circumstances are deprived of a certain semantic independence and uniqueness that leads to additional interpretations. Different types of ambiguous data from various sources need in specific ways of processing.

Ambiguity of data caused by the use of abbreviations: it often happens that the use of abbreviations leads to confusion in the data interpretation. In this case, data can be transform into CD by abbreviation expansion of the stored value instead of truncation, if it possible.

Data ambiguity caused by incomplete context: examples of such ambiguity are the use of different units of measurement (without an explicit definition in which measurement units the value is given). For example, the temperature can be indicated in degrees Celsius or Fahrenheit, and the price—in Ukrainian hryvnas or Euro. Another variant of the ambiguity of the data is that it is not clearly indicated what number of data units is given (grams or kilograms, meters or kilometers). In most cases, such ambiguity is easily resolved by an expert by connecting the information with an external source of knowledge. But there are quite common situations where it is impossible to determine the units of measurement from the values themselves (for example, wind speed in meters per second or kilometers per hour).

Ambiguity of data caused by different order of words: such data arise in cases where the same information is represented semantically correctly, but in different ways. An example of such ambiguity is the date format—in Ukraine the first digit usually indicates the day, and the second one—the month, while in some other countries it is the other way around. Another quite common example of the ambiguity is the definition of the name and surname in personal data for foreigners.

Fuzzy or vague data. Vagueness or fuzziness implies deals with such degree of reflection where the value and meaning of the data cannot be clearly and precisely determined. Quite often, the reason for the emergence of such data is the ambiguity of the concepts of NL, which describe vague sets of objects and are determined quite subjectively. Special mathematical mechanisms (for example, fuzzy logic) can be applied to process such data.

Fuzzy data can contains fuzzy predicates (for example, "old" and "young", "short" and "tall") that are modeled by fuzzy *linguistic variables.*

A linguistic variable takes the value from some non-empty fuzzy set of words or phrases of some natural or artificial language [9]. If any linguistic variable is modeled as a fuzzy subset of values in the interval $[0, \infty)$ with the membership function on the interval [0, 1], then the projection $[0, \infty) \rightarrow [0, 1]$ is a mathematical description of the value of the linguistic variable. These are the so-called fuzzy sets of Zadeh, who introduced one of the basic concepts of fuzzy logic—the linguistic variable. A linguistic variable is a variable whose value is determined by a set of verbal characteristics [6]. It should be noted that determining the values of a linguistic variable can use, if possible, restrictions of orderliness, completeness, consistency, and normality.

A set of values of a linguistic variable is called a *set of terms* where *term* is any element of a set formalized with the help of a fuzzy membership function: some instances of sets belong to some class unambiguously, while others can be assigned to two or more classes with different probabilities (the sum of such probabilities is equal to 1). For example, if age of a person's is 99 years, then she/he belongs to the set "old" with probability 1, but if the age of a person is 50 years, then she/he belongs to the set "old" with probability 0.6, and to "young" with 0.4.

Imprecise data are not false or erroneous data and do not violate the integrity of the information system if their characteristics are caused by the existence of value that cannot be measured with sufficient precision. In such cases, availability at least some information about the range or restrictions of possible values can be valuable. For example, it is not possible to measure the temperature of the air in the absence of a thermometer, but from the fact that the water turned into ice, it can be concluded that it is below 0.

Disjunctive imprecise data are divided into always-true imprecise data, probably imprecise data, imprecise interval data and imprecise data due to errors.

First kind of imprecise data takes one alternative value from a fixed set of values, and although the probability of choosing a particular value can vary, the precision and reliability of the values of their population is always be 1. For example, certain building has three available doors. That is, from the information that a person is inside the building, data is generated that she/he used one of these enters, but we do not know which one was used.

Second kind of data is similar to always-true imprecise data because the values have to be chosen from some fixed discrete set or from interval of integers. The difference is that the precision and reliability of all data values can not be equal to 1, and therefore has different mathematical probability. This is caused by the fact that individual options of values are not alternatives and can have intersection. For example, values "*the temperature is higher than +5*" and "*the temperature is in the range between +2 and +10*" have non-empty intersection.

Imprecise interval data means that their values are true on a certain interval, not for specific values. Such data does not represent probabilities of separate values from interval. For example, "*The temperature yesterday was from +3 to +15*".

Like imprecise interval data, imprecise data caused by errors can only take on values within a certain interval. But the main difference is that the interval has to be a fuzzy singleton. A fuzzy singleton means that only one value from a finite range of values is true and exact. The membership function returns only 1 for this value, while the membership function of other values are always less than one. For example, "My dog was born around 1.07.2000."

NULL data: This data is a critical case of imprecise data. A NULL value usually indicates absence of information. The main problem with such data is that this absence can be interpreted in different ways. The most common interpretations:

- the value is unknown (it exists, but unknown);
- the value does not exist.

A broader approach to classifying and analyzing NCD is proposed in [10] that consider dirty data, their sources and types. As a result, such NCD groups are divided into incorrect data, incomplete and unusable. This paper develops a comprehensive classification of dirty data that can be used as a basis for understanding how dirty data is generated, detected and can be cleaned to enable better data analysis. NCD can be divided into erroneous and those that are not erroneous, but unsuitable for analysis. For example, authors distinguish the subclass "Integrity constraints not supported in relational database systems today" of wrong data subclass of "Non-enforcement of automatically enforceable integrity constraints" and distinguish its subclasses "Wrong categorical data" and "Outdated temporal data". This classification is very detailed and multi-level, but, unfortunately, it is too difficult for perception and practical use. They analyze the interrelations between certain types of dirty data and the reasons why these data are classified as NCD. This approach is proposed for considering the differences of NCD from CD represented on the basis of relational databases, but it can be extended easily to the analysis of other data, including Big Data, NL texts and semi-structured documents. This approach substantiates criteria for dividing NCD into subgroups and means that allow converting raw data into usable data. Thus, the classification of raw data is one of the preconditions for the application of Smart data methods. Without such classification, it is difficult to determine the quality of data analysis results and the quality of decisions made on the basis of these results.

In a broad sense, dirty data obtained from various data sources can be divided into missing data, incorrect data and non-standard representations of the same data. The results of intelligent analysis of such NCD can be unreliable and incorrect. In some cases data needs in automated cleaning. In other situations it is appropriate to apply various models of soft computing to them, and sometimes data need explicit verification and correction. But in all these variants the basis for obtaining useful results is to determine the type of data difference from classical data.

New information can be obtained using various data analysis tools for processing of combining data from distributed open sources. But the quality of new information depends not only on analysis algorithms, but also on the quality of data. Despite the large number of various software products that help to clean dirty data, the results of such pre-processing are not always suitable for the purposes of further analysis and require some additional specialized operations instead of excessive "cleaning".

3 Problem Definition

NCD taxonomies provide a framework for understanding the impact of dirty data on data mining, as well as help to select methods for dealing with dirty data and metrics for measuring data quality. Some of NCD taxonomies are limited to some subtypes of data (for example, taxonomy proposed in [10] considers only numerical and text dates, and multimedia data and metadata are not analyzed).

But in more general cases of NCD analysis, multimedia data, as well as data in more specific formats (for example, streaming data from various equipment) and partially structured information (for example, metadata without full standardization or knowledge representation formats) have to be considered too.

Q1 is an example query in a medical image database. For example, geographic multimedia *information objects* (IOs) can be represented by fuzzy templates and be linked by fuzzy operators such as "partially-surrounded-by" [11].

In this work, we propose to use a more advanced classification to determine the sources of dirty data about IOs, formalize their types for more correct processing by means of soft computing, and ways to prevent their occurrence, if possible. Important aspect of proposed approach deals with semantic technologies and data semantization [12]: they define need in processing of NCD data that are used as elements of semantic markup and metadata.

4 Taxonomy of NCD

The great heterogeneity of representation and processing of semantics in information systems causes a great variety of approaches to finding and solving such inaccuracies. For example, some aspects of data inaccuracy related to metadata processing and integration of software engineering components, which are related to the application of Data Mining techniques to real-time multimedia data, are discussed in [13].

We propose classification that contains an additional class of NCD—*semantically incorrect data.* In contrast to dirty data, the difference between such data and classic data can be detected only at the stage of their semantic interpretation, if the stored values do not meet the domain restrictions, and such detection requires the analysis of knowledge about this domain from external sources. For example, it data refers to the age of the employee, then the minimum or maximum values are not only greater than zero but determined by certain characteristics of her/his profession and the requirements of the legislation of a particular country (Fig. 1).

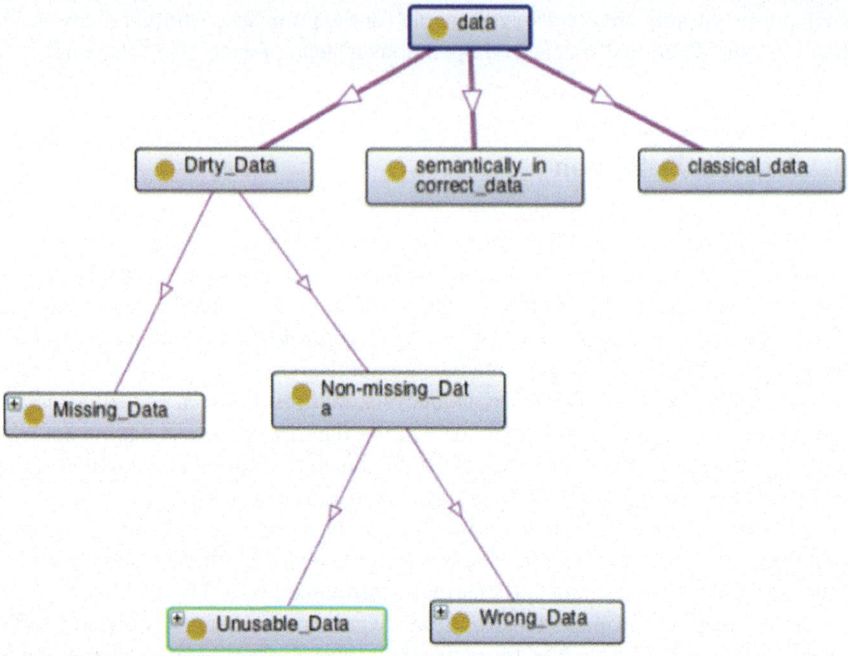

Fig. 1 Taxonomy of classical and non-classical data (upper level)

Semantically incorrect data can have the same sources as ordinary dirty but we can distinguish some additional subclasses of data, which are NCD due to the incorrectness of the choice of domain concepts associated with those attributes whose values are incorrect, with incomplete semantic similarity of the selected attributes and the selection of the range of values of these attributes, etc. For example, if instead of the concept of "employee" the concept of "person" is chosen, then it does not allow to correctly display the data on service dogs (such as the dog Patron), which are part of a certain unit.

All classes of data (both classical data and NCD, as well as semantically incorrect data) have many subclasses, and their hierarchy and level of knowledge refinement depend on the purpose of classification (Fig. 2).

It is important to understand that the same piece of data can be assigned to different subclasses at the same time if it contains several different incorrectnesses at the same time. For example, data for a person's phone number can simultaneously contain an invalid "@" symbol, consist of a longer sequence of characters than required, and begin with a code that does not correspond to any country. Therefore, such data may be an example of three different types of NCD from taxonomy. Each such incorrectness requires different processing methods. Therefore we need to identify exactly what types of incorrectness is present because it may not be clear from the data itself.

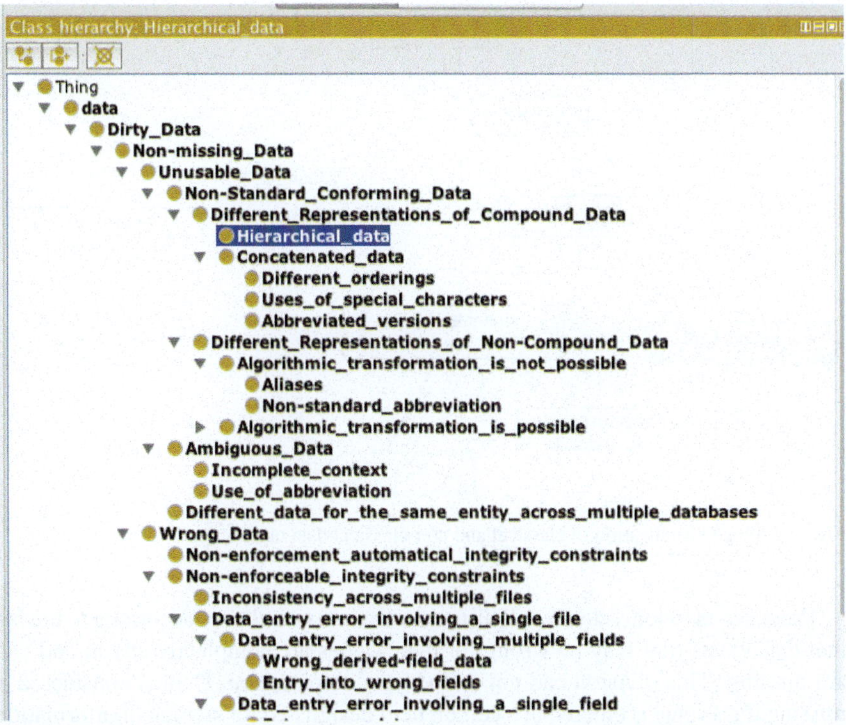

Fig. 2 Basic classes and subclasses of the taxonomy of classical and non-classical data

This NCD taxonomy is implemented as a special case of ontology with a single type of "*class-subclass*" relation and formalized by free, open-source ontology editor and Protégé (protege.stanford.edu) that provides framework for building intelligent systems [14].

The classes of this ontology connected by one hierarchical relation, which can be visualized using the OntoGraf plugin (Fig. 3), and the instances are various examples of NCD and those classic data into which they can be transformed—manually or automatically. The main goal of creating such taxonomy is to provide an unambiguous identification of the NCD type in order to solve questions about the possibility and ways of their transformation into CD.

Usually, the taxonomy of dirty data is based on a hierarchical decomposition of its main manifestations—the absence of data, its incorrectness (in various understandings) and its unsuitability for further analysis and use. Such taxonomy includes only atomic types of dirty data and does not consider their various combinations. The upper levels of the taxonomy of data proposed in this paper cannot contain other subclasses, because they take into account all possible alternatives (but not their combinations). If the taxonomy is used for a more specific domain or applies to a certain subset of data, some of its subclasses may be deleted, and others may be extended with additional lower-level subclasses.

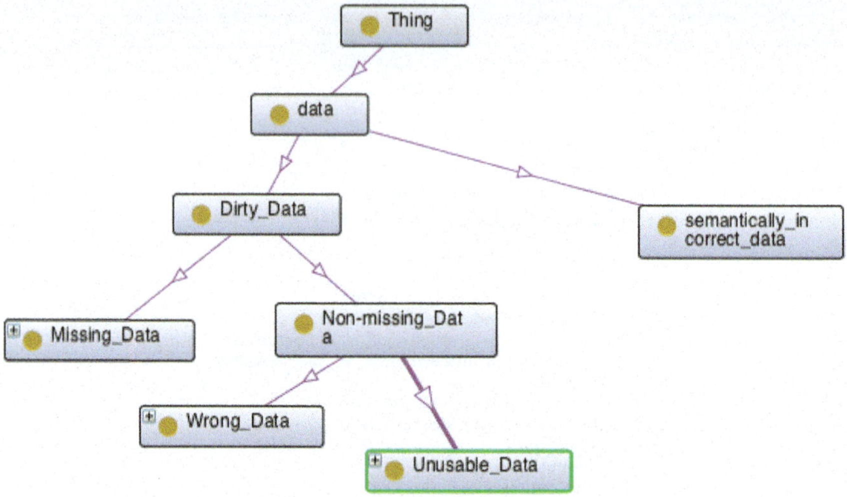

Fig. 3 Protégé visualization of classical and non-classical data taxonomy

Proposed taxonomy divides all dirty data are into two classes according to the fact that data exists (but may be wrong) or their values are completely absent and data are missing. This division can not include any third option. Data is considered as missing if no value is entered in a certain field designated for storage of information. Otherwise, the data is entered and is considered dirty for other reasons.

Missing data are also called null data. Data can be missed for various reasons: 1. if it is allowed according to data meaning (null data)—the values are unknown or unimportant, or 2. if data input is not allowed.

In the first case, the data can be missing due to the fact that it is not yet known, but already exists (for example, we known that some person has an e-mail address, but this address is unknown to us), due to the fact that it is not yet available (the person has not yet register e-mail, but she/he is going to do it) or because their meaning is missing in principle (a person died many centuries before the Internet appeared).

It is clear that these are different types of null data, and therefore various soft computing systems associate with them different special values (with meanings "not known", "does not exist", "undefined"). Logical inference based on such data uses multi-valued logics with sets of the special inference rules and axioms for each value. The simplest approach to null data processing is replacing of the value "not known" by the set of all possible values, and "does not exist" value by a value that does not coincide with any existing one.

5 Methods of NCD Processing

Each type of NCD requires different methods of detection and processing. Sometimes we first need to discover exactly the type of inaccuracy in the data, because it is not clear from the data itself. Then we have to answer the following questions: does such data need correction, can it be transformed into classic data (and into what CD subclasses), and, if such transformation is possible, can it be performed automatically or with human assistance, and does it require the use of additional knowledge sources or analysis tools.

It is advisable to link such answers directly with the classes and subclasses of the NCD taxonomy. For this purpose we propose an extended ontological model of the NCD that contains the following classes for values of object properties of the NCD subclasses:

- method of NCD detection;
- method of transforming NCD into CD;
- external sources of information about NCD.

Instances of these classes are specific methods of Data Mining, machine learning, logical inference, as well as references to external domain ontologies.

In order to speed up and simplify their NCD processing, we propose to add the following classes used as data properties of NCD subclasses:

- possibility of automatic NCD detection;
- possibility of automatic NCD transformation (Fig. 4);
- need for external sources of information (Fig. 5).

These data properties can be treated as Boolean variables with the values Yes and No, or as fuzzy Boolean variables with probabilistic values ranging from 0 to 1. Such data properties in a certain way duplicate the information implicitly represented by NCD taxonomy (because these parameters are the basis for the top level taxonomic division), but their use greatly facilitates the processing of information—these data can be used as query conditions for retrieval of NCD processing means.

Populating the ontological model with such instances is a complex process that requires a detailed analysis of current research from various areas of data analysis and is beyond the scope of this study. But the proposed ontological model sets the structure of how relevant information can be presented and connected with other elements (Fig. 6).

Vagueness is characteristic of many types of information processed by humans and information systems. As it was said above, one of the sources of NCD deals with elements of ambiguity that have not a probabilistic nature and are caused by use of linguistic variables. Inaccuracies reflect various deviations from the classic properties of formal data models—completeness, certainty, consistency, etc. Such deviations arise due to a fundamental difference between objective reality and its model.

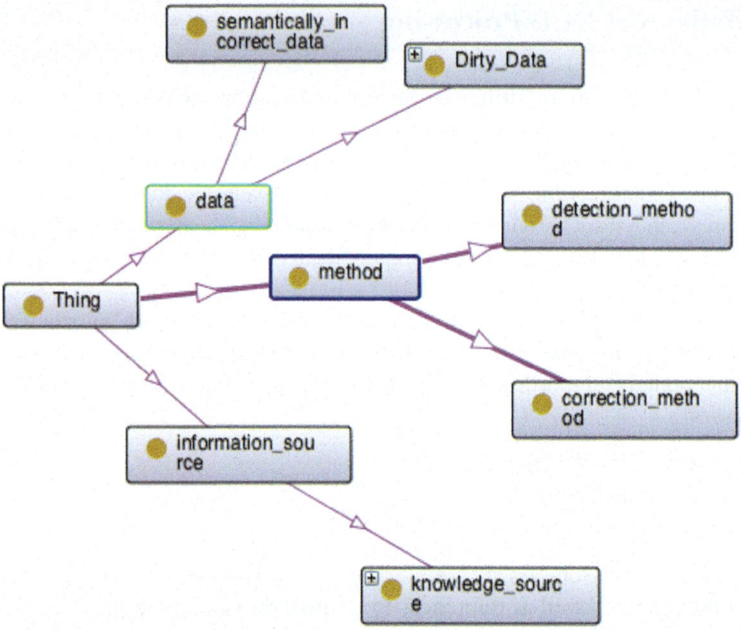

Fig. 4 Methods used to detect and correct NCD

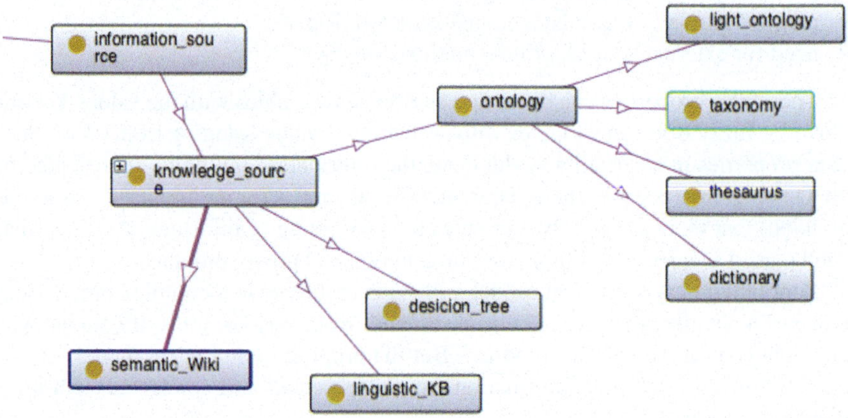

Fig. 5 External sources of information used to find and correct NCDs

Most NCD can be matched with different types of fuzzy knowledge. For example, fragment of NL text (data) corresponds with value of linguistic variable (knowledge) where NL texts can be considered as containers for fuzzy statements. Other non-classical forms of data (for example, probability estimates) can be matched with

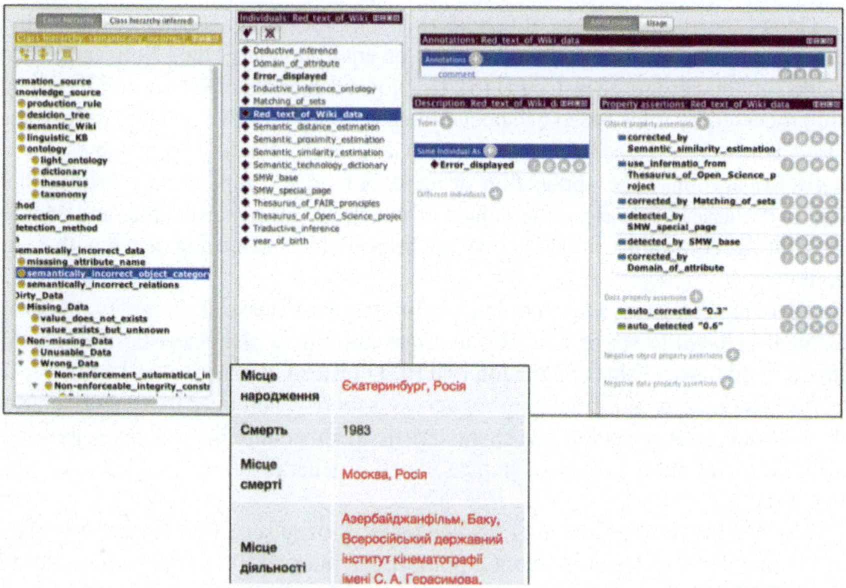

Fig. 6 An instance of the semantically incorrect data subclass for Semantic MediaWiki environment

fuzzy production rules where two or more fuzzy statements are combined by logical operators.

The theory of fuzzy sets ("fuzzy logic") is a tool for integrating the accuracy of classical mathematics with the inaccuracy of the surrounding real world. The main idea of the theory of fuzzy sets is that the way of NL reasoning used by humans cannot be described within the framework of traditional mathematical formalisms, which are characterized by a strictly single-valued interpretation, while the use of NLs implies a multi-valued interpretation. Fuzzy set theory provides a rigorous mathematical description of fuzzy statements, thereby bridging the linguistic barrier between human approximate and fuzzy statements and evaluations and computers that follow precise instructions.

Theory of fuzzy sets generalizes the achievements of such areas of classical mathematics as:

- multi-valued logic (three-valued and N-valued Lukasevich logics, n-valued logic of Post, infinite-valued logic, which made it possible to move from two—"true" or "false"—to an arbitrary number of evaluations of the truth of statements: for example, "true", "unknown" or "false");
- probability theory that products a significant number of methods of statistical processing of experimental data;
- discrete mathematics (theory of matrices, theory of graphs, theory of automata, etc.) that provides tools for building models of multidimensional and multilevel systems.

One of the common ways of formalizing and processing incomplete data, which can be applied to data in an open information environment, is the "Null Values" (A-marks) method proposed by Codd [15]. Unknown values denoted by special constants are processed by rules based on special three-digit logic with epistemic truth-values (T—"True", F—"False", W—"Possible") and corresponding truth tables for all logical operations. The application of such logic to incomplete data divides them into two classes: True-data, the values of which are always available, and Maybe-data, the specific values of which may not be available and are associated with a later definition.

There are some ways to overcome the limitation of classical set theory that every element belongs to set or not. The first one introduces characteristic membership function that takes values on the interval [0, 1] instead of exact value 1 or 0.

The second way of formalizing vagueness is more generalized in comparison with the first one. It assumes that the characteristic membership function takes its values in finite or infinite distributive lattices. Such a generalization is called Gauguin's fuzzy set.

The third way of formalizing vagueness is P-fuzzy sets. In this case, any value of the membership function is not a point in the interval [0, 1], but a subset of this interval. The algebra of P-sets can be reduced to the algebra of classes.

The fourth method uses heterogeneous fuzzy sets. Different elements of the universal set correspond to values in different distributive lattices. Each element is associated with the most suitable assessment for it. The estimates themselves can be vague and are defined in the form of functions. Heterogeneous fuzzy sets and the associated high-order linguistic variables allow modeling data about multi-criteria decision-making situations that use properties with both quantitative and ordinal scales.

There is a close relation between CI, types of fuzzy data and the theory of fuzzy sets, which follows from L. Zadeh's thesis that "a person think with fuzzy concepts, but does not with numbers". This statement means that objects of real world described by data are fuzzy by their nature.

Let's define the information unit I: $I = \{ O,A,a,K,k\}$, where O is the object described by the data; A is an attribute whose value is data; a is a value of attribute A; K is a confidence in choosing an attribute A; k is a confidence in the value of the attribute. In this context, uncertainty is a characteristic of the information meaning represented by a and A, and unreliability is a characteristic of the truth of information represented by k and K that refers to their conformity with reality.

The information is unreliable, if information unit I contains confidence k and K that cannot be represented by two values: 1 ("*true*") and 0 ("*false*"). One of the forms of unreliability is inaccuracy. It refers to the quality of the values of the facts. To process such data, confidence coefficients are used, which quantify the degree of confidence that the attribute has exactly this value, and this value refers to exactly this attribute. Estimates of the likelihood of k and K significantly depend on the conditional probabilities subjectively defined for each rule. The sources of fuzzy information lie within the very interaction of a person with the surrounding world, that is, due to the nature of the reflection of objective reality.

6 Semantic Similarity Estimations of Data

Estimating semantic relatedness with use of various network representations of domain is a problem in artificial intelligence and psychology that can be useful for CI purposes. *Semantic similarity* (SS) is a special case of semantic relatedness [16]: for example, cars and gasoline may seem more closely related than cars and bicycles that are actually more semantically similar. SS estimations can be used as an instrument for detecting and transformation of semantically incorrect data. Use of the domain concepts as semantic markup tags requires to make sure that the chosen concept corresponds to selected domain and is not used in a different sense as a concept of another domain.

Therefore, one of the important steps of checking the semantic consistency of Wiki-data is to determine the quantitative estimations of the semantic similarity between the tags of a certain Wiki-page. This similarity can estimated as a function of semantic distance between the concepts of the corresponding domain ontology.

Some researchers suggest that the estimation of semantic similarity should be considered with the involvement of only taxonomic ("*is-a*") connections [17], excluding other types of relations between concepts; but relations such as "*part-whole*" or "*synonym*" can also be seen as attributes that influence the definition of similarity.

Semantically close domain concepts can be considered as fuzzy sets that include concepts with quantitative value of semantic closeness with the selected concept above some given threshold. Measures for determining the semantic proximity of concepts based on ontologies can use various other semantic characteristics of these concepts such as their properties (attributes and relations with other concepts, values of these attributes), relations between other concepts, mutual position in ontological hierarchies, etc.

Similarity of concepts is related to their meaning. Let C is the set of concepts in the "is-a" taxonomy that allows multiple inheritance. One of the key factors in the similarity of two concepts is the degree to which they share information specified in this taxonomy by a highly specific concept that applies to both of these concepts. The edge-counting method takes this into account indirectly, because if the minimum path of "*is-a*" links between two nodes is long, then it means that it is necessary to rise in the taxonomy to more abstract concepts to find the smallest upper bound defined as a concept to which both concepts under analysis belong.

According to the standard argumentation of information theory, the information content of the concept c can be quantified as $-\log p(c)$. . Such quantitative definition of information content corresponds to an intuitive idea: if the probability of the appearance of a concept increases, then its informativity decreases. Thus, more abstract concept (that is situated higher is in the taxonomy) has lower information content. Moreover, if taxonomy has a unique top concept, then its information content is equal to 0.

This quantitative characteristic of information provides a new way of measuring semantic similarity [18]. If two concepts c_1 and c_2 jointly use more information then concepts c_1 and c_3, then concepts c_1 and c_2 are more similar then c_1 and c_3.

Information jointly used by two concepts is determined by the informational content of the concepts included in the taxonomy.

Formally, such semantic similarity is defined as follows:

$$\mathrm{sim}(C_1, C_2) = \max_{c \in S(c_1, c_2)} [-\log p(c)], \quad (1)$$

where $S(c_1, c_2)$ is the set of concepts included in both c_1 and c_2.

Estimation of the similarity between tags of semantic Wiki resource needs to measure the similarity of the words used as tag names (that is, the names of the page's semantic properties), and not the concepts to which these tags correspond. Such similarity allows to separate quite similar names of different concepts from different names of similar or identical concepts. This makes it possible to resolve semantic inaccuracy arising from the interoperable parallel work of specialists from various fields to improve the structure of the Wiki resource: quite often they create semantic properties with similar names that have different meanings in different fields of knowledge.

Estimation of SS of words can be defined, for example, with use the function s(w), such that $s : W \rightarrow C$ for the representation of NL words from the set W to the set of domain taxonomy concepts C. This function transform NL words into their meanings: $s(w \in W) = \{c_k \in C, k = \overline{1, m}\}$. Then semantic similarity of words w_1 and w_2 based on (1) is defined as:

$$\mathrm{sim\_w}(w_1, w_2) = \max \mathrm{sim}(c_i, c_j), \quad (2)$$

where $c_i \in s(w_1)$, $c_j \in s(w_2)$. If concepts consist of more then one NL word we use various functions of these estimates (sum, arithmetic average, etc.).

This approach is consistent with the definition of "disjunctive concepts" that uses edge counting: they define the distance between two disjunctive sets of concepts as the minimum path length from any element in the first set to any element in the second. The similarity of words is evaluated by finding the maximum information content for all concepts for which both words can be an instance.

Another point of view on the estimation of concept similarity differed from (2) is based on the use of the probability of the concepts instead of their information content:

$$\mathrm{sim\_p}(c_1, c_2) = \max_{c \in S(c_1, c_2)} (1 - p(c)). \quad (3)$$

With the help of estimation (3) we can compare the similarity of words used as semantic markup tags:

$$\text{sim\_ p\_ w}(\text{w}_1, \text{w}_2) = \max_{\text{c}_1, \text{c}_2}(\text{sim\_p}(c_1, c_2)). \quad (4)$$

Probability-based similarity estimation allows accounting frequency of concept occurrence into informational content. Many ontological measures of closeness are based on Tversky's set-theoretical approach [19], which determines the measure of closeness of two objects by matching of their features. The measure of proximity S(a,b) between objects a and b is a function of three arguments of sets of properties of these objects A and B: their intersection A ∩ B and complements A–B and B–A. This approach can be useful for comparison of semantic Wiki pages and their tags.

These measures allow to evaluate the proximity between Wiki semantic markup elements in order to detect and correct semantic incorrectness of the data used in the corresponding resource. But for this we need to choose clearly the pertinent domain ontology, which represents the knowledge and rules of this area. Such estimations can be integrated with estimations that use taxonomic relations between Wiki categories and semantic properties that can be represented by several independent taxonomies. The simplest measure of closeness between them is the length of the shortest paths in these taxonomies between the corresponding concepts [17]: $S(a, b) = \log_2 N \big/ d(a, b)$ where N is the depth of the taxonomic tree, and $d(a, b)$ is the length of the shortest path between concepts a and b.

Various similarity estimations (choice of them is based on task characteristics, features of domain model and specification of data, etc.) can be used for detection of semantic incorrect data. For example, if semantic distance between attributes of data is more then pre-defined constant, then expert has to check the correctness of selected attribute names (such as "*mouse*" instead of "*month*").

Other situation where this approach can indicate semantic incorrectness of data is a semantic similarity of attribute and value (such as "city" is semantically similar to "*Dnipro (city)*" and not to "*Dnipro (river)*").

7 Dirty Data and Semantic Wikis

Manifestations of dirty and incorrect data significantly depend on the technological environment that supports creation, storage and processing of this data.

The most important factor is the expressiveness that the environment provides both to store information (data and metadata) and to retrieve data elements relevant to some current problem: only something that is generated and represented can be distorted.

In addition, various tools contain different types of automated data reconciliation. It should be taken into account that in some cases the source of dirty data is precisely the wrong choice of a data representation model or the use of an information structure that is not pertinent to the solved problem. E.g., if a certain type of data is not provided into the information system, but it is necessary to enter data of this type, then the use of another type of data can lead to incorrect processing. An example of this situation: use of text data type instead of a numeric type causes incorrect data sorting.

We propose to consider the NCD taxonomy on the example of the technological environment of semantic Wikis (namely, MediaWiki and its extension Semantic MediaWiki) where a wide range of dirty data from various sources is represented. Semantization of Wiki-resources significantly expands the expressiveness of this technology, but also makes it necessary to analyze the semantic correctness of information.

Wiki technologies are aimed on distributed processing of information in the Web open environment. The specific feature of this technology is that users independently create and edit the content of the pages. On the one hand, this ensures the rapid development of the Wiki-resources and an increase in their volume, and on the other hand, it causes a large number of various errors and ambiguities in the data.

Therefore the semantic Wiki environment becomes an interesting illustration for the means of dirty data classification and for finding ways to transform it into classic data.

This process requires the development of additional models and methods for verifying raw data, determining the sources of what makes them NCD. It is important to understand that the reasons for vagueness and incompleteness of data in Wiki-resources are not always the result of errors or lack of reliable information. In many cases, Wiki resource turns into a NCD due to a bad choice of content structure that is not relevant to the real world. Unfortunately, in many cases, this situation is determined only in the process of accumulating of heterogeneous content and generation of complex information objects [20] with big number of properties, restrictions and relations.

The main content element of a Wiki resource is a Wiki page that has unique name and an arbitrary set of links, properties and categories that can be considered as metadata.

Now one of the most popular Wiki software is MediaWiki [21] used by many projects such as Wikipedia, Wikidata, and Wikibooks. Data entry in MediaWiki is supported by a user-friendly content editor and provides the following data structuring elements:

- *categories* that allow to group Wiki pages (each page can belong to an arbitrary number of categories at the same time, and the relation of partial ordering allows to create sets of hierarchies of these categories);
- *links* between Wiki-pages (the meaning of the link is not formally described by MediaWiki instruments and can be determined only by context);
- page *namespaces* that provide additional instrument relationship (without subgroups and intersections);
- *templates* that unify data elements of page.

MediaWiki core software does not propose any means of checking the consistency of usage of these elements and cannot represent the semantics of the relations between pages and data.

To solve these problems, now various semantic extensions are developed to expand the expressiveness of the Wiki resource with the help of semantic markup. Use of the semantic markup allows to associate certain elements of the Wiki content with domain concepts. Such markup helps to structure information and makes data more accessible for automatic analysis. For example, Semantic MediaWiki (SMW) [22] is a MediaWiki plugin that allows to associate links between Wiki-pages and data with arbitrary domain concepts, which are used as tags. SMW provides semantic search by names and values of these tags to integrate information from different Wiki-pages and generate ontological structures for Wiki-pages [23] that can be used by other systems [24].

SMW allows to extend the content of MediaWiki by the following data structuring elements:

- *semantic properties* of Wiki pages that are represented by "property-value" pairs with a data type definition, for example, text, number, date, or a link to another Wiki page (in this case, these properties can be considered as meaningful relations between pages that extend links between Wiki-pages);
- *templates* of typical information objects represented by relevant Wiki-page (semantic properties can be used as template parameters, and their values are entered with the help of template as structured data);
- *semantic queries* that can generate new data by processing of the values of semantic properties and categories of Wiki pages that meet the query conditions.

Semantic Wiki-resources need in check the semantic consistency of data with domain rules and requirements. The built-in possibilities of Semantic MediaWiki do not provide all possibilities to detect fuzzy and semantically incorrect data, and therefore it is advisable to use external means of distributed knowledge management based on ontological analysis and develop specific ones oriented on aims of some practical task. One of the instruments for this is the use of metrics of semantic similarity and semantic proximity that allow to make quantitative estimates of the pertinence of used semantic properties.

We chose this example because this information resource is built on MediaWiki and Semantic MediaWiki, it has a complex structure, a large volume, and reflects different semantic connections between the concepts of various domains. Information is entered by specialists from various domains that have different experience in Wiki technology and knowledge processing. It causes a lot of various fuzzy data that need in immediate detection and corrections. Use of proposed above taxonomy can become an instrument of such data processing.

The use of the described above taxonomy of NCD allows to identify more precisely the causes of inaccuracy and incorrectness of data in this Wiki-resource and to recommend ways of their transformation, if possible. Depending on the type of incorrectness, you need to change the data itself or make additions and changes to their model. Therefore we need in external knowledge bases and methods of their matching with semi-structures Wiki-content (such as semantic similarity estimations) for solving these semantic problems.

Fig. 7 e-VUE portal main page

Let's consider this in more detail using examples related to the development and replenishment of the portal version of the Great Ukrainian Encyclopedia e-VUE (vue.gov.ua) [25]. The main tool for organizing of the e-VUE portal knowledge is provided by Semantic MediaWiki (categories and semantic properties). Structured representation of information uses a set of independent taxonomies, which are associated with appropriate sets of categories and semantic properties (Fig. 7).

Semantic markup of the content of the Wiki-pages is the basis for intelligent search and integration of data, but problems of its creating and processing are complicated in the process of e-VUE development, both due to the increase in the number of information objects represented in the encyclopedia, and due to the complexity of the structure of the content itself and the increase in the number of those semantic relations that determine the content of hyperlinks between pages. Ambiguity of meaning and types of markup tags causes semantically incorrect content (Fig. 8).

Typical information objects (IOs) are distinguished for pages with similar sets of properties, categories and content structure.

Null data can appear in IO template if certain properties are unknown (currently or in general) for some instances. This situation has to be taken into account during the creation of the template: we need to check that the value is not empty, because otherwise trying to display non-existent information will lead to errors. Such check requires additional calculations, and therefore it is necessary to consult with the domain specialists about its necessity for every semantic property of every IO.

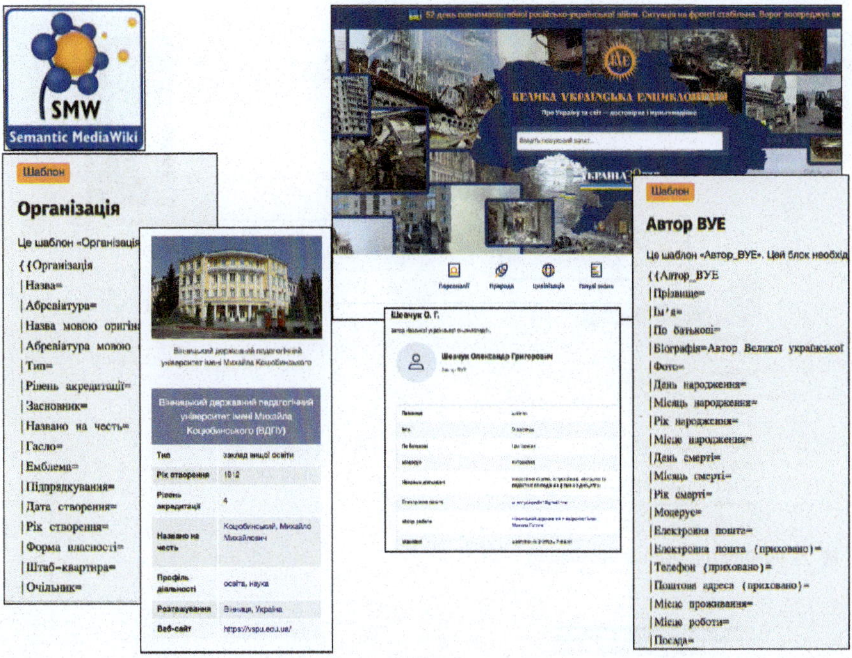

Fig. 8 e-VUE templates and typical IOs (fragments)

Every correctly designed IO template provides for the possibility of entering all the main attributes, but takes into account the verification of their presence. For example, such information as a surname or first name of outstanding personalities of antiquity can be missing because it is not exist, and year of birth can be unknown exactly. In such situations we don't display values of relevant properties because entering of text data instead of number is processed as error (and displayed by red color). SMW template language [26] is expressive enough to program such checks. Another variant of missing data in a Wiki resource is a link to a page that does not exist. This type of NCD is controlled by MediaWiki tools—the link is displayed in red. In this case, it is advisable to choose one of two possible solutions—to create a corresponding referral page or to convert the data type of such an attribute to text instead of a link (Fig. 9). Other MediaWiki plug-ins can be useful for detection of other types of data incorrectness [27].

Recognizing semantically incorrect data for the Wiki environment includes the following situations (Fig. 10):

- the not-existing name of attribute is used;
- the value category does not correspond to the attribute in terms of content;
- the complex information object formed with the help of semantic properties cannot exist in the real world.

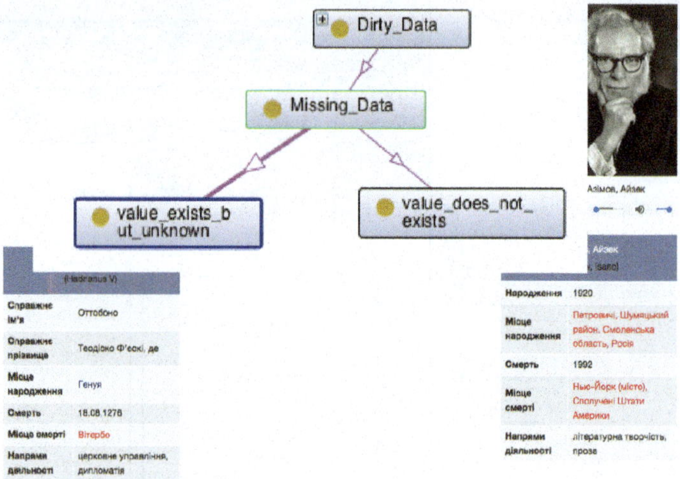

Fig. 9 Displaying of missing data in e-VUE

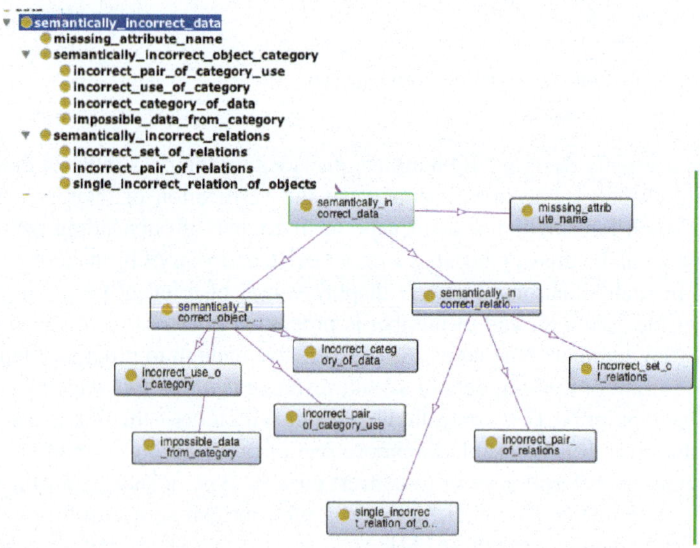

Fig. 10 Taxonomy of semantically incorrect data (fragment)

A separate element of incorrectness is the use of similar names of templates and properties, but SMW partially checks this problem (Fig. 11). Unfortunately, such detection provides only statistical estimates of similarity and provides only the basis for further analysis: it is necessary to check whether both similar terms are used in this domain (and then the situation is not erroneous), or whether only one of them

```
[
    {
        "property": [
            "Використовується",
            "Використовується в"
        ],
        "similarity": 95.52
    },
    {
        "property": [
            "Автор ВУЕ",
            "Актор ВУЕ"
        ],
        "similarity": 94.12
    },
    {
        "property": [
            "Відзнака",
            "Відзнаки"
        ],
```


Fig. 11 Search for usage of similar property names in Semantic MediaWiki

is used (then the other one is incorrect), are there are no such or similar concepts at all, and both need to be removed. For this purpose, it is advisable to apply queries to the domain ontology with the help of external knowledge management tools.

In the first case, the values of property by default are assigned to the "link" type. The probability that Wiki page with such name exists is very low, therefore entered values are displayed by red color, which in this technological environment indicates an error. Processing of such semantic incorrectness is quite simple: the user has to choose one of the options—to create a corresponding property or to replace the used property name with the name of existing one.

Semantic errors are not detected automatically. For example, the value "Dnipro (river)" is selected for the "Place of birth" attribute instead of "Dnipro (city)". Only experts can recognize such incorrectness, because only they can separate the wrongly used values from special situations (for example, a person was really born in the ocean on a ship). Other way deals with development of semantic requests. But we can develop special API-based requests for searching of semantically similar pages. Such similarity can use arbitrary subset of the Wiki categories, properties and their values.

It should be noted that all semantic checks in Wiki environment concerning the use of categories require writing additional software code, in contrast to the processing of semantic properties by SMW queries. Therefore, it is advisable to duplicate information about categories using the apparatus of semantic properties provided by Semantic MediaWiki.

An example of the last situation: Person A page refers to Person B page with the help of semantic property "research predecessor", but Person B page contains link to Person A page with the help of semantic property "research predecessor" too.

This is a semantic mistake, because "research predecessor" relation in this domain is not symmetric. Another example: Person A page links to Person B page with the help of semantic property "father", but Person B page links to Person A page with the help of semantic property "brother". This is a semantic mistake as well, because according to domain rules father of person is not his brother.

Automated verification of these situations in Semantic MediaWiki is not possible due to the fact that the expressiveness of the environment does not allow to formally define such characteristics of properties as transitivity, symmetry, antisymmetry, etc. More complex combinations require derivation in multi-valued logic.

Such semantic inconsistencies can be detected on base of logical inference, which is not supported by SMW, but can be executed by external means of ontology analysis.

Therefore, we can propose the following steps in detection of semantic incorrectness of Wiki data:

- Define semantic properties of IOs that can be incorrect (for example, if they have similar names or if experts have different definitions of their domain space, have intersecting sets of categories, have similar but not identical values);
- Construct the semantic request that contain these properties based on the set of Wiki pages for which we need to check on the NCD;
- Generate query result as an RDF file;
- Perform check on this generated set of RDF data by external tools.

Now a large number of tools for testing various aspects of the quality of ontologies are created. For example, OOPS! [28] is an open source software that allows to detect transitive and symmetric properties of objects. The choice of verification tools depends on type of semantic inaccuracies that is checked. Taxonomy of NCD proposed above can help in selection of relevant instruments.

Alternative is manual search for semantic inaccuracies with the help of domain specialists and ontologists that takes much more time for large volumes of data and is less reliable. Therefore, it is advisable to use it only at the initial stages of creating the structure of knowledge base for Wiki resource, when the expert is still looking for correct correspondences between data models and the real world.

8 Conclusion

The methods used by CI are close to the human fuzzy way of thinking and are able to adaptively produce management actions. Intelligent information systems that use fuzzy data and knowledge bases can become a powerful instrument of decision making in various fields of science, business, and manufacturing.

Enriching fuzzy data with semantic models that use domain knowledge can help to overcome data incompleteness.

The approaches to the classification and transformation of dirty data proposed in the work can be useful for practical tasks where it is important to separate the fuzziness and falsity of the data itself from the semantic incorrectness of the models of their representation and interpretation. This difference is especially important for problems where fuzzy data (received from measuring devices, acquired from natural language texts or incorrectly entered by users) are analyzed on the basis of fuzzy rules and multi-valued logic in a dynamic environment where both the conditions themselves and the decision-making criteria change.

Now we use in practice proposed methods of NCD detection and their transforming into CD in process of development of knowledge structure and content of e-VUE. Some of these methods are automates, and other ones are applied manually or semi-automatically (for example, by informing of content manager about possible errors or data incompleteness). Such support of content management causes more high-quality data representation and more efficient knowledge acquisition from this data.

Examples of such tasks where we use external ontological sources of knowledge and developed data integration methods (including based on Semantic MediaWiki) are the assessment of teamwork of military units with multi-level hierarchies, the analysis of competencies and the validation of the results of distance learning [29], Smart Home semantic support and formation of groups of unmanned aerial vehicles that interact by of swarm intelligence methods [30].

In all these tasks we detect some elements of dirty data that demand additional means of their modeling and processing. But sources of them were not defined formally (it was beyond the scope of research projects), and methods of processing were not unified. Therefore, we consider, that taxonomy of NCD expanded by means of their detection and transforming can be usable for solving these practical tasks.

References

1. RDF Web Ontology Language. Overview, W3C, 2012. https://www.w3.org/RDF/, last accessed 2023/02/15.
2. OWL Web Ontology Language. Overview, W3C Recommendation: W3C, (2009). URL: http://www.w3.org/TR/owl-features/, last accessed 2023/02/16.
3. Kruse, R., Borgelt, C., Braune, C., Mostaghim, S., Steinbrecher, M., Klawonn, F., & Moewes, C. Computational intelligence. Vieweg+ Teubner Verlag. (2011). DOI: https://doi.org/10.1007/978-1-4471-7296-3_1.
4. Konar, A. Artificial intelligence and soft computing: behavioral and cognitive modeling of the human brain. CRC press. (2018).https://doi.org/10.1201/9781420049138.
5. Kaburlasos, V.G. Towards a unified modeling and knowledge-representation based on lattice theory: computational intelligence and soft computing applications. In: Springer Science & Business Media, V. 27 (2007).
6. Zadeh, L. A. (1979). Fuzzy sets and information granularity. Fuzzy sets, fuzzy logic, and fuzzy systems: selected papers, 433–4. www2.eecs.berkeley.edu/Pubs/TechRpts/1979/ERL-m-79-45.pdf
7. Motro A. Uncertainty Management in Information Systems: From Needs to Solutions / Motro, A., Smets, P. – Springer, 464 p. (1997). DOI: https://doi.org/10.1007/978-1-4615-6245-0.
8. Parsons S. Current Approaches to Handling Imperfect Information in Data and Knowledge Bases. In: Knowledge and Data Engineering IEEE, Vo.8, No. 3, 483–488. (1996). DOI: https://doi.org/10.1109/69.506705.

9. Zadeh, L. A. The concept of a linguistic variable and its application to approximate reasoning—I. Information sciences, 8(3), 199-249, (1975).https://doi.org/10.1016/0020-0255(75)90036-5.
10. Kim, W., Choi, B. J., Hong, E. K., Kim, S. K., & Lee, D. A taxonomy of dirty data. Data mining and knowledge discovery, 7, 81–99 (2003). URL: http://biosoft.kaist.ac.kr/bisl_2018/paperPDF/2003TaxonomyOfDirty.pdf.
11. Aygun, R. S., Yazici, A. Modeling and management of fuzzy information in multimedia database applications. In: Multimedia Tools and Applications, 24, 29–56 (2004). DOI: https://doi.org/10.1023/B:MTAP.0000033982.50288.14.
12. Gladun A., Khala K., Martinez-Bejar R., Development of Object's Structured Information Field with Specific Properties for Its Semantic Model Building. CEUR Workshop Proceedings, Vol-3241, 102–111 (2021).
13. Kim, W., Chae, K. J., Cho, D.S., et al. The Chamois component-based knowledge engineering framework. In: Computer, 35(5), 45–54. (2002).
14. Gennari, J. H., Musen, M. A., Fergerson, et al. The evolution of Protégé: an environment for knowledge-based systems development. In: International Journal of Human-computer studies, 58(1), 89–123 (2003). DOI: https://doi.org/10.1016/S1071-5819(02)00127-1.
15. Codd, E. F. Missing information (applicable and inapplicable) in relational databases. In: ACM Sigmod Record, 15(4), 53–53 (1986).
16. Collins, A. M., & Loftus, E. F. A spreading-activation theory of semantic processing. In: Psychological review, 82(6), 407, (1975). DOI:https://doi.org/10.1037/0033-295X.82.6.407.
17. Rada, R., Mili, H., Bicknell, E., Blettner, M. Development and application of a metric on semantic nets. IEEE transactions on systems, man, and cybernetics, 19(1), 17-30 (1989).
18. Resnik P. Semantic Similarity in a Taxonomy: An Information-Based Measure and its Application to Problems of Ambiguity in Natural Language. In: Journal of Artificial Intelligence Research 11, 95–130 (1999).
19. Tversky, A. Features of similarity. Psychological review, 84(4), 327 (1977).
20. Rogushina, J., Gladun, A. Use of ontological knowledge for multi-criteria comparison of complex information objects. In: Proc. of the 13th International Scientific and Practical Conference of Programming (UkrPROG 2020), CEUR Workshoop Proceedings, Vol-2866, 222–231 (2022).
21. Koren, Y. Working with MediaWiki. San Bernardino, CA, USA: WikiWorks Press. 157–159(2012). URL: uplooder.net.
22. Semantic MediaWiki. – https://www.semantic-mediawiki.org/wiki/Semantic_MediaWiki, last accessed 2023/02/18.
23. Guarino N. Formal Ontology in Information Systems. Formal Ontology in Information Systems. In: Proc. of FOIS'98, 3–15 (1998).
24. Rogushina J.V., Grishanova I.J. Ontological methods and tools for semantic extension of the media WIKI // Problems of programming, No. 2–3, 61–73 (20200. http://pp.isofts.kiev.ua/ojs1/article/download/398/437. DOI:https://doi.org/10.15407/pp2020.02-03.061.
25. Andon P., Rogushina J., Grishanova I. et al. Experience of semantic technologies use for development of intelligent web encyclopedia. In: Proc. of the 12th International Scientific and Practical Conference of Programming (UkrPROG 2020),CEUR Workshoop Proceedings, Vol-2866, 246–259 (2021). URL: http://ceur-ws.org/Vol-2866/ceur_246-259andon24.pdf.
26. Semantic templates. URL: www.semantic-mediawiki.org/wiki/Help:Semantic_templates, last accessed 2023/02/20.
27. Product Analytics/Data Products. URL: https://www.mediawiki.org/wiki/Product_Analytics/Data_Products, last accessed 2023/02/15.
28. OntOlogy Pitfall Scanner! URL: http://oops.linkeddata.es, last accessed 2023/02/23.
29. Pryima, S., Rogushina, J., Strokan, O. Use of semantic technologies in the process of recognizing the outcomes of non-formal and informal learning. In: Proc. of the 11th International Conference of Programming UkrPROG, 226–235 (2018). URL: http://ceur-ws.org/Vol-2139/226-235.pdf.
30. Gladun, A., Rogushyna, J., Lesage, M. Ontological approach to aggregated evaluation of the work of teams with multiple levels of hierarchy. In: Information Technology and Security, 10(2), 126–140. (2022). DOI: https://doi.org/10.20535/2411-1031.2022.10.2.270284

GPSR Compliance

The European Union's (EU) General Product Safety Regulation (GPSR) is a set of rules that requires consumer products to be safe and our obligations to ensure this.

If you have any concerns about our products, you can contact us on ProductSafety@springernature.com

In case Publisher is established outside the EU, the EU authorized representative is:

Springer Nature Customer Service Center GmbH
Europaplatz 3
69115 Heidelberg, Germany

The manufacturer's authorised representative in the EU is Springer Nature Customer Service Centre GmbH, Europaplatz 3, 69115 Heidelberg, Germany. If you have any concerns regarding our products, please contact ProductSafety@springernature.com

Printed and bound by CPI Group (UK) Ltd, Croydon, CR0 4YY
28/09/2026
02224085-0011